Current Progress in Microbiology

Current Progress in Microbiology

Editor: Dean Watson

Callisto Reference
www.callistoreference.com

Callisto Reference,
118-35 Queens Blvd., Suite 400,
Forest Hills, NY 11375, USA

Visit us on the World Wide Web at:
www.callistoreference.com

ISBN: 978-1-64116-246-3 (Hardback)

The publisher's policy is to use permanent paper from mills that operate a sustainable forestry policy. Furthermore, the publisher ensures that the text paper and cover boards used have met acceptable environmental accreditation standards.

Printed in China.

Cataloging-in-Publication Data

Current progress in microbiology / edited by Dean Watson.
p. cm.
Includes bibliographical references and index.
ISBN 978-1-64116-246-3
1. Microbiology. 2. Microorganisms. 3. Biology. I. Watson, Dean.
QR41.2 .C87 2020
579--dc23

Table of Contents

Preface

This book aims to highlight the current researches and provides a platform to further the scope of innovations in this area. This book is a product of the combined efforts of many researchers and scientists, after going through thorough studies and analysis from different parts of the world. The objective of this book is to provide the readers with the latest information of the field.

Microbiology is the science concerned with the study of microorganisms, whether acellular, unicellular or multicellular. Unicellular organisms fall within the categories of prokaryotes and eukaryotes. Prokaryotes include bacteria and archaea, while eukaryotes include unicellular algae, fungi and protozoa. The primary candidates for acellular life are viruses. All species of plants, animals, algae and fungi are multicellular organisms. Some of the various sub-disciplines of microbiology are parasitology, bacteriology, virology and mycology. While many microbes are causative agents of various human diseases, many microorganisms have potential for human benefit. They are useful in industrial fermentation, particularly in the production of vinegar, alcohol and dairy products. They are also utilized for antibiotic production. Microorganisms are also used for the biosynthesis of cellulose, xanthan, cyanophycin, alginate, levan, etc. The ability of microorganisms to degrade toxic waste is of utmost importance for microbial biodegradation or bioremediation of agricultural, domestic and industrial wastes, and subsurface pollution in sediments, soils and marine environments. This book brings forth some of the most innovative concepts and elucidates the unexplored aspects of microbiology. Most of the topics introduced herein cover new techniques and applications of this field. It is a complete source of knowledge on the present status of this important field.

I would like to express my sincere thanks to the authors for their dedicated efforts in the completion of this book. I acknowledge the efforts of the publisher for providing constant support. Lastly, I would like to thank my family for their support in all academic endeavors.

Editor

Deletion of *AIF1* but not of *YCA1/MCA1* protects *Saccharomyces cerevisiae* and *Candida albicans* cells from caspofungin-induced programmed cell death

Christopher Chin[1,2,#], Faith Donaghey[1,#], Katherine Helming[1,3,#], Morgan McCarthy[1,#], Stephen Rogers[1], and Nicanor Austriaco[1,*]

[1] Department of Biology, Providence College, Providence, RI 02918, U.S.A.

[2] Current address: University of Massachusetts School of Medicine, 55 Lake Ave. N., Worcester, MA 01655, U.S.A.

[3] Current address: Dana-Farber Cancer Institute, 44 Binney St., Boston, MA 02115, U.S.A.

[#] These authors contributed equally to this work, and are listed in alphabetical order.

* Corresponding Author: Rev. Nicanor Pier Giorgio Austriaco, O.P., Ph.D., Department of Biology, Providence College, Providence, RI 02918, USA; E-mail: naustria@providence.edu

ABSTRACT Caspofungin was the first member of a new class of antifungals called echinocandins to be approved by a drug regulatory authority. Like the other echinocandins, caspofungin blocks the synthesis of β(1,3)-D-glucan of the fungal cell wall by inhibiting the enzyme, β(1,3)-D-glucan synthase. Loss of β(1,3)-D-glucan leads to osmotic instability and cell death. However, the precise mechanism of cell death associated with the cytotoxicity of caspofungin was unclear. We now provide evidence that Saccharomyces cerevisiae cells cultured in media containing caspofungin manifest the classical hallmarks of programmed cell death (PCD) in yeast, including the generation of reactive oxygen species (ROS), the fragmentation of mitochondria, and the production of DNA strand breaks. Our data also suggests that deleting *AIF1* but not *YCA1/MCA1* protects *S. cerevisiae* and *Candida albicans* from caspofungin-induced cell death. This is not only the first time that *AIF1* has been specifically tied to cell death in *Candida* but also the first time that caspofungin resistance has been linked to the cell death machinery in yeast.

***Keywords**: caspofungin, AIF1, MCA1/YCA1, programmed cell death, Saccharomyces cerevisiae, Candida albicans.*

INTRODUCTION

Defined as any cell death that results from the activation of a genetic program, programmed cell death (PCD) has been classified by the Nomenclature Committee on Cell Death (NCCD) into twelve different functional categories based on measurable biochemical features [1]. In recent years, numerous laboratories have shown that a wide range of antifungal drugs leads to PCD in yeast that is reminiscent of apoptosis in mammalian cells [2]. This cell death is accompanied by several distinctive phenotypical markers, including the generation of reactive oxygen species (ROS), the fragmentation of mitochondria, and the dissipation of the mitochondrial membrane potential [3-5]. It also involves intrinsic pathways of cell death defined by several genes, including *AIF1, BIR1, BXI1, NMA111,* and *YCA1,* which have homologs in the metazoa [6-11].

Caspofungin was the first member of the class of antifungals called the echinocandins to be approved by a drug regulatory authority [12, 13]. Like the other echinocandins, caspofungin blocks the synthesis of β(1,3)-D-glucan of the fungal cell wall by inhibiting the enzyme, β(1,3)-D-glucan synthase [14]. However, the precise mechanism of cell death associated with the cytotoxicity of caspofungin was unclear.

In this report, we provide evidence that *Saccharomyces cerevisiae* cells cultured in media containing caspofungin manifest the classical hallmarks of PCD in yeast, including the generation of reactive oxygen species (ROS), the fragmentation of mitochondria, and the activation of caspase-like activity. Moreover, our data also suggests that this cell death scenario requires the pro-apoptotic gene, *AIF1*, but not the metacaspase, *YCA1,* in both *S. cerevisiae* and *Candida albicans.* Our findings reveal an additional dimension of caspofungin's mechanism of action as a general antifungal agent that points to a possible novel pathway to drug resistance.

RESULTS AND DISCUSSION

To determine if caspofungin triggers PCD in budding yeast, we cultured W303-1A cells in YPD containing 0.02µg/ml caspofungin for three hours, a drug concentration below the MIC of 0.04µg/ml for this strain (data not shown). With this protocol, propidium iodide staining (50µg/ml) revealed that just over half of the cells (58±10%; $p<0.005$) were viable as compared to cells grown in the absence of drug (99±0.1%). The cells were also treated with 2.5µM CellROX Green (Molecular Probes, Inc.) and imaged with a Zeiss

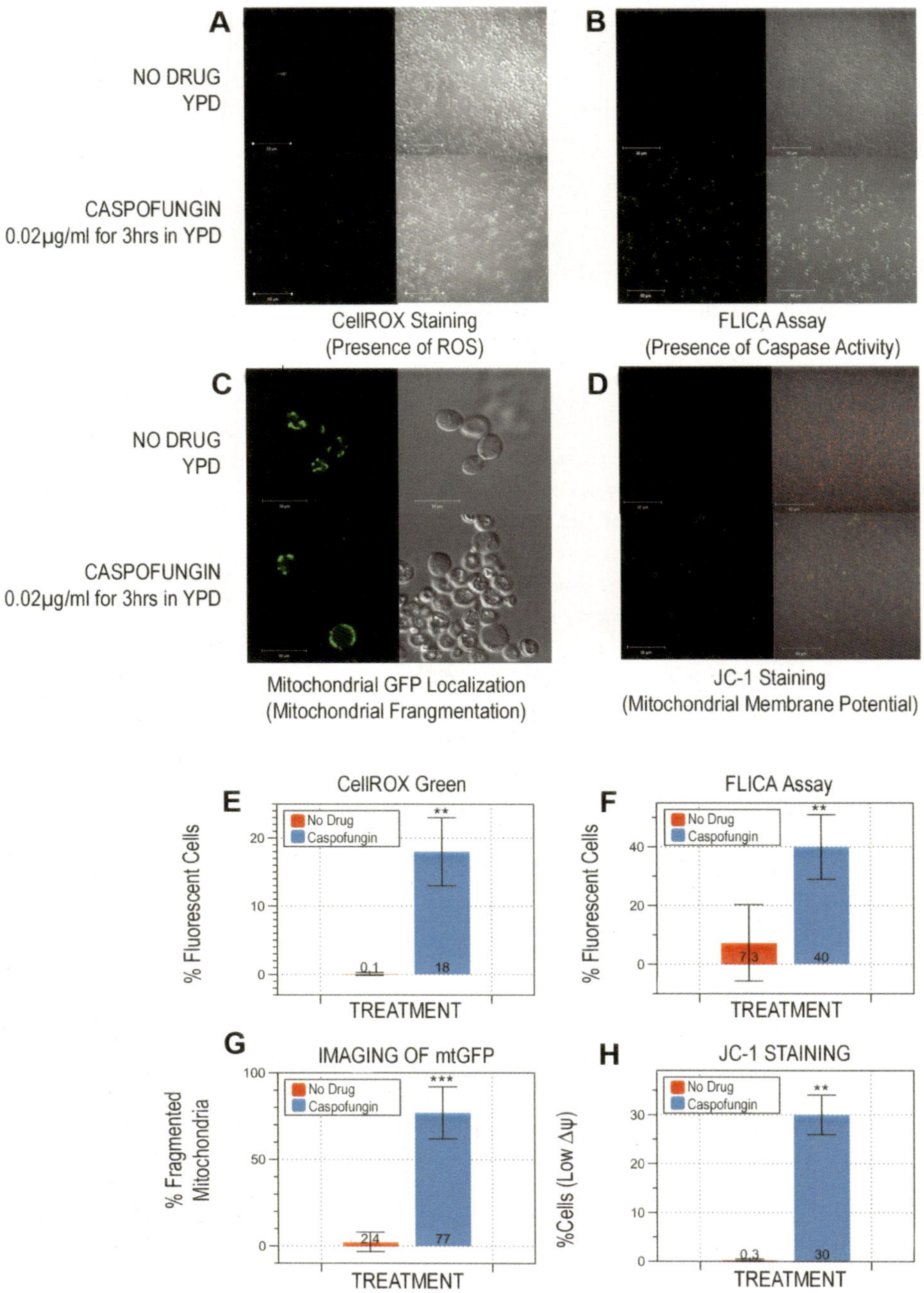

FIGURE 1: *Saccharomyces cerevisiae* cells cultured in caspofungin manifest numerous phenotypic markers associated with yeast programmed cell death. Exponentially-growing wildtype cells from the W303-1A strain background were cultured in YPD containing 0.02µg/ml caspofungin for three hours. The cells were stained with 2.5µM CellROX Green to detect ROS **(A, E)**; with a FLICA kit to detect activated caspase activity **(B, F)**; and with the mitochondrial vital dye, JC-1, to assess their mitochondrial membrane potential **(D, H)**. To visualize the structure of their mitochondria, the cells were transformed with a p416 GPD-mtGFP plasmid constitutively expressing a mitochondria-localized GFP marker and their mitochondria were visualized after they had been treated with caspofungin as described above **(C, G)**. Representative images are shown **(A-D)**. Statistical significance for the graphs **(E-H)** was determined with the unpaired Student's t-test (**: $p<0.05$; ***: $p<0.005$), and a minimum of 500 cells was counted for each assay.

LSM700 confocal microscope. We discovered that including caspofungin in the culture media generates significantly more fluorescent cells (18±5%; $p<0.05$; Figure 1A, E) as compared to controls (0.1±0.2%) suggesting that the drug induces the production of ROS, a characteristic hallmark of apoptotic cell death in yeast [3, 15].

Next, we used a FLICA apoptosis detection kit (ImmunoChemistry Technologies, LLC) according to the manufacturer's specifications to determine if the caspofungin treatment described above also induced caspase activity in W303-1A cells. Numerous studies have associated caspase activation with apoptotic cell death, not only in yeast, but also in mammalian cells [16-18]. We discovered that drug treatment leads to significantly more cells with activated caspase-like activity (40±11%; $p<0.05$; Figure 1B, F) as compared to no-drug controls (7.3±13%).

Third, we imaged the same yeast cells, this time transformed with the p416 GPD-mtGFP plasmid constitutively expressing a mitochondria-localized GFP marker to visualize the structure of their mitochondria [19], after they had been treated with caspofungin as described above. Several reports have shown that apoptosis in *Saccharomyces cerevisiae* is accompanied by the fragmentation of mitochondria [4, 20]. Our experiments revealed that the majority of mitochondria of the caspofungin-treated cells had lost their tubular architecture after three hours (77±15%; $p<0.005$; Figure 1C, G). In contrast, only a few cells cultured in media without the drug (2.4±5.6%) displayed the punctate mitochondrial-staining pattern associated with fragmented mitochondria.

Fourth, to determine the functional state of the mitochondria in cells treated with caspofungin as described above, we incubated the cells with the mitochondrial vital dye, JC-1 (5mM; Molecular Probes, Inc.) for 15min at 30°C to assess their mitochondrial membrane potential ($\Delta\psi_m$). In healthy cells with high $\Delta\psi_m$, JC-1 forms aggregates that exhibit intense red fluorescence, while in apoptotic cells with low $\Delta\psi_m$, the dye remains in the monomeric form, which has green fluorescence [21-23]. Our experiments revealed that caspofungin-treated W303-1A cells had significantly more unhealthy cells with low $\Delta\psi_m$ (30±4%; $p<0.05$; Figure 1D, H) than untreated controls (0.3±0.2%).

In toto, all of our data suggests that caspofungin induces PCD in *S. cerevisiae* cells. Notably, while our studies were in progress, a paper was published that reported that caspofungin induces PCD in the yeast, *Candida albicans*, accompanied with phenotypic markers characteristic of apoptosis and necrosis, including some of the markers described above [24]. We have repeated and confirmed these results suggesting that caspofungin kills with a similar

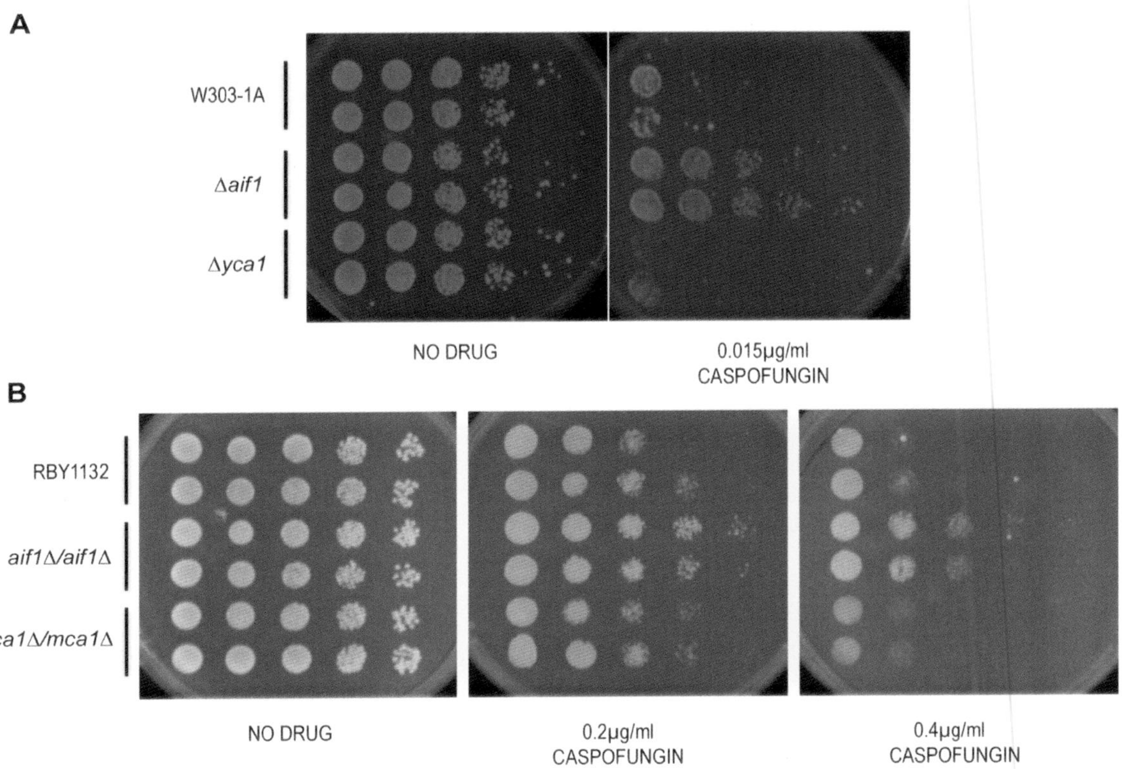

FIGURE 2: Deletion of *AIF1* but not *YCA1/MCA1* protects *Saccharomyces cerevisiae* and *Candida albicans* cells from caspofungin-induced programmed cell death. (A) A series of ten-fold dilutions of two cultures each of wildtype, *Δaif*, and *Δyca1* W303-1A *Saccharomyces* cells were plated on YPD plates and YPD plates with 0.015 µg/ml caspofungin for two days. **(B)** A series of ten-fold dilutions of two independent cultures each of wildtype, *aifΔ/aifΔ* and *mca1Δ/mca1Δ* RBY1132 *Candida* cells were plated on YPD plates and YPD plates with the indicated concentrations of caspofungin for two days. Representative images are shown.

mechanism of action in both *Saccharomyces cerevisiae* and *Candida albicans* (data not shown).

Finally, ROS-mediated PCD in yeast is often distinguished into two different types depending upon their mechanism of killing, either caspase-dependent or caspase-independent yeast apoptosis [18, 25, 26]. Caspase-dependent cell death can be bypassed, at least in part, by the deletion of the yeast metacaspase, *YCA1/MCA1* [6, 27, 29]. In contrast, caspase-independent cell death cannot be suppressed by a *YCA1/MCA1* deletion [29]. Importantly, the yeast homolog of the mammalian apoptosis-inducing factor, called *AIF1* in both *Saccharomyces cerevisiae* and *Candida alibcans,* has often been linked to caspase-independent cell death [7, 25].

While the recent paper showed that caspofungin induces PCD in *Candida*, it did not determine the mechanism of killing for the drug [24]. Indeed, the authors proposed that future studies should determine if either the caspase-dependent or the independent pathway is also involved in regulating yeast apoptosis upon exposure of caspofungin. To see if caspofungin induces either caspase-dependent or caspase-independent cell death, we spotted a series of ten-fold dilutions of two cultures each of wildtype, *Δaif1*, and *Δyca1* W303-1A cells on YPD plates and YPD plates with caspofungin. After two days, we discovered that the viability of cells lacking the metacaspase was indistinguishable from that of wildtype cells (Figure 2A). This suggests that the caspase-like activity that we observed with our FLICA assay described above is associated with caspases and/or proteases other than Yca1p [18]. However, *Δaif1* cells were relatively more resistant to caspofungin killing suggesting that this anti-fungal agent triggers *AIF1*-dependent PCD in *Saccharomyces cerevisiae* (Figure 2A). We repeated these experiments with wildtype, *aif1Δ/aif1Δ*, and *mca1Δ/mca1Δ* mutants of *Candida albicans*, generously provided by Richard Bennett (Brown University). *MCA1* is the *Candida albicans* homolog of the *Saccharomyces cerevisiae* metacaspase gene, *YCA1* [16], while *AIF1* (orf19.3362) is the single homolog of the *Saccharomyces AIF1* gene [7, 25]. The assay was repeated with two independent isolates for each *Candida* null mutant to ensure that the observed phenotype is due to the knockout of the ORF and not to some spurious mutation. Once again, after two days incubation, we discovered that *aif1Δ/aif1Δ* cells were more resistant to caspofungin than either wildtype or *mca1Δ/mca1Δ* cells (Figure 2B).

In conclusion, our study suggests that caspofungin leads to caspase-independent cell death requiring *AIF1* in both *Saccharomyces cerevisiae* and *Candida albicans*. Of significance, though caspase-independent cell death has been observed in *Candida* before [29], this is the first time that *AIF1* has been implicated in this process for this pathogenic yeast. In sum, our findings reveal an additional dimension of this echinocandin's mechanism of action as an anti-fungal agent. They also point to a novel scenario for the acquisition of caspofungin resistance since this is the first time that drug resistance in *Candida* has been linked to a gene involved in cell death.

MATERIALS AND METHODS

Yeast Strains and Growth Conditions

All experiments were done with isogenic *Saccharomyces cerevisiae* strains in the W303-1A background (MAT**a** *ade2, his3, leu2, trp1, ura3, ssd1-d2*). The *Δyca1* mutant (SKY2880) was previously described [30]. The *Δaif1* mutant was created by disrupting the ORF with the kanR marker using a PCR-based knock-out strategy [31] and verified both by PCR and by phenotypic analysis. Experiments with *Candida albicans* were done with wildtype, *aif1Δ/aif1Δ*, and *mca1Δ/mca1Δ* mutants of *Candida albicans*, generously provided by Richard Bennett (Brown University). Mutants were created in the wildtype RBY1132 strain background (MTL **a**/**a**, *leu2/leu2 his1/his1 arg4/arg4*) by disrupting the ORF with a disruption cassette generated by fusion PCR and verified as previously described [32]. For all the experiments described in this paper, cells were cultured using standard protocols [33]. Caspofungin (*Cancidas*) was generously provided by Merck (Rahway, NJ) and by Timothy Flanigan (Rhode Island Hospital/Warren Alpert School of Medicine at Brown University). Unless noted otherwise, all other drugs and reagents were purchased from SIGMA-Aldrich.

Caspofungin Activity Assay

Wildtype budding yeast cells in the W303-1A background were cultured in YPD overnight. In the morning, they were reinoculated in fresh media and allowed to grow for several hours until they reached the exponential phase (OD_{600} 0.2-0.4). They were then resuspended at the same cell density in fresh media or in fresh media containing 0.02µg/ml caspofungin for three hours, a drug concentration below the MIC of 0.04µg/ml for this strain (data not shown). Next, the cells were washed and resuspended in a solution of YPD and 50µg/ml propidium iodide. Propidium iodide is a membrane impermeable fluorescent molecule that is excluded from viable cells. Fluorescent cells were counted with either a Zeiss Axiophot Fluorescence Microscope or a Zeiss LSM 700 Laser Confocal Microscope fitted with Filter Cube 43 Cy3/Rhod/RFP (EX BP 545/25, BS FT 565, EM BP 606/70). Measurements were repeated with at least three independent cell cultures.

Apoptosis Assays

Wildtype budding yeast cells in the W303-1A background were cultured in YPD overnight. In the morning, they were reinoculated in fresh media and allowed to grow for several hours until they reached the exponential phase (OD_{600} 0.2-0.4). They were then resuspended at the same cell density in fresh media or in fresh media containing 0.02µg/ml caspofungin for three hours. To detect the presence of reactive oxygen species (ROS), they were stained with 2.5µM CellROX Green (Molecular Probes, Inc.) for 10min at room temperature. This cell-permeable dye exhibits bright fluorescence upon oxidation by ROS, which was visualized with a Zeiss LSM700 confocal microscope fitted with Filter Cube 38 GFP (EX BP 470/40, BS FT 495, EM BP 525/50). Next, to determine the presence of activated caspase-like activity, they were stained with a FLICA apoptosis detection kit (ImmunoChemistry Technologies, LLC) according to the manufacturer's specifications, and imaged with a Zeiss LSM700 confocal microscope fitted with Filter Cube 38 GFP. Third, to determine the functional state of the mitochondria in cells treated with caspofungin as described above, they were incubated with the mitochondrial vital dye, JC-1 (5mM; Molecular Probes, Inc.) for 15min at 30°C to assess their mitochondrial membrane potential ($\Delta\psi_m$). In healthy cells with high $\Delta\psi_m$, JC-1 forms aggregates that exhibit intense red fluorescence, while in apoptotic cells with low $\Delta\psi_m$, the dye remains in the monomeric form, which has green fluorescence [21-23]. Both were imaged with the Zeiss LSM700 fitted either with Filter Cube 43 Cy3/Rhod/RFP or with Filter Cube 38 GFP. Finally, to detect the fragmentation of mitochondria, the cells were transformed with a p416 GPD-mtGFP plasmid constitutively expressing a mitochondria-localized GFP marker [19], treated with caspofungin as described above, and imaged with a Zeiss LSM 700 Laser Confocal Microscope fitted with Filter Cube 38 GFP. For all the apoptosis assays, the experiment was repeated with at least three independent cultures. Representative images are shown with consistent parameters applied to comparable panels in each figure. Statistical significance was determined with the unpaired Student's t-test, and a minimum of 500 cells was counted for each assay.

Spot Assays

Wildtype and mutant cells of the indicated genotypes were grown overnight in YPD at 30°C (*Saccharomyces*) or at 37°C (*Candida*) respectively. For each yeast strain, a series of 10-fold dilutions was then prepared in water with cells in the exponential phase (OD_{600} 0.2-0.4) over a range of concentrations from 10^{-1} to 10^{-5} relative to the initial culture. Spots of 5 µl from each dilution series were then plated on the indicated media and cultured at 30°C (*Saccharomyces*) or at 37°C (*Candida*) for 2 days. Plates supplemented with caspofungin were poured and used on the same day. All spot assays were repeated at least three times and a representative experiment is shown.

ACKNOWLEDGMENTS

We thank Richard Bennett (Brown University) for the *Candida* strains, Stephen Kron (University of Chicago) for SKY2880, May Godoy and her colleagues (Merck Research Laboratories, Rahway, NJ) for providing us with caspofungin, and for helpful comments on an earlier version of this paper, Timothy Flanigan (Rhode Island Hospital/Warren Alpert School of Medicine at Brown University) for additional caspofungin, John Yang (Providence College) for technical support, and Antoninus Niemiec, O.P., for help with preparation of the figures. Our laboratory is supported by the following grants awarded to N. Austriaco: NIGMS R15 GM094712, NSF MRI-R2 0959354, and NIH/Rhode Island INBRE Program Grant 8 P20 GM103430-12 for undergraduate student training. *Non nisi te, Domine.*

CONFLICT OF INTEREST

The authors declare no conflict of interest.

REFERENCES

1. Galluzzi L, Vitale I, Abrams JM, Alnemri ES, Baehrecke EH, Blagosklonny MV, Dawson TM, Dawson VL, El-Deiry WS, Fulda S, Gottlieb E, Green DR, Hengartner MO, Kepp O, Knight RA, Kumar S, Lipton SA, Lu X, Madeo F, Malorni W, Mehlen P, Nunez G, Peter ME, Piacentini M, Rubinsztein DC, Shi Y, Simon HU, Vandenabeele P, White E, Yuan J, et al. (**2012**). Molecular definitions of cell death subroutines: recommendations of the Nomenclature Committee on Cell Death 2012. **Cell Death Differ** 19(1): 107-20.

2. Almeida B, Silva A, Mesquita A, Sampaio-Marques B, Rodrigues F, Ludovico P (**2008**). Drug-induced apoptosis in yeast. **Biochim Biophys Acta** 1783(7): 1436-48.

3. Madeo F, Frohlich E, Ligr M, Grey M, Sigrist SJ, Wolf DH, Frohlich KU (1999). Oxygen stress: a regulator of apoptosis in yeast. J Cell Biol 145(4): 757-67.

4. Fannjiang Y, Cheng WC, Lee SJ, Qi B, Pevsner J, McCaffery JM, Hill RB, Basanez G, Hardwick JM (2004). Mitochondrial fission proteins regulate programmed cell death in yeast. Genes Dev 18(22): 2785-97.

5. Madeo F, Frohlich E, Frohlich KU (**1997**). A yeast mutant showing diagnostic markers of early and late apoptosis. **J Cell Biol** 139(3): 729-34.

6. Madeo F, Herker E, Maldener C, Wissing S, Lachelt S, Herlan M, Fehr M, Lauber K, Sigrist SJ, Wesselborg S, Frohlich KU (**2002**). A caspase-related protease regulates apoptosis in yeast. **Mol Cell** 9(4): 911-7.

7. Wissing S, Ludovico P, Herker E, Buttner S, Engelhardt SM, Decker T, Link A, Proksch A, Rodrigues F, Corte-Real M, Frohlich KU, Manns J, Cande C, Sigrist SJ, Kroemer G, Madeo F (**2004**). An AIF orthologue regulates apoptosis in yeast. **J Cell Biol** 166(7): 969-74.

8. Sharon A, Finkelstein A, Shlezinger N, Hatam I (**2009**). Fungal apoptosis: function, genes and gene function. **FEMS Microbiol Rev** 33(5): 833-54.

9. Walter D, Wissing S, Madeo F, Fahrenkrog B (**2006**). The inhibitor-of-apoptosis protein Bir1p protects against apoptosis in *S. cerevisiae* and is a substrate for the yeast homologue of Omi/HtrA2. **J Cell Sci** 119(Pt 9): 1843-51.

10. Fahrenkrog B, Sauder U, Aebi U (**2004**). The *S. cerevisiae* HtrA-like protein Nma111p is a nuclear serine protease that mediates yeast apoptosis. **J Cell Sci** 117(Pt 1): 115-26.

11. Cebulski J, Malouin J, Pinches N, Cascio V, Austriaco N (**2011**). Yeast Bax inhibitor, Bxi1p, is an ER-localized protein that links the unfolded protein response and programmed cell death in *Saccharomyces cerevisiae*. **PLoS One** 6(6): e20882.

12. Letscher-Bru V, Herbrecht R (**2003**). Caspofungin: the first representative of a new antifungal class. **J Antimicrob Chemother** 51(3): 513-21.

13. Onishi J, Meinz M, Thompson J, Curotto J, Dreikorn S, Rosenbach M, Douglas C, Abruzzo G, Flattery A, Kong L, Cabello A, Vicente F, Pelaez F, Diez MT, Martin I, Bills G, Giacobbe R, Dombrowski A, Schwartz R, Morris S, Harris G, Tsipouras A, Wilson K, Kurtz MB (**2000**). Discovery of novel antifungal (1,3)-beta-D-glucan synthase inhibitors. **Antimicrob Agents Chemother** 44(2): 368-77.

14. Douglas CM, D'Ippolito JA, Shei GJ, Meinz M, Onishi J, Marrinan JA, Li W, Abruzzo GK, Flattery A, Bartizal K, Mitchell A, Kurtz MB (**1997**). Identification of the *FKS1* gene of *Candida albicans* as the essential target of 1,3-beta-D-glucan synthase inhibitors. **Antimicrob Agents Chemother** 41(11): 2471-9.

15. Perrone GG, Tan SX, Dawes IW (**2008**). Reactive oxygen species and yeast apoptosis. **Biochim Biophys Acta** 1783(7): 1354-68.

16. Shirtliff ME, Krom BP, Meijering RA, Peters BM, Zhu J, Scheper MA, Harris ML, Jabra-Rizk MA (**2009**). Farnesol-induced apoptosis in *Candida albicans*. **Antimicrob Agents Chemother** 53(6): 2392-401.

17. Zhu J, Krom BP, Sanglard D, Intapa C, Dawson CC, Peters BM, Shirtliff ME, Jabra-Rizk MA (**2011**). Farnesol-induced apoptosis in *Candida albicans* is mediated by Cdr1-p extrusion and depletion of intracellular glutathione. **PLoS One** 6(12): e28830.

18. Wilkinson D, Ramsdale M (**2011**). Proteases and caspase-like activity in the yeast *Saccharomyces cerevisiae*. **Biochemical Society transactions** 39(5): 1502-8.

19. Okamoto K, Kondo-Okamoto N, Ohsumi Y (**2009**). Mitochondria-anchored receptor Atg32 mediates degradation of mitochondria via selective autophagy. **Developmental cell** 17(1): 87-97.

20. Kitagaki H, Araki Y, Funato K, Shimoi H (**2007**). Ethanol-induced death in yeast exhibits features of apoptosis mediated by mitochondrial fission pathway. **FEBS Lett** 581(16): 2935-42.

21. Cossarizza A, Salvioli S (**2001**). Flow cytometric analysis of mitochondrial membrane potential using JC-1. **Current protocols in cytometry / editorial board, J Paul Robinson, managing editor [et al]** Chapter 9(Unit 9 14).

22. Salvioli S, Ardizzoni A, Franceschi C, Cossarizza A (**1997**). JC-1, but not DiOC6(3) or rhodamine 123, is a reliable fluorescent probe to assess delta psi changes in intact cells: implications for studies on mitochondrial functionality during apoptosis. **FEBS Lett** 411(1): 77-82.

23. Cossarizza A, Salvioli S (**2001**). Analysis of mitochondria during cell death. **Methods in cell biology** 63(467-86).

24. Hao B, Cheng S, Clancy CJ, Nguyen MH (**2013**). Caspofungin kills *Candida albicans* by causing both cellular apoptosis and necrosis. **Antimicrob Agents Chemother** 57(1): 326-32.

25. Madeo F, Carmona-Gutierrez D, Ring J, Buttner S, Eisenberg T, Kroemer G (**2009**). Caspase-dependent and caspase-independent cell death pathways in yeast. **Biochem Biophys Res Commun** 382(2): 227-31.

26. Carmona-Gutierrez D, Eisenberg T, Buttner S, Meisinger C, Kroemer G, Madeo F (**2010**). Apoptosis in yeast: triggers, pathways, subroutines. **Cell Death Differ** 17(5): 763-73.

27. Mazzoni C, Falcone C (**2008**). Caspase-dependent apoptosis in yeast. **Biochim Biophys Acta** 1783(7): 1320-7.

28. Fu Z, Lu H, Zhu Z, Yan L, Jiang Y, Cao Y (**2011**). Combination of baicalein and Amphotericin B accelerates *Candida albicans* apoptosis. **Biological & pharmaceutical bulletin** 34(2): 214-8.

29. Aerts AM, Carmona-Gutierrez D, Lefevre S, Govaert G, Francois IE, Madeo F, Santos R, Cammue BP, Thevissen K (**2009**). The antifungal plant defensin RsAFP2 from radish induces apoptosis in a metacaspase independent way in *Candida albicans*. **FEBS Lett** 583(15): 2513-6.

30. Wysocki R, Kron SJ (**2004**). Yeast cell death during DNA damage arrest is independent of caspase or reactive oxygen species. **J Cell Biol** 166(3): 311-6.

31. Brachmann CB, Davies A, Cost GJ, Caputo E, Li J, Hieter P, Boeke JD (**1998**). Designer deletion strains derived from *Saccharomyces cerevisiae* S288C: a useful set of strains and plasmids for PCR-mediated gene disruption and other applications. **Yeast** 14(2): 115-32.

32. Alby K, Schaefer D, Sherwood RK, Jones SK, Jr., Bennett RJ (**2010**). Identification of a cell death pathway in *Candida albicans* during the response to pheromone. **Eukaryot Cell** 9(11): 1690-701.

33. Amberg DC, Burke D, Strathern J (**2005**). Methods in Yeast Genetics. Cold Spring Harbor Laboratory Press, Cold Spring Harbor, NY. ISBN 978-087969728-0

2

Tracking autophagy during proliferation and differentiation of *Trypanosoma brucei*

William R. Proto[1], Nathaniel G. Jones[1], Graham H. Coombs[2], and Jeremy C. Mottram[1,*]
[1] Wellcome Trust Centre for Molecular Parasitology, Institute of Infection, Immunity and Inflammation, College of Medical, Veterinary and Life Sciences, University of Glasgow, Glasgow, G12 8TA, UK.
[2] Strathclyde Institute of Pharmacy and Biomedical Sciences, University of Strathclyde, Glasgow, G4 0RE, UK.
* Corresponding Author: Jeremy C. Mottram, 120 University Place; Glasgow G12 8TA, UK;
E-mail: jeremy.mottram@glasgow.ac.uk

ABSTRACT Autophagy is a lysosome-dependent degradation mechanism that sequesters target cargo into autophagosomal vesicles. The *Trypanosoma brucei* genome contains apparent orthologues of several autophagy-related proteins including an ATG8 family. These ubiquitin-like proteins are required for autophagosome membrane formation, but our studies show that ATG8.3 is atypical. To investigate the function of other ATG proteins, RNAi compatible *T. brucei* were modified to function as autophagy reporter lines by expressing only either YFP-ATG8.1 or YFP-ATG8.2. In the insect procyclic lifecycle stage, independent RNAi down-regulation of ATG3 or ATG7 generated autophagy-defective mutants and confirmed a pro-survival role for autophagy in the procyclic form nutrient starvation response. Similarly, RNAi depletion of ATG5 or ATG7 in the bloodstream form disrupted autophagy, but did not impede proliferation. Further characterisation showed bloodstream form autophagy mutants retain the capacity to undergo the complex cellular remodelling that occurs during differentiation to the procyclic form and are equally susceptible to dihydroxyacetone-induced cell death as wild type parasites, not supporting a role for autophagy in this cell death mechanism. The RNAi reporter system developed, which also identified TOR1 as a negative regulator controlling YFP-ATG8.2 but not YFP-ATG8.1 autophagosome formation, will enable further targeted analysis of the mechanisms and function of autophagy in the medically relevant bloodstream form of *T. brucei*.

***Keywords**: trypanosome, protozoan, autophagy, differentiation, cell death.*

INTRODUCTION

Autophagy is a primary intracellular mechanism for the degradation of long lived proteins and organelles. It describes multiple distinct pathways that all utilise the lysosome (or vacuole) to degrade and recycle unwanted cellular components. Of these pathways, the major form is macroautophagy (hereafter referred to as autophagy), which is characterised by the engulfment of cargo destined for the lysosome into a double membrane bound vesicle called an autophagosome [1]. Autophagy was once regarded solely as a stress response mechanism enabling cells to withstand nutrient starvation; however it is now clear that it participates in wide ranging cellular processes influencing aspects of growth, development, differentiation and immunity [2]. The dramatic expansion in the known functions of autophagy stems largely from the improved definition of the pathway's molecular mechanisms achieved through extensive study of model systems such as yeast. This has led to the identification of multiple autophagy related genes (ATGs) that control and execute a complex process broadly summarised by the following principle stages - induction, autophagosome formation and degradation.

Autophagy is subject to tight regulation to ensure that it functions optimally. The signalling pathways that control autophagy induction are organism specific, but typically function through the protein kinase, Atg1, and its interacting partners [3]. Post induction, the isolation membrane expands to engulf and eventually surround the cargo in a mature autophagosome [4]. This process requires the coordinated recruitment of multiple factors, including components of two ubiquitin-like conjugation systems (Atg8-phosphatidylethanolamine (PE) and Atg12-Atg5) that participate in membrane elongation and autophagosome completion [5–7]. Atg12 is a small ubiquitin-like protein (UBL) that covalently links to Atg5 in a process requiring Atg7 and Atg10. Post conjugation, Atg16 associates with Atg5, stimulating homo-multimer formation, which facilitates membrane binding and Atg8 recruitment [8,9]. During membrane expansion the UBL Atg8 is uniquely conjugated to PE, an abundant membrane phospholipid [6]. This process requires the concerted action of Atg4, Atg7 and Atg3, and produces a lipidated protein that tightly associates with membrane. The Atg8-PE that decorates the external surface of the mature autophagosome is removed by the proteolytic action of Atg4, whereas internal Atg8-PE remains associated [10]. To recycle the autophagosome contents, the external autophagosome membrane fuses with the lysosome and its cargo, still bound by the inner membrane, is then degraded.

In silico screening of genome databases for ATG genes has indicated that autophagy is widely conserved amongst eukaryotes, and this has been confirmed experimentally with some species including evolutionary divergent parasitic protozoa, such as the kinetoplastids and apicomplexans [11–14]. *Trypanosoma brucei* is a parasitic protozoon, transmitted by the tsetse fly, which causes Human and Animal African Trypanosomiasis (HAT and AAT). During its life cycle, the parasite progresses through a number of developmental stages within the insect and mammalian hosts. Long slender bloodstream form (BSF) trypomastigotes replicate within the mammalian host, whilst procyclic form (PCF) trypomastigotes replicate within the tsetse mid gut. The short stumpy bloodstream trypomastigote is an intermediate transition life cycle stage that is transmission competent. The *T. brucei* genome contains identifiable genes encoding for approximately half of the yeast Atg proteins, including components of both the ATG8 and ATG12 conjugation pathways [11–13].

Investigation of autophagy in protozoan parasites has revealed that autophagy is central to fundamental processes such as cellular differentiation, virulence and the starvation response (reviewed in [15]). In addition to these pro-survival functions, autophagy has also been linked to a form of regulated cell death, known as autophagic cell death, in some protozoan parasites [16,17]. These findings are controversial, however, and detailed molecular analyses are required to identify whether activation of autophagy has occurred to orchestrate cell death or is merely a response to cell stress coincident with cell death [18]. Accordingly, it is important to elucidate whether or not autophagy is an active death mechanism; for instance, by using targeted genetic or chemical manipulation of the autophagy pathway and determining if this delays or abolishes the cell death phenotype under investigation.

Eukaryotic Target of Rapamycin (TOR) regulates diverse cellular events, most notably, in the context of this study the induction of autophagy. TOR functions through two multi-component complexes to control distinct aspects of cellular growth; TOR complex 1 (TORC1), which regulates temporal cell growth and TOR complex 2 (TORC2), which controls spatial cell growth [19]. Rapamycin is a macrolide widely used to induce autophagy and the compound's selective inhibition of TORC1 suggests this is the complex responsible for autophagy control [20,21]. *T. brucei* has both TORC1 and TORC2 complexes, but it has been reported that only TORC2 is sensitive to rapamycin and thus that rapamycin is not suitable for inducing autophagy in *T. brucei* [22].

In order to resolve some of these confounding issues we attempted a comprehensive analysis of *T. brucei* autophagy, focusing on both mechanism and physiological function. Confirming *T. brucei* ATG8.1 and ATG8.2 as *bone fide* ATG8-like proteins informed the subsequent development of transgenic *T. brucei* that combined autophagy-specific fluorescent reporter systems with RNAi silencing technology. Through RNAi targeting of genes predicted to function in the *T. brucei* autophagy pathway we were able to generate loss of function autophagy mutants. In PCF this identified a role for autophagy facilitating survival during nutrient starvation, but not in executing specific autophagic cell death events. In contrast, disruption of BSF autophagy did not lead to detectable phenotypes in cellular proliferation or differentiation.

RESULTS

Autophagy in procyclic form *T. brucei*

T. brucei has 3 *ATG8*-like genes, *ATG8.1*, *ATG8.2* and *ATG8.3* (TritrypDB gene ID: Tb927.7.5900, Tb927.7.5910 and Tb927.7.3320 respectively). Structural analysis of ATG8.2 showed the protein contains a classical ubiquitin-like fold and displays significant structural homology to *bone fide* ATG8s, including conservation of key functional residues required for Atg4 interaction and lipidation as identified in yeast [23]. The *T. brucei* ATG8.1 and ATG8.2 proteins share 82% amino acid identity (the only divergence occurring at their N-termini (Fig. S1)) and are both incorporated into autophagosomes in response to nutrient depletion of cultured procyclic form (PCF) *T. brucei* [16]. However the role of ATG8.3 remains unclear, as despite good sequence identity to ATG8.1 and ATG8.2, a large insertion is predicted to interfere with the ubiquitin fold prohibiting interaction with ATG4 (Fig. S1)[23].

To investigate the 3 *T. brucei ATG8*-like genes we engineered PCF cell lines to conditionally express the proteins as N-terminal yellow fluorescent protein (YFP) fusions under the control of a tetracycline (tet) inducible promoter. The tet-induced expression of each fusion protein was verified by western blotting using anti-ATG8.1, an affinity purified polyclonal antibody raised against recombinant *T. brucei* ATG8.1 and anti-GFP, showing that the proteins were expressed at their predicted molecular masses (Fig. 1A). Only YFP-ATG8.1 and YFP-ATG8.2 were recognised by anti-ATG8.1, whilst YFP-ATG8.3 expression was only detected with anti-GFP antibody. While the anti-ATG8.1 confirmed that endogenous ATG8 expression occurred in both BSF and PCF wild type cell lines (Fig.1B), the similarity in the ATG8 predicted molecular masses (ATG8.1, 13.3 kDa; ATG8.2, 13.8 kDa; ATG8.3, 15.3 kDa), and the selective cross reactivity of anti-ATG8.1, prohibited identification of the endogenous ATG8 isoforms. Anti-ATG8.1 antibody also detected an unknown protein of approximately 40 kDa in both lifecycle stages (Fig. 1B).

Having established tight conditional expression of the ATG8 fusion proteins fluorescent microscopy was used to determine whether any characteristic ATG8 protein properties were evident. During standard nutrient rich culture YFP-ATG8.1 and YFP-ATG8.2 were predominantly distributed throughout the cytoplasm with punctate structures observed in a minority of cells. However, subjecting the tet induced cells to nutrient starvation, a widely used inducer of autophagy, significantly increased the number of observed puncta; the mean number of YFP-ATG8.1 puncta per cell increased from 0.22 (±0.03) to 1.30 (±0.25) and YFP-ATG8.2 puncta per cell increased from 0.13 (±0.07) to 0.87 (±0.08) (Figs 1C and 1D). These findings are in good agreement with data published previously showing autophagosome formation in PCF trypanosomes [16]. In contrast to the study by Li *et al.* [16] glucose supplementation of the starvation media was not required for puncta formation.

YFP-ATG8.3 was markedly different to YFP-ATG8.1 and YFP-ATG8.2, with comparatively low levels of puncta formation observed (Figs 1C and 1D). During standard culture conditions the mean puncta per cell was 0.02 (± 0.01), more than 10- and 5-fold lower than YFP-ATG8.1 and YFP-ATG8.2, respectively. Nutrient starvation did increase the observed puncta per cell to 0.04 (±0.03), however the

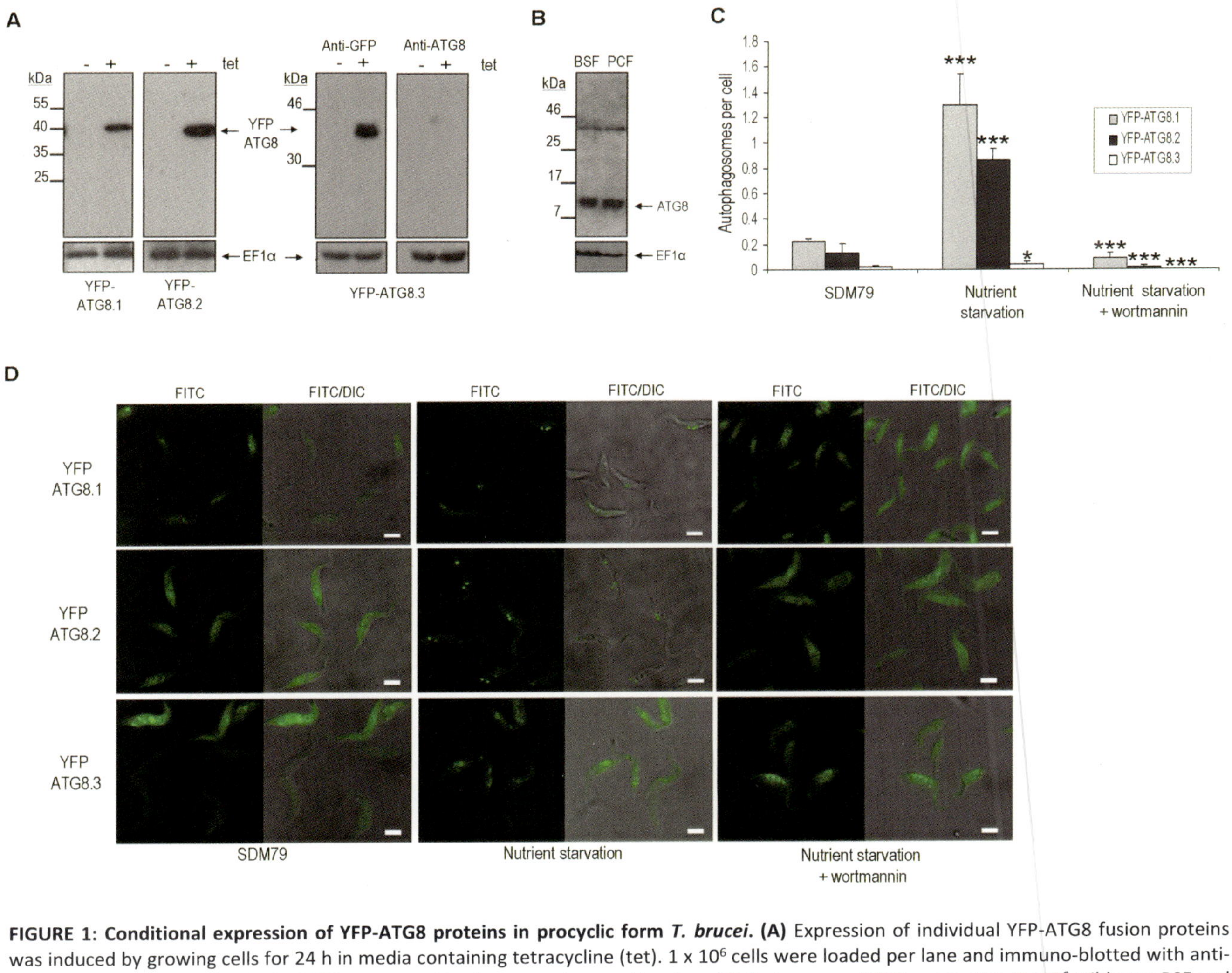

FIGURE 1: Conditional expression of YFP-ATG8 proteins in procyclic form *T. brucei*. (A) Expression of individual YFP-ATG8 fusion proteins was induced by growing cells for 24 h in media containing tetracycline (tet). 1 x 10^6 cells were loaded per lane and immuno-blotted with anti-ATG8.1 or anti-GFP antibody. Anti-EF1α was used to demonstrate equal loading. **(B)** Endogenous ATG8 expression. 5 x10^6 wild type BSF and PCF cells loaded per lane, were probed with anti-ATG8.1. Anti-EF1α was used to demonstrate equal loading. **(C)** Fluorescent microscopy was used to monitor the presence of autophagosomes in PCF cells conditionally expressing YFP fusion proteins following growth in SDM79 media, starvation in PBS for 2.5 h or 2.5 h nutrient starvation in PBS supplemented with 10 μM wortmannin. The mean number of autophagosomes per cell was determined by counting >200 cells with data displayed as a mean of three replicate experiments from three independent clones. Error bars represent standard deviation and asterisks indicate where data differed significantly from the mean of the non-starved controls *p<0.05, ***p<0.001. **(D)** Representative images of cells assessed as for C. Left hand images FITC filter set, right hand images DIC FITC merge. Scale bar 5 μm.

magnitude of change was far lower than for YFP-ATG8.1 and YFP-ATG8.2. To provide more evidence for a link between puncta formation and the autophagy pathway, the effect of supplementing the nutrient starvation buffer with wortmannin, a PI3 kinase inhibitor and established autophagy inhibitor [16], was monitored (Figs 1C and 1D). As expected this reduced the appearance of puncta; effectively abolishing YFP-ATG8.1 and YFP-ATG8.3 puncta formation and significantly decreasing YFP-ATG8.2 puncta. These findings, in combination with those of Li *et al.* [16], provide convincing evidence that ATG8.1 and ATG8.2 are incorporated into autophagosomes and are therefore true ATG8-like proteins.

RNAi ablation of autophagy in procyclic form *T. brucei*

To provide functional insights into *T. brucei* autophagy the YFP-ATG8 expression constructs were modified to operate in conjunction with RNAi compatible *T. brucei* cell lines [24,25]. To simplify the *T. brucei* autophagy reporter system, a plasmid designed for constitutive ectopic expression in *T. brucei* [26] was generated by switching the drug resistance to BSD and then inserting the YFP-ATG8.1 and –ATG8.2 ORFs into the expression site. The new constructs, were individually transfected into PCF 29-13 cells [24] already containing 2T7^{ti} RNAi constructs [27] targeting the *T. brucei* orthologues of either *ATG3* (Tb927.2.1890) or *ATG7* (Tb927.10.11180). These genes were chosen as initial proof of concept targets based on presumed essentiality to the autophagy pathway and high confidence bioinformatic predictions indicating single copy orthologues in the *T. brucei* genome [11,13,15].

Microscopic analysis of the PCF cells confirmed that the constitutively expressed YFP-ATG8.1 and YFP-ATG8.2 fusion proteins were functional and formed autophagosomes

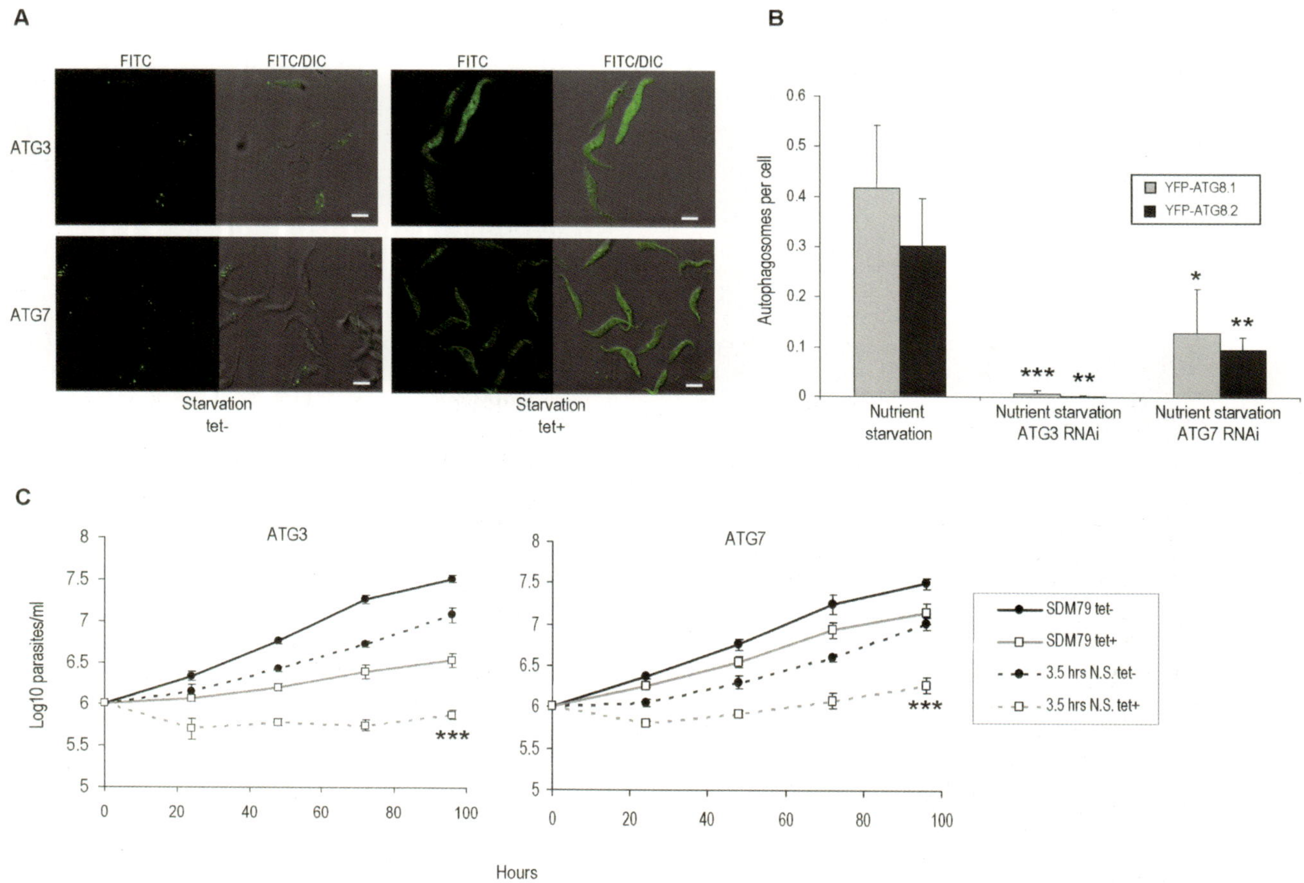

FIGURE 2: Procyclic form *ATG3* and *ATG7* RNAi mutants. Constitutive expression of YFP-ATG8.1 and YFP-ATG8.2 in PCF was visualised by fluorescent microscopy following RNAi of either *ATG3* or *ATG7* for 72 h before nutrient starvation in PBS for 2.5 h. **(A)** Representative images, left hand images FITC filter set, right hand images DIC FITC merge. Scale bar 5 µm. **(B)** The mean number of autophagosomes per cell was determined by counting >200 cells with data displayed as a mean of three replicate experiments. Error bars represent standard deviation and asterisks indicate where data differed significantly from the means of uninduced controls *$p<0.05$, ** $p<0.01$, ***$p<0.001$. **(C)** *ATG3* and *ATG7* RNAi was induced for 72 h in cells expressing YFP-ATG8.1 before nutrient starvation for 3.5 h in PBS (dashed lines) or control treatment in SDM79 media (solid lines). Cells were re-seeded at 0 h in SDM79 + tet (empty boxes grey line, tet+) and SDM79 –tet (filled boxes black line, tet-) and growth was determined daily for 96 h. Data represent mean of three replicate experiments. Error bars show standard deviation and asterisks indicate where data differed significantly from the means of tet induced starved and non-starved lines at the 96 h time point ***$p<0.001$.

during nutrient starvation (Figs 2A and 2B). RNAi depletion of both ATG3 and ATG7 orthologues for 72 h significantly decreased the number of autophagosomes observed per cell following starvation (Figs 2A and 2B). Although cells depleted of ATG3 were almost entirely devoid of puncta after starvation, the effect of *ATG7* RNAi was less severe. This difference could be linked to RNAi penetration, with real time PCR confirming post RNAi depletion of *ATG7* transcript occurs to a lesser extent than *ATG3* (Fig. S2). However, it could also reflect varying half-lives of the target proteins, their different roles in the autophagy pathway or the presence of additional, and as yet unknown, ATG7 orthologues. Nevertheless, these findings provided additional experimental evidence supporting the identification of the *T. brucei* ATG3 and ATG7 orthologues [13,16] and created PCF autophagy mutants for detailed loss of function analysis.

Independent induction of *ATG3* or *ATG7* RNAi impaired cell growth of PCF cultured in normal media (Fig. 2C). The expression of YFP-ATG8.1 or YFP-ATG8.2 did not appear to influence the RNAi growth phenotypes, as RNAi in the PCF 29-13 cells produced similar results (data not shown). In *T. brucei* PCF RNAi autophagy mutants have previously been used to implicate the autophagy machinery in an active cell death process during prolonged *in vitro* PCF nutrient deprivation [16]. However, in this study disruption of PCF autophagy by 72 h RNAi of *ATG3* or *ATG7*, significantly diminished the capacity of the parasites to withstand and recover from acute nutrient starvation (Fig. 2C).

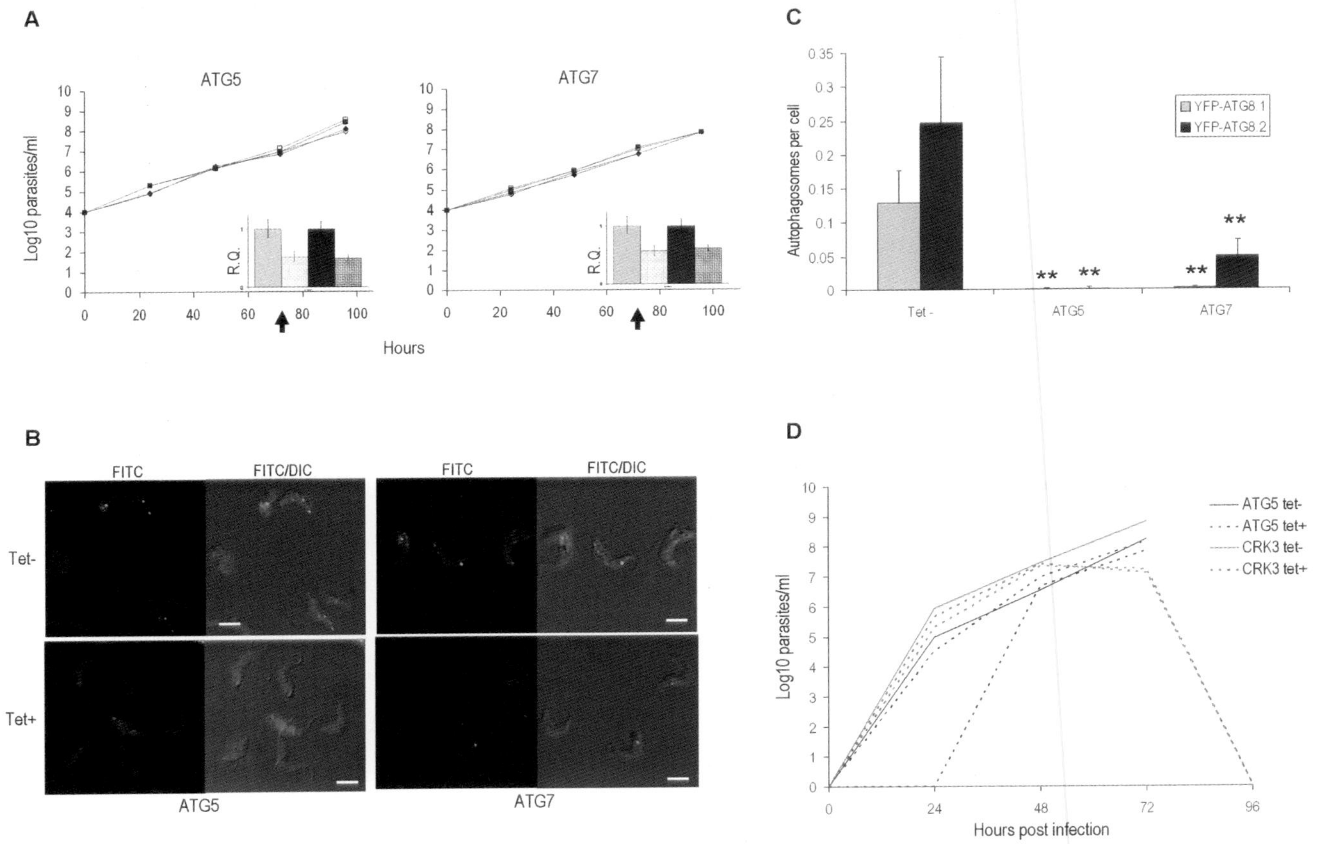

FIGURE 3: RNAi of autophagy genes in bloodstream form. (A) Growth analysis of BSF RNAi mutants *in vitro*. RNAi of *ATG5* and *ATG7* genes in BSF 2T1 constitutively expressing YFP-ATG8.1 (diamonds) and YFP-ATG8.2 (squares). RNAi induced with tetracycline (filled symbols) and growth compared to controls (empty symbols) for 96 h. Cells were reseeded to 1 x 10^4 ml^{-1} at 48 h as required, with cumulative values shown. Arrows indicate time point selected for microscopy analysis. Inset: qPCR of *ATG5* or *ATG7* cell lines 72 h after tetracycline induction (hatch) or control in BSF 2T1 clones expressing either YFP-ATG8.1 (grey) or YFP-ATG8.2 (black). Error bars represent one standard deviation derived from three replicates. **(B)** RNAi depletion of *ATG* genes influences autophagy. Constitutive expression of YFP-ATG8.1 and YFP-ATG8.2 in BSF 2T1 cells was visualised by fluorescent microscopy following individual RNAi of *ATG5* and *ATG7*. Left hand images FITC filter set, right hand images FITC DIC merge. Scale bar 5 µm. **(C)** The mean number of autophagosomes per cell was determined in *ATG5* and *ATG7* RNAi lines by counting >200 cells 72 h after induction, with data displayed as a mean of three replicate experiments. Error bars represent standard deviation and asterisks indicate where data differed significantly from the mean of the un-induced controls $**p<0.01$. **(D)** Growth analysis of BSF *ATG5* and *CRK3* RNAi mutants *in vivo*. 1 x 10^5 (*ATG5*) and 5 x 10^5 (*CRK3*) trypanosomes were inoculated in 3 mice and RNAi induced with doxycycline (Dox) in 2 mice 24 h (*ATG5*) or 48 h (*CRK3*) later. Parasitaemia was monitored by tail bleed and counting on a haemocytometer.

A B

ATG5 RNAi	IC50 (mM)	Standard Deviation
Tet-	1.32	0.29
Tet+	1.29	0.04

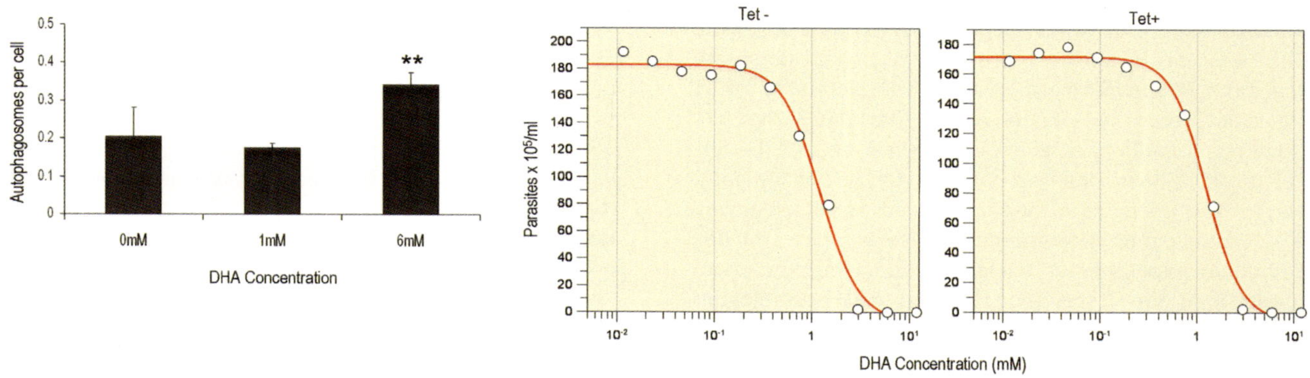

FIGURE 4: Autophagy is not required for dihydroxyacetone (DHA) induced bloodstream form cell death. (A) DHA increases autophagosome numbers in BSF 2T1 cells. BSF 2T1 cells expressing YFP-ATG8.2 were grown with 1 mM and 6 mM DHA for 20 h. The mean number of autophagosomes per cell was determined by counting >200 cells with data displayed as a mean of three replicate experiments and error bars represent standard deviation and asterisks indicate where data differed significantly from the mean of the untreated controls **$p<0.001$. **(B)** The IC_{50} of DHA was determined for BSF 2T1 YFP-ATG8.2 following 48 h *ATG5* RNAi induction (tet+) and un-induced control cells (tet-). Parasites were cultured with serially diluted DHA for 48 h before determining cell density. Graphs are representative of three replicate experiments.

RNAi ablation of autophagy in bloodstream form *T. brucei*

Having effectively combined the YFP-ATG8 reporters with RNAi in PCFs we extended our analyses to the BSF of the parasite. The constitutive expression constructs for YFP-ATG8.1 and YFP-ATG8.2 were used to create RNAi autophagy reporter lines in the 2T1 BSF cell line [25]. The resulting parental 2T1 cell lines individually expressing YFP-ATG8.1 or YFP-ATG8.2, were therefore compatible with the stem loop RNAi vector pTL that had been specifically designed to facilitate rapid construct generation [28]. Initially, RNAi constructs were designed to target the *T. brucei* ATG5 and ATG7 orthologues. RNAi depletion of either ATG5 or ATG7 did not impact cell growth over 96 h (Fig. 3A), and therefore 72 h post induction was validated as an appropriate time point for autophagosome analysis with RNAi specificity confirmed by quantitative RT-PCR (Fig. 3A). As predicted, RNAi depletion of both ATG5 and ATG7 significantly decreased the number of autophagosomes observed (Fig. 3B and 3C). However, RNAi of *ATG5* was the most effective at inhibiting puncta formation and was therefore selected for the subsequent loss of function analyses (Figs 3D, 4 and 5). To determine whether autophagy might contribute to parasite growth or survival in a more stringent *in vivo* environment, mice were infected with YFP-ATG8.2 *ATG5* RNAi parasites and RNAi was induced 24 h later by supplementing a subset of the mice's drinking water with doxycycline. This protocol allows clearance of parasites from the blood within 48 h for essential genes, such as the kinase *CRK3* (Figs 3D and S3) [28]. However, no difference in parasitaemia was observed between control and induced *ATG5* RNAi cell lines (Fig. 3D), suggesting that under the conditions tested autophagy was dispensable for the short term growth of monomorphic 2T1 BSF.

Having established that autophagy disruption does not impact BSF growth, we sought to investigate the response of ATG5 autophagy mutants to dihydroxyacetone (DHA). DHA is a trypanocidal compound that has been suggested to function by stimulating autophagic cell death; evidence for this is electron micrograph detection of vesicles reminiscent of autophagosomes post DHA treatment [29]. 1 mM DHA did not affect the number of observed YFP-ATG8.2 labelled autophagosomes, whereas 6 mM produced a mild yet significant increase in the number of autophagosomes per cell (Fig. 4A). To see whether the autophagy machinery was actively required for DHA-induced cell death, the IC_{50} for DHA against cells with and without a functional autophagy pathway were determined by inducing *ATG5* RNAi 48 h pre-DHA incubation. This showed that there is no significant difference between the IC_{50} values for the un-induced control and ATG5 autophagy mutants (Fig. 4B), indicating that the autophagy machinery is not required for DHA-induced cell death in trypanosomes.

In *Leishmania* species autophagy facilitates lifecycle differentiation and thereby contributes to parasite virulence [12,30]. Monomorphic BSF 2T1 parasites are not able to progress through the entire *T. brucei* lifecycle, however, stimulation with citrate/cis-aconitate and reduced incubation temperature to 27°C initiates differentiation to viable PCF parasites [31]. Accordingly, autophagy was disrupted with 48 h of *ATG5* RNAi induction before stimulation of differentiation and analysis of characteristic PCF markers. The dynamics of EP-procyclin expression and repositioning

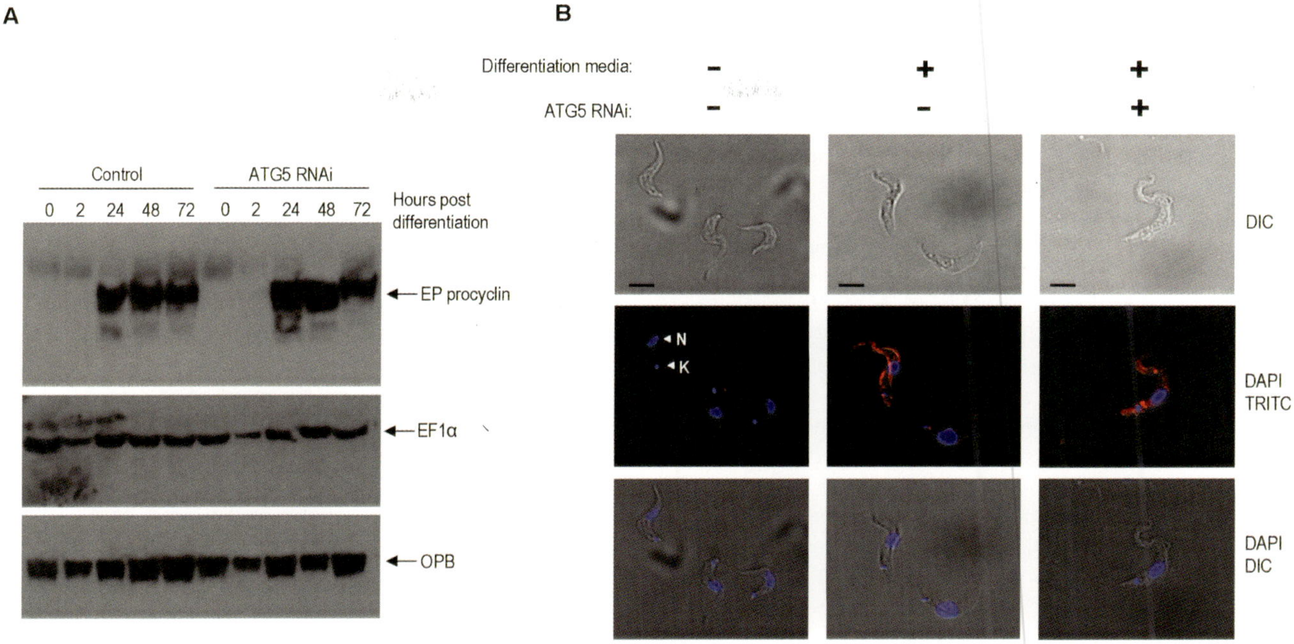

FIGURE 5: Differentiation in bloodstream form RNAi autophagy mutants. Differentiation of BSF YFP-ATG8.2 *ATG5* RNAi mutants (48 h after tet induction) was stimulated by switching cells to differentiation media and incubating at 27°C. Cells were analysed for characteristic PCF markers such as EP-procyclin expression and nuclear/kinetoplast re-configuration. **(A)** Western blot of whole cell extracts taken at specific times after induction of differentiation and probed with EP-procyclin, anti-EF1α and anti-OPB. **(B)** Fluorescent microscopy of cells 24 h post induction, fixed and labelled with anti-EP procyclin (red) and DAPI (blue) to show changes in the nuclear configuration (nucleus, N and kinetoplast, K). Scale bar 5 μm.

of the kinetoplast in respect to the nucleus was unaffected by *ATG5* RNAi (Fig. 5) indicating that autophagy is not required for BSF to PCF differentiation.

Regulation of autophagy in BSF *T. brucei* by TOR

Although *T. brucei* contains an expanded TOR family (TOR1-4 [32]), rapamycin treatment of both BSF and PCF failed to significantly increase the number of YFP-ATG8.1 or YFP-ATG8.2 puncta (Fig. S4). This is in good agreement with findings for the PCF [16] and supports the atypical resistance displayed by the *T. brucei* TORC1 complex to rapamycin [22]. Nevertheless electron micrographs indicate that selective disruption of TORC1 through *TOR1* RNAi can stimulate autophagy in BSF parasites [22]. To investigate this observation, the 2T1 autophagy reporter lines were used to analyse YFP-ATG8.1 and YFP-ATG8.2 labelled autophagosome formation following independent RNAi depletion of the four *T. brucei TOR* genes (Figs 6A and 6B). RNAi of the *TOR* orthologues produced differing effects on cell growth over the 96 h time course; depletion of TOR2 caused rapid cell death, while TOR1 and TOR4 depletion lead to substantial albeit less severe growth defects, and TOR3 depletion did not affect cell growth (Fig. 6A). Given such diversity, the time point selected to assess autophagosome formation was determined individually for each RNAi target, balancing the maximum time for target protein depletion with cell integrity. Autophagosome analysis was therefore carried out ahead of detected growth defects or at 72 h if no growth defect was observed (arrow heads Fig. 6A). To ensure valid assessment and comparison, the down-regulation of RNAi target transcript and YFP-ATG8 fusion protein expression post induction were confirmed for each autophagy analysis time point (Figs 6A and S5).

In this study the only *T. brucei* TOR orthologue for which RNAi downregulation significantly increased the number of autophagosomes detected was TOR1, with RNAi increasing the YFP-ATG8.2 labelled autophagosomes per cell from 0.24 (±0.07) to 0.35 (±0.04) (Fig. 6B). However, only YFP-ATG8.2 was sensitive to *TOR1* RNAi, as no significant change to YFP-ATG8.1 autophagosome formation was observed, pointing towards the presence of independent mechanisms of regulation for the two ATG8 orthologues. Furthermore, depletion of TOR2, TOR3 and TOR4 did not induce autophagosome formation; in fact, the only other significant change in comparison to the basal autophagosome levels was a slight decrease in YFP-ATG8.2 labelled autophagosomes following *TOR2* RNAi (Fig. 6A). However, given the strong growth phenotype associated with *TOR2* RNAi further detailed investigation is required to reveal the biological significance of this finding.

DISCUSSION

The widely established role of autophagy in the nutrient starvation response of many eukaryotes led us to initially investigate PCF *T. brucei,* which unlike the BSF are able to withstand periods of nutrient starvation. Our findings support the assertion that autophagy participates in a nutrient starvation response [16], with both YFP-ATG8.1 and YFP-ATG8.2 showing dramatic reorganisation into autophagosomes post starvation (Fig. 1). ATG8.3, however, despite significant similarity to ATG8.1 and ATG8.2 proteins, does not appear to play a functionally analogous role as during both normal growth and nutrient starvation YFP-ATG8.3 remains almost entirely dispersed in the cytosol. It is possible that ATG8.3 functions as a *bone fide* ATG8-like protein, but only in processes outside of the PCF nutrient starvation response. Such functional selectivity has been reported for ATG8.1 and ATG8.2 [16] and also occurs for the expanded *L. major* ATG8 family, with specific proteins linked to distinct autophagy processes [33]. However, an alternative hypothesis is that ATG8.3 functions as an ATG12. In support of this hypothesis, ATG8.3 is the syntenic orthologue of LmjF22.1300, a protein that despite some ATG8-like features, functions as an *L. major* ATG12 [30]. Furthermore, in yeast, the Atg12-Atg5 complex is predominantly located in the cytoplasm and associates only to the isolation membranes and not with mature autophagosomes [4]. Accordingly, the limited YFP-ATG8.3 puncta formation that occurs in response to PCF nutrient starvation in comparison to YFP-ATG8.1 or YFP–ATG8.2 could reflect a similar phenomenon. In the absence of definitive *in vivo* and *in vitro* biochemical data caution is still required with regard to definitively ascribing a function to ATG8.3 in *T. brucei.*

Although the application of molecular level analyses has begun to shed light on the mechanisms of *T. brucei* autophagy, its physiological role(s) remain largely unknown [16,18]. To help address this issue we have developed and now validated BSF *T. brucei* autophagy reporter cell lines, in an RNAi compatible genetic background, that can be used to interrogate genes predicted to be involved in the autophagy pathway. The effectiveness of this system was

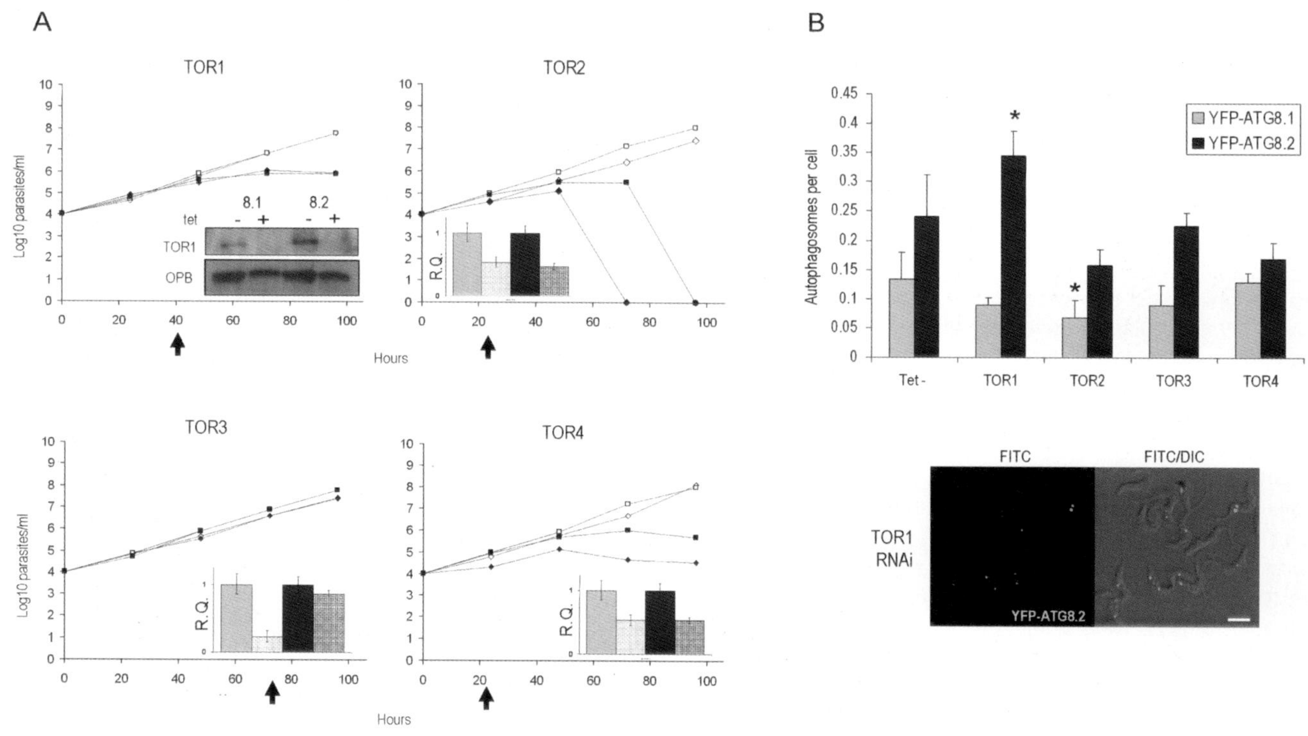

FIGURE 6: RNAi of *TOR* genes in bloodstream form. (A) Growth analysis of BSF RNAi mutants in vitro. RNAi of *TOR* genes in BSF 2T1 constitutively expressing YFP-ATG8.1 (diamonds) and YFP-ATG8.2 (squares). RNAi induced with tetracycline (filled symbols) and growth compared to controls (empty symbols) for 96 h. Cells were reseeded to 1 x 10^4 ml^{-1} at 48 h as required, with cumulative growth shown. Arrows indicate time point selected for microscopy analysis following. Inset: confirmation of RNAi specificity by western blot (where antibody available, TOR1 [22]) or by qPCR of TOR RNAi cell lines at selected analysis time points after tetracycline induction (hatch) or control in BSF 2T1 clones expressing either YFP-ATG8.1 (grey) or YFP-ATG8.2 (black). Error bars represent one standard deviation derived from three replicates. **(B)** The mean number of autophagosomes per cell was determined in *TOR* RNAi lines by counting >200 cells at selected times after induction (arrows in A), with data displayed as a mean of at least three replicate experiments. Error bars represent standard deviation and asterisks indicate where data differed significantly from the mean of the un-induced controls *$p<0.05$. Constitutive expression of YFP-ATG8.2 in BSF 2T1 cells was visualised by fluorescent microscopy following individual RNAi of *TOR1*. Left hand images FITC filter set, right hand images FITC DIC merge. Scale bar 5 μm.

demonstrated through the *in vivo* study of the *T. brucei* ATG5, ATG7 and the TOR family. Depletion of ATG5 following induction of *ATG5*-specific RNAi led to disruption of autophagosome formation, but this did not impact cell growth or the ability to differentiate under the conditions tested. The absence of detectable growth or differentiation phenotypes for the *ATG5* RNAi mutant could, in part, be due to the highly virulent monomorphic BSF 2T1 line. The rapid growth rate of these cells might obscure the detection of subtle autophagy-linked defects under the experimental conditions tested. Future studies would benefit from creating *ATG5* gene deletion mutants, or *ATG5* RNAi in pleomorphic cell lines to enable the role of autophagy to be studied over the complete parasite life-cycle.

Analysis of the *T. brucei* TOR family successfully identified TOR1 as a negative regulator of autophagy in monomorphic BSF; controlling YFP-ATG8.2 puncta formation but not YFP-ATG8.1. However, further investigation is required to identify whether this discrimination relates to the potential hierarchical relationship of ATG8.1 and ATG8.2 in *T. brucei* autophagosome formation proposed to occur in the procyclic form [16] or reflects the existence of different signalling pathways that coordinate discreet subsets of autophagosomes. Interestingly, TOR4 has recently been shown to negatively regulate the differentiation of long slender BSF into the short stumpy stage pre-adapted for transmission to the insect host [32]. However, in this analysis RNAi of *TOR4* did not increase the formation of autophagosomes (Fig. 6), although further analysis at additional time points might reveal a role for autophagy in this differentiation process.

The participation of autophagy in the PCF nutrient starvation response is consistent with the complex and prolonged differentiation processes that occur within the tsetse fly vector [34]. To ensure successful transmission to a mammalian host, the developing PCF parasites must be able to survive fluctuating nutrient stress attributed to inconsistent feeding patterns of tsetse flies in the wild [35]. It has been speculated that programmed cell death pathways, such as autophagic cell death, could function in altruistic mechanisms intended to limit the parasite burden placed upon the host, thereby promoting transmission of surviving parasites [16]. However, in the absence of conclusive evidence describing such pathways in *T. brucei*, it remains a distinct possibility that, as for many other eukaryotes, autophagy's primary function is pro-survival [18]. Indeed, the reduced capacity of PCF autophagy mutants to recover from acute periods of nutrient starvation (Fig. 2), hints at how such a process could facilitate PCF survival during nutrient stress in the tsetse fly. Furthermore, the redundancy of the established BSF autophagy pathway during DHA-induced death (Fig. 4), highlights the difficulties in confidently attributing the autophagic machinery to an active cell death process and we believe that such an involvement has not been demonstrated to date for trypanosomatids [18].

MATERIALS AND METHODS

Parasite culture and genetic manipulation

Trypanosoma brucei Lister (strain 427) were grown in culture as bloodstream form (BSF) and procyclic form (PCF) parasites. BSF were maintained at 37°C in 5 % carbon dioxide, with wild type grown in HMI-9 media supplemented with 10 % heat-inactivated foetal calf serum (FCS), 10 % serum plus and 0.5 µg ml^{-1} penicillin-streptomycin (PS) (Sigma) and BSF 2T1 [25] grown in HMI-11 supplemented with tetracycline free FCS and 0.5 µg ml^{-1} PS. Wild type PCF and PCF 29-13 [24] *T. brucei* were grown at 27°C in 5 % carbon dioxide in SDM-79 media supplemented with 10 % FCS and 0.5 µg ml^{-1} PS (Sigma). Cell densities were determined by counting the number of cells in 10 µl of culture using an Improved Neubauer haemocytometer counting chamber (Weber Scientific).

For transfection, mid-log BSF (1 x 10^{7}) or PCF (3 x 10^{7}) cells were resuspended in 100 µl Human T Cell Nucleofector Solution (Lonza), transferred to a cuvette loaded with 10 µg of linearised DNA and pulsed on programme X-001 using the Human T Cell Nucleofector machine (Amaxa). The parasites were recovered overnight in HMI9, HMI11 or SDM79, respectively, before limited dilution cloning in appropriate media containing antibiotics for selection. BSF *T. brucei* 427: 5 µg ml^{-1} hygromycin B (Calbiochem), 2.5 µg ml^{-1} G418 (Calbiochem), 2.5 µg ml^{-1} phleomycin (InvivoGen) and 10 µg ml^{-1} blasticidin (Calbiochem). PCF *T. brucei* 427: 50 µg ml^{-1} hygromycin B (Calbiochem), 10 µg ml^{-1} G418 (Calbiochem), 10 µg ml^{-1} zeocin (Calbiochem) and 20 µg ml^{-1} blasticidin (Calbiochem).

RNAi was induced by adding 1 µg ml^{-1} tetracycline to cells at 1 x10^{4} ml^{-1} (BSF) or 5 x 10^{5} ml^{-1} (PCF) and the growth rate was compared to identical non-induced control lines. Cells were passaged every 24 h or 48 h (BSF) or 72 h (PCF) with tetracycline added to induced cells. For *in vivo* growth analysis 1 × 10^{5} BSF parasites (*ATG5*) or 5 x 10^{5} BSF parasites (*CRK3*) from a donor mouse were inoculated into 3 ICR mice by intraperitoneal injection. After 24 h (*ATG5*) or 48 h (*CRK3*), the drinking water of two mice was supplemented with 0.2g L^{-1} doxycycline with 5% sucrose. The parasitaemia in all mice was monitored daily by haemocytometer cell count of mouse blood taken from a tail vein diluted in 0.83% ammonium chloride. At parasitaemias above 1x10^{8} cells per ml mice were humanely culled.

To monitor autophagy the number of YFP-ATG8 labelled autophagosomes per cell was determined by fluorescence microscopy of fixed cells. For each condition tested >200 cells were analysed, with data presented as a mean of at least three replicate experiments. Where required the conditional expression of YFP-ATG8 in PCF was induced by overnight incubation with 1 µg ml^{-1} tetracycline before further analysis. For nutrient starvation of PCF, the cells were washed twice and incubated in PBS warmed to 27°C, supplemented with wortmannin (Sigma) or 1 µg ml^{-1} tetracycline, as required.

For PCF starvation recovery assays, *ATG3* and *ATG7* RNAi were induced in cell lines expressing YFP-ATG8.1 for 72 h. The induced and control cells were washed twice in PBS and then either starved in PBS for 3.5 h or returned to SDM79, both supplemented with tetracycline as required. Cells were seeded at 1 x 10^{6} ml^{-1} into fresh SDM79, with tetracycline as required, and growth was determined every 24 h by haemocytometer cell count for 96 h post starvation or mock SDM79 treatment.

For the phenotypic analysis of BSF autophagy mutants, *ATG5* RNAi was induced in cells expressing YFP-ATG8.2 for 48 h to disrupt autophagy. To determine the IC_{50} of dihydroxyacetone (Sigma D107204) induced and control cells were counted, adjusted to 3 x 10^4 ml^{-1} in HMI-11 supplemented as required with 2 tetracycline μg ml^{-1} and 100 μl was added across a 96 well plate. Dihydroxyacteone was prepared to 24 mM in HMI-11 and 100 μl was serially diluted across the plate, with media only added to the last column. After 48 h cells were resuspended by gentle pipetting and growth determined by haemocytometer cell count of at least two replicate rows per treatment. The mean IC_{50} values were derived from three experimental replicates. To assess differentiation induced and control cells were counted, washed twice and adjusted to in 5 x 10^6 ml^{-1} in PCF differentiation media (SMD79, 10 mM glycerol, 6 mM cis-aconitate) and incubated at 27°C.

Plasmid Construction

The plasmids used for the ectopic expression of epitope tagged genes are described in detail by [36]. For the conditional expression of YFP-ATG8 fusion proteins *ATG8.1* (Tb927.7.5900) was amplified from genomic DNA (strain 427) by PCR using primers OL2621 (*Hind*III) and OL2622 (*Bam*HI); see table S1 for oligonucleotide sequences. The resulting fragment was cloned into p2628 [36], pre-digested with *Bam*-HI and *Hind*III. This created an N-terminal yellow fluorescent protein (YFP) tagged ATG8.1 expression construct, named pGL1843. Additional N-terminally tagged YFP expression constructs were prepared by following the same procedure for *T. brucei ATG8.2* (Tb927.7.5910), using primers OL2623 (*Hind*III) and OL2624 (*Bam*HI), producing pGL1835 and *T. brucei ATG8.3* (Tb927.7.3320), using primers OL2625 (*Hind*III) and OL2626 (*Bam*HI), producing pGL1836. Constructs were linearised with *Not*I for transfection. For constitutive ectopic expression of YFP-ATG8s, ORFs from pGL1843 and pGL1835 were cloned with *Hind*III and *Xho*I into a version of the *T. brucei* expression construct pGL2113 [26], that had been previously modified to change the drug resistance marker to BSD.

RNAi constructs. RNAi target fragments were identified using the TrypanoFAN: RNAit target selection script (http://trypanofan.path.cam.ac.uk/software/RNAit). For 2T7^{ti} [27] constructs products were cloned with *Bam*HI and *Hind*III using oligo pairs OL3668/OL3669 for *ATG3* (Tb927.2.1890) creating pGL2065 and OL3670/OL3671 for *ATG7* (Tb927.10.11180) creating pGL2066. To prepare the RNAi plasmids for transfection they were digested with *Not*I. For the stem-loop RNAi constructs, PCR amplified products were Gateway® cloned into pTL [28], an attP modified version of pRP^{iSL} [25]. This method was used to clone RNAi target fragments for: *ATG5* (Tb927.6.2430) with oligos TL0413 and TL0414 creating pTL209; *ATG7* (Tb927.10.11180) with oligos OL3810 and OL3811 creating pGL2171; and *TOR1* (Tb927.10.8420) with oligos TL0297 and TL0298 creating pTL149, *TOR2* (Tb927.4.420) with oligos TL0295 and TL0296 creating pTL148, *TOR3* (Tb927.4.800) with oligos TL0294 and TL0295 creating pTL147 and *TOR4* (Tb927.1.1930) with oligos TL0291 and TL0292 creating pTL146 [28]. The *CRK3* (Tb927.10.4990) stem-loop RNAi construct pGL1987 was created by cloning the PCR product amplified with oligos OL3332 and OL3333 into pRPa^{iSL} [25].

Antibodies

Antibodies were raised to recombinant *T. brucei* ATG8.1. *ATG8.1* was amplified from genomic DNA (strain 427) using primers OL2775 (*Nhe*I) and OL2776 (*Xho*I). The product was cloned into the pET-28a(+) *E. coli* protein expression vector, pre-digested with *Nde*I and *Xho*I, generating pGL2115. pGL2155 was transformed into *E. coli* BL21 Rosetta (Stratagene) and ATG8.1 was expressed in 300 ml Overnight Express TB Medium (Novagen). The pellet was lysed with Bacterial Protein Extraction Reagent (B-PER) (Pierce) supplemented with 10 μg ml^{-1} DNAse 1 (Sigma). Recombinant HIS tagged protein was purified from the soluble fraction by immobilised metal ion affinity chromatography on a column packed with Metal Chelate-20 (Poros). Peak fractions were pooled and buffer exchanged into 50 mM Tris pH 8.0 then adjusted to pH 5.5 and 150 mM NaCl added. Purified recombinant ATG8.1 was used as an antigen to produce polyclonal antiserum in a rabbit (Scottish National Blood Transfusion Service, Penicuik, Midlothian). Antibodies were affinity purified using AminoLink Coupling Gel (Pierce) as per manufacturer's guidelines.

Immunoblotting

Whole cell extracts were denatured by heating at 100°C for 4 min in LDS sample buffer (Invitrogen), and then electrophoresed on 12% or 15% (w/v) SDS-PAGE. Proteins were transferred to nitrocellulose membrane (Hybond-C, Amersham Biosciences) for western blotting. Primary antibodies rabbit anti-TbATG8.1, mouse anti-GFP (Living Colours), mouse anti-EF-1α (Millipore), rabbit anti-TbTOR1 [22] and sheep anti-oligopeptidase B [37] were used at 1:400, 1:800, 1:15000, 1:5000, respectively. HP-conjugated anti-mouse (Promega), anti-rabbit (Promega), and anti-sheep (Santa Cruz Biotechnology, Inc., Santa Cruz, CA) were used at 1:5000. Chemiluminescent detection was with the SuperSignal West Pico substrate or West Femto Maximum Sensitivity Substrate (Pierce).

Microscopy

For each slide ~1 x 10^5 *T. brucei* cells were washed twice in trypanosome dilution buffer (TDB) (20 mM Na_2HPO_4, 2 mM NaH_2PO_4, 80 mM NaCl, 5 mM KCl, 1 mM $MgSO_4$, 20 mM glucose, pH 7.4) before fixation in 1 % (w/v) paraformaldehyde (Sigma) on ice for 15 min (BSF) or 30 min (PCF). After two washes in TDB cells were resuspended in 20 μl TDB and evenly spread onto slides coated in 0.1 % poly-l-lysine (Sigma) and allowed to sediment. Where indicated DNA was visualised by staining with 1 μg ml^{-1} 4,6-Diamindino-2-phenylindole (DAPI) in mounting solution (PBS, 50 % (v/v) glycerol, 2.5 % (w/v) DABCO). All slides were fitted with a coverslip sealed with nail varnish.

The number of autophagosomes per cell was determined using a Zeiss Axioplan microscope. Mean autophagosomes per cell were derived from >200 cells from at least 3 replicate experiments. Representative fluorescent microscopy images were obtained using an Applied Precision DeltaVision Deconvolution microscope system fitted with a CoolSnap HQ camera. Image capture and microscope operation was performed using the software package SoftWoRx. Fluorescence was viewed using the FITC filter (λ_{Ex} 490 nm / λ_{Em} 528 nm), RD-TR-PE filter (λ_{Ex} 555 nm / λ_{Em} 617 nm), DAPI filter (λ_{Ex} 360 nm / λ_{Em} 457 nm) and reference images were obtained using the differential interference contrast (DIC) filter. Exposure time was set to 1 second with the transmission level of the neutral filter optimised for each sample and 60 x or 100 x oil immersion objectives were used. The Z-stacks obtained were deconvolved using the automatic conservative ratio setting before projection into one representative image. Merged images were created using Adobe Photoshop.

RNA isolation and quantitative PCR

RNA was isolated from 1.5 x 10^7 cells using the RNAeasy kit (Qiagen) processed by the QIAcube (Qiagen) according to manufacturer's instructions, with an added on column DNase - 1 (Sigma) digestion step. cDNA was produced from 1 µg of isolated RNA using random hexamer primers (Invitrogen) and SuperScript Reverse Transcriptase III (Invitrogen) according to manufacturers' instructions. Genomic DNA contamination was controlled for by omitting reverse transcriptase (-RT). Quantitative real time PCR (qPCR) analysis was performed on cDNA samples in reactions comprised of following: 1 µl of cDNA, 12.5 µl Power SYBR Green PCR Master Mix (Applied Biosystems), 3 µM primer 1, 3 µM primer 2 and 6.5 µl of water. PCR reactions were carried out in triplicate in sealed 96 well plates, using an Applied Biosystems 7500 Real Time PCR machine, with resulting data analysed on Applied Biosystems 7500 System Software. Primers used for real time PCR were designed using the Applied Biosystems Primer Express 3.0 programme and designed to amplify ~50 bp product specific to each target ORF outside of the RNAi target region.

Statistical analysis of data

Where indicated data were expressed as means ± standard deviation from the mean. P-values were calculated using an unpaired, two-tailed Student's T-test on Microsoft Excel. Differences were considered significant at a p-value of <0.05.

ACKNOWLEDGMENTS

This work was supported by an MRC studentship to WRP and MRC programme grants (G0700127 and MR/K019384/1) to JCM and GHC. The Welcome Trust Centre for Molecular Parasitology is supported by core funding from the Welcome Trust (085349). We thank Elaine Brown and Alana Hamilton for assistance culturing parasites and Miguel Navarro for the anti-TOR1 antibody.

CONFLICT OF INTEREST

The authors declare no conflict of interest.

REFERENCES

1. Klionsky DJ, Abdalla FC, Abeliovich H, Abraham RT, Acevedo-Arozena A, Adeli K, Agholme L, Agnello M, Agostinis P, Aguirre-Ghiso JA, Ahn HJ, Ait-Mohamed O, Ait-Si-Ali S, Akematsu T, Akira S, Al-Younes HM, Al-Zeer MA, Albert ML, Albin RL, Alegre-Abarrategui J, Aleo MF, Alirezaei M, Almasan A, Almonte-Becerril M, Amano A, Amaravadi R, Amarnath S, Amer AO, Andrieu-Abadie N, Anantharam V et al. (**2012**) Guidelines for the use and interpretation of assays for monitoring autophagy. **Autophagy** 8: 445-544.

2. Yang Z, Klionsky DJ (**2009**) An overview of the molecular mechanism of autophagy. **Curr Top Microbiol Immunol** 335: 1-32.

3. He C, Klionsky DJ (**2009**) Regulation mechanisms and signaling pathways of autophagy. **Annu Rev Genet** 43: 67-93.

4. Mizushima N, Yamamoto A, Hatano M, Kobayashi Y, Kabeya Y, Suzuki K, Tokuhisa T, Ohsumi Y, Yoshimori T (**2001**) Dissection of autophagosome formation using Apg5-deficient mouse embryonic stem cells. **J Cell Biol** 152: 657-667.

5. Mizushima N, Noda T, Yoshimori T, Tanaka Y, Ishii T, George MD, Klionsky DJ, Ohsumi M, Ohsumi Y (**1998**) A protein conjugation system essential for autophagy. **Nature** 395: 395-398.

6. Ichimura Y, Kirisako T, Takao T, Satomi Y, Shimonishi Y, Ishihara N, Mizushima N, Tanida I, Kominami E, Ohsumi M, Noda T, Ohsumi Y (**2000**) A ubiquitin-like system mediates protein lipidation. **Nature** 408: 488-492.

7. Suzuki K, Kubota Y, Sekito T, Ohsumi Y (**2007**) Hierarchy of Atg proteins in pre-autophagosomal structure organization. **Genes to Cells** 12: 209-218.

8. Mizushima N, Noda T, Ohsumi Y (**1999**) Apg16p is required for the function of the Apg12p-Apg5p conjugate in the yeast autophagy pathway. **EMBO J** 18: 3888-3896.

9. Romanov J, Walczak M, Ibiricu I, Schuchner S, Ogris E, Kraft C, Martens S (**2012**) Mechanism and functions of membrane binding by the Atg5-Atg12/Atg16 complex during autophagosome formation. **EMBO J** 31: 4304-4317.

10. Ohsumi Y, Mizushima N (**2004**) Two ubiquitin-like conjugation systems essential for autophagy. **Sem Cell & Dev Biol** 15: 231-236.

11. Rigden DJ, Michels PA, Ginger ML (**2009**) Autophagy in protists: Examples of secondary loss, lineage-specific innovations, and the conundrum of remodeling a single mitochondrion. **Autophagy** 5: 784-794.

12. Williams RAM, Tetley L, Mottram JC, Coombs GH (**2006**) Cysteine peptidases CPA and CPB are vital for autophagy and differentiation in *Leishmania mexicana*. **Mol Microbiol** 61: 655-674.

13. Herman M, Gillies S, Michels PA, Rigden DJ (**2006**) Autophagy and related processes in trypanosomatids: insights from genomic and bioinformatic analyses. **Autophagy** 2: 107-118.

14. Tomlins AM, Ben-Rached F, Williams RAM, Proto WR, Coppens I, Ruch U, Gilberger TW, Coombs GH, Mottram JC, Muller S, Langsley G (**2013**) *Plasmodium falciparum* Atg8 implicated in both autophagy and apicoplast formation. **Autophagy** 9: 1-13.

15. Duszenko M, Ginger ML, Brennand A, Gualdron-lopez M, Colombo MI, Coombs GH, Coppens I, Jayabalasingham B, Langsley G, Lisboa de Castro S, Menna-Barreto R, Mottram JC, Navarro M, Rigden DJ, Romana PS, Stoka V, Turk B, Michels PAM (**2011**) Autophagy in protists. **Autophagy** 7: 127-158.

16. Li FJ, Shen Q, Wang C, Sun Y, Yuan AY, He CY (**2012**) A role of autophagy in *Trypanosoma brucei* cell death. **Cell Microbiol** 14: 1242-1256.

17. Ghosh D, Walton JL, Roepe PD, Sinai AP (**2012**) Autophagy is a cell death mechanism in *Toxoplasma gondii*. **Cell Microbiol** 14: 589-607.

18. Proto WR, Coombs GH, Mottram JC (**2013**) Cell death in parasitic protozoa: regulated or incidental? **Nat Rev Microbiol** 11: 58-66.

19. Wullschleger S, Loewith R, Hall MN (**2006**) TOR Signaling in Growth and Metabolism. **Cell** 124: 471-484.

20. Zheng XF, Fiorentino D, Chen J, Crabtree GR, Schreiber SL (**1995**) TOR kinase domains are required for two distinct functions, only one

of which is inhibited by rapamycin. **Cell** 82: 121-130.

21. Kamada Y, Funakoshi T, Shintani T, Nagano K, Ohsumi M, Ohsumi Y (**2000**) Tor-Mediated Induction of Autophagy via an Apg1 Protein Kinase Complex. **J Cell Biol** 150: 1507-1513.

22. Barquilla A, Crespo JL, Navarro M (**2008**) Rapamycin inhibits trypanosome cell growth by preventing TOR complex 2 formation. **Proc Natl Acad Sci U S A** 105: 14579-14584.

23. Koopmann R, Muhammad K, Perbandt M, Betzel C, Duszenko M (**2009**) *Trypanosoma brucei* ATG8: Structural insights into autophagic-like mechanisms in protozoa. **Autophagy** 5: 1085-1091.

24. Wirtz E, Leal S, Ochatt C, Cross GA (**1999**) A tightly regulated inducible expression system for conditional gene knock-outs and dominant-negative genetics in *Trypanosoma brucei*. **Mol Biochem Parasitol** 99: 89-101.

25. Alsford S, Horn D (**2008**) Single-locus targeting constructs for reliable regulated RNAi and transgene expression in *Trypanosoma brucei*. **Mol Biochem Parasitol** 161: 76-79.

26. Myburgh E, Coles JA, Ritchie R, Kennedy PGE, McLatchie AP, Rodgers J, Taylor MC, Barrett MP, Brewer JM, Mottram JC (**2013**) In vivo imaging of trypanosome-brain interactions and development of a rapid screening test for drugs against CNS stage trypanosomiasis. **PLoS Neg Trop Dis** 7: e2384.

27. LaCount DJ, Bruse S, Hill KL, Donelson JE (**2000**) Double-stranded RNA interference in *Trypanosoma brucei* using head-to-head promoters. **Mol Biochem Parasitol** 111: 67-76.

28. Jones N.G, Thomas E.B, Brown E, Dickens NJ, Hammarton TC, Mottram JC (**2013**) Regulators of *Trypanosoma brucei* cell cycle progression and differentiation identified using a kinome-wide RNAi screen. **PLoS Path**, *in press*.

29. Uzcategui NL, Denninger V, Merkel P, Schoenfeld C, Figarella K, Duszenko M (**2007**) Dihydroxyacetone induced autophagy in African trypanosomes. **Autophagy** 3: 626-629.

30. Williams RAM, Smith TK, Cull B, Mottram JC, Coombs GH (**2012**) ATG5 is essential for ATG8-dependent autophagy and mitochondrial homeostasis in *Leishmania*. **PloS Path** 8: e1002695.

31. Ziegelbauer K, Quinten M, Schwarz H, Pearson TW, Overath P (**1990**) Synchronous differentiation of *Trypanosoma brucei* from bloodstream to procyclic forms *in vitro*. **Eur J Biochem** 192: 373-378.

32. Barquilla A, Saldivia M, Diaz R, Bart JM, Vidal I, Calvo E, Hall MN, Navarro M (**2012**) Third target of rapamycin complex negatively regulates development of quiescence in *Trypanosoma brucei*. **Proc Natl Acad Sci U S A** 109: 14399-14404.

33. Williams RAM, Woods KL, Juliano L, Mottram JC, Coombs GH (**2009**) Characterisation of unusual families of ATG8-like proteins and ATG12 in the protozoan parasite *Leishmania major*. **Autophagy** 5: 159-172.

34. Van den Abbeele J, Claes Y, van Bockstaele D, Le Ray D, Coosemans M (**1999**) *Trypanosoma brucei spp*. development in the tsetse fly: characterization of the post-mesocyclic stages in the foregut and proboscis. **Parasitol** 118: 469-478.

35. Aksoy S, Gibson WC, Lehane MJ (**2003**) Interactions between tsetse and trypanosomes with implications for the control of trypanosomiasis. **Adv Parasitol** 53: 1-83.

36. Kelly S, Reed J, Kramer S, Ellis L, Webb H, Sunter J, Salje J, Marinsek N, Gull K, Wickstead B, Carrington M (**2007**) Functional genomics in *Trypanosoma brucei*: A collection of vectors for the expression of tagged proteins from endogenous and ectopic gene loci. **Mol Biochem Parasitol** 154: 103-109.

37. Munday JC, McLuskey K, Brown E, Coombs GH, Mottram JC (**2011**) Oligopeptidase B-deficient mutants of *Leishmania major*. **Mol Biochem Parasitol** 175: 49-57.

3

Exogenous addition of histidine reduces copper availability in the yeast *Saccharomyces cerevisiae*

Daisuke Watanabe, Rie Kikushima, Miho Aitoku, Akira Nishimura, Iwao Ohtsu, Ryo Nasuno, and Hiroshi Takagi*

Graduate School of Biological Sciences, Nara Institute of Science and Technology, 8916-5 Takayama, Ikoma, Nara 630-0192, Japan.
* Corresponding Author: Hiroshi Takagi, Graduate School of Biological Sciences, Nara Institute of Science and Technology, 8916-5 Takayama, Ikoma; Nara 630-0192, Japan; E-mail: hiro@bs.naist.jp

ABSTRACT The basic amino acid histidine inhibited yeast cell growth more severely than lysine and arginine. Overexpression of *CTR1*, which encodes a high-affinity copper transporter on the plasma membrane, or addition of copper to the medium alleviated this cytotoxicity. However, the intracellular level of copper ions was not decreased in the presence of excess histidine. These results indicate that histidine cytotoxicity is associated with low copper availability inside cells, not with impaired copper uptake. Furthermore, histidine did not affect cell growth under limited respiration conditions, suggesting that histidine cytotoxicity is involved in deficiency of mitochondrial copper.

***Keywords**: yeast Saccharomyces cerevisiae, basic amino acids, histidine cytotoxicity, copper transporter Ctr1, mitochondrial respiration.*

INTRODUCTION

Free amino acids play pivotal roles as building blocks of proteins, as intermediates in metabolism, and also as regulators of a wide variety of cellular functions. In our previous studies, several amino acids including proline and arginine show cryoprotective activity in the budding yeast *Saccharomyces cerevisiae* [1-3]. We recently showed that, under oxidative stress conditions, increased conversion of proline into arginine led to the flavoprotein Tah18-dependent synthesis of nitric oxide, which confers stress tolerance on *S. cerevisiae* cells [4-6]. Although other charged amino acids, such as lysine and glutamate, also effectively enhance the freeze tolerance for yeast cells [1], the mechanism underlying the freezing stress tolerance by these amino acids remained unclear. In contrast, intracellular excessive levels of basic amino acids (lysine, arginine, and histidine) in mammals were reported to induce toxicity leading to gastrointestinal diseases, such as hepatomegaly and acute pancreatitis [7-9]. Although pleiotropic disorders in lipid metabolism, protein synthesis, and mitochondrial functions have been observed in cells damaged by basic amino acids, their primary cellular effects are not fully understood.

Basic amino acids are incorporated via three similar permeases, Can1, Lyp1, and Alp1, which share 60-65% sequence identity with each other [10]. Can1 was originally identified as an arginine permease [11], though it has been also reported to transport lysine, histidine, and ornithine with lower affinities [12]. An ortholog of Can1 in *Candida albicans* actively transports lysine, arginine, and histidine [13]. Lyp1 efficiently mediates only lysine transport [14, 15]. Overexpression of *ALP1* leads to specific uptake of arginine, although it is unclear whether this gene is expressed under physiological conditions [15]. While these permeases transport amino acids with substrate preferences, the general amino-acid permease Gap1 is a transporter for all of 20 L-forms and also D-forms of the common α-amino acids, as well as other related compounds, such as citrulline, ornithine, γ-aminobutyric acid (GABA), and polyamines [16-19]. Gap1 is most closely related to Hip1 in terms of amino acid sequence, although Hip1 seems to be a rather specific permease for histidine [15, 20, 21]. Recent comprehensive studies revealed that single overexpression of these permeases decreases the growth rate [22, 23], suggesting that *S. cerevisiae* can be utilized as a model to analyze the cytotoxicity caused by excess basic amino acids.

Among basic amino acids, histidine is especially related to copper transport [24]. Since the discovery of copper(II)-bis(L-histidinato) complex in human blood [25], extensive research has been performed to determine its physiological roles. Consequently, histidine was found to facilitate copper uptake in hepatic, placental, and brain cells [26-28] by removing copper from albumin, which physically inhibits incorporation of copper ions [29]. The copper(II)-bis(L-histidinato) complex has thus been applied for the treatment of Menkes disease and hypertrophic cardioencephalomyopathy, both of which are closely associated with copper deficiency [30, 31]. In *S. cerevisiae*, copper uptake is mediated by the high-affinity transporters Ctr1 and Ctr3 and a ferric/cupric reductase Fre1, which oxidizes copper(II) into usable copper(I) ions in advance of their uptake

[32-34]. To maintain copper homeostasis, the *CTR1*, *CTR3*, and *FRE1* genes are upregulated or downregulated under copper starvation or excess copper conditions, respectively, via the action of the copper-sensing transcription factor Mac1 [35, 36]. Intriguingly, the mutations in the histidine biosynthetic genes of *S. cerevisiae* increase sensitivity to the excess amounts of copper, which is suppressed by addition of histidine [37]. This finding supports the idea that histidine might directly interact with copper ions in yeast cells to alleviate the copper toxicity.

To explore novel roles of free amino acids, we analyzed here the cytotoxicity caused by exogenous addition of excess basic amino acids in *S. cerevisiae*.

RESULTS AND DISCUSSION

To understand the mechanism by which excess of basic amino acids mediate cytotoxicity, we examined cell growth of *S. cerevisiae* under culture conditions supplemented with an elevated concentration of lysine, arginine, or histidine. As shown in Figure 1A, 5 mM of histidine severely impaired the growth of yeast cells, although a higher concentration of lysine or arginine (25 mM) did not affect growth. The concentrations for three basic amino acids were selected based on the intracellular contents of these amino acids in L5487 cells (Figure 1B). When the *CAN1* gene, encoding basic amino acids permease on the plasma membrane [12, 15, 38], was overexpressed, there was little effect on growth on SCGal medium that contained no excess of basic amino acids. In contrast, the overexpression of *CAN1* markedly inhibited growth under elevated levels of basic amino acids. In particular, the growth of yeast cells that overexpress *CAN1* was relatively slow on the medium in the presence of 5 mM of histidine. We confirmed that the overexpression of *CAN1* increased the intracellular levels of basic amino acids in SCGal medium (approximately 1.6- to 4.1-fold increase) (Figure 1B). Thus, these results suggest that the excess amount of intracellular basic amino acids exerts toxic effects on yeast cells. Considering that Can1 preferentially transports lysine and arginine [15], overexpression of the high-affinity histidine transporter gene *HIP1* [15, 21] might enhance histidine uptake and effectively confer more severe toxicity to yeast cells in the presence of excess histidine. As histidine conferred more sensitivity to yeast cells than lysine and arginine despite having the lowest intracellular level, we further analyzed histidine cytotoxicity.

To identify multicopy suppressor genes that alleviate histidine toxicity, a yeast genomic library YEp51B [39] was introduced into L5487 cells overexpressing *CAN1*, and the transformants were screened for growth on SCGal medium containing 10 mM histidine. Sixteen independent genomic DNA fragments were isolated from the transformant colonies, and 25 full-length open reading frames were included in these fragments. After subcloning into pYES2, each gene was tested for its effect on the growth of L5487 cells in the presence of excess histidine. Consequently, the *CTR1* gene, which encodes a high-affinity copper transporter that predominantly mediates copper uptake under low copper conditions [32], exhibited the most significant suppression of the histidine-caused growth defect (Figure 1C). Although the overexpression of *CTR1* reduced cell growth in the presence of excess lysine by unknown mechanism(s), it is suggested that Ctr1 functions in alleviating histidine toxicity. Another high-affinity copper transporter gene, *CTR3* [33], and a cupric reductase gene, *FRE1*, the latter of which is required for conversion of copper(II) to copper(I) ions prior to uptake [34], also suppressed the growth defect under histidine-excess conditions when overexpressed. In addition, deletion of the *CTR1* or *FRE1* gene slightly increased sensitivity to excess histidine (data not shown). These results indicate that the histidine cytotoxicity in *S. cerevisiae* is alleviated by enhancement of copper uptake. In agreement with these results, the growth defect caused by 5 mM histidine became significantly more severe when the concentration of $CuSO_4$ in SD medium was decreased (2.5 and 25 nM), although higher concentrations of $CuSO_4$ (2.5 and 25 μM) suppressed the histidine toxicity (Figure 1D). It has been well studied that histidine directly binds to copper(II) to form copper(II)-bis(L-histidinato) complex under physiological conditions as in human blood [24]. However, considering that the overexpression of *CAN1* enhanced histidine toxicity (Figure 1A), it seems unlikely that the elevated level of histidine simply chelates copper(II) ions outside of the cells to inhibit copper uptake. Instead, excess histidine might interact with copper(I) ions to reduce the availability of copper after incorporation into yeast cells. To verify this hypothesis, we quantified the intracellular level of copper ions. Although cell growth was delayed starting after 4-hour incubation with 5 or 10 mM histidine (data not shown), intracellular copper ions were slightly increased in the histidine-treated cells (Figure 1E). Therefore, our data consistently demonstrate that an excess level of histidine enhances copper uptake but impairs copper availability in yeast cells. Regarding the *CTR1*-overexpressing cells (Figure 1C), sufficient copper ions might be incorporated into cells in bioavailable forms and thus contributed to relieving histidine toxicity. In a similar manner as clioquinol [40], histidine may act as both a chelator and an ionophore: histidine chelates copper(II) outside of the cell and is taken in as a complex, which may facilitate uptake of copper ions, though copper ions in this complex are not bioavailable. Additionally, it is possible that histidine-induced copper deficiency upregulates expression of copper transporters, which may increase copper uptake.

What functions of copper does excess histidine inhibit? Intracellular copper is distributed to distinct target proteins via specific cytosolic copper chaperons, such as Atx1, Ccs1, and Cox17. Atx1 assists in the transport of copper to the cell-surface iron uptake protein Fet3 through the function of Ccc2, which has copper-transporting ATPase activity, on the post-Golgi vesicle [41]. Ccs1 delivers copper specifically to the superoxide dismutase Sod1, which scavenges reactive oxygen species in the intermembrane space of mitochondria [42, 43]. Another copper chaperon Cox17 transfers copper to the mitochondrial inner membrane proteins

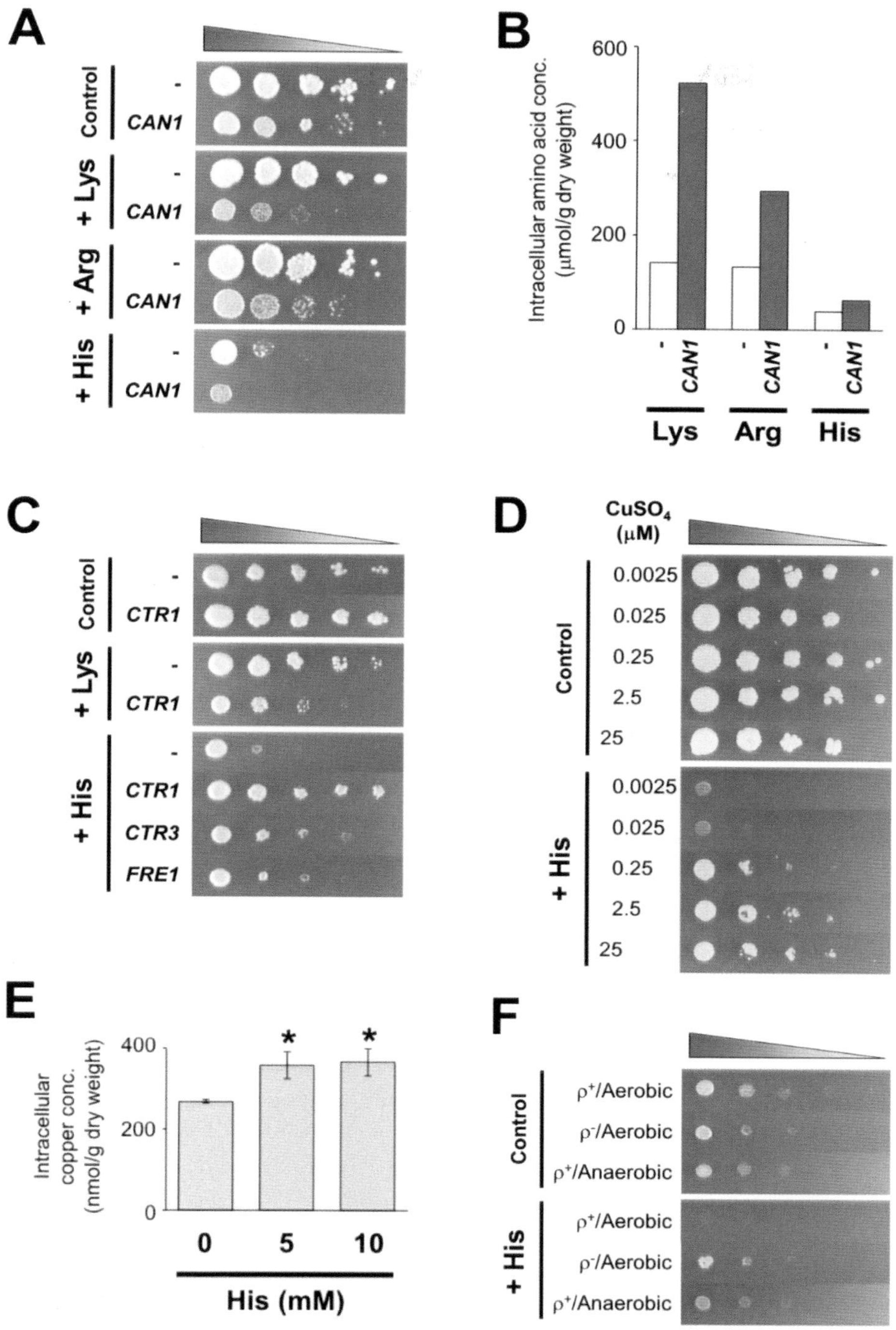

FIGURE 1: The effect of excess histidine on yeast cell growth and copper uptake. (A) Growth phenotypes of *S. cerevisiae* Σ1278b wild-type strain L5487 (complemented with pRS415 (*LEU2*)) carrying an empty vector pYES2 (*URA3*) or pYES2-CAN1 (*URA3*). After overnight cultivation in SC-Leu-Ura liquid medium, approximately 10^6 cells of each strain, and serial dilutions of 10^{-1} to 10^{-4} (from left to right) were spotted and incubated onto SCGal-Leu-Ura-Lys-Arg-His agar medium in the absence (Control) and presence of excess basic amino acids (25 mM lysine (+ Lys), 25 mM arginine (+ Arg), or 5 mM histidine (+ His)). **(B)** Intracellular contents of basic amino acids in strain L5487 (complemented with pRS415) carrying an empty vector pYES2 (white) or pYES2-CAN1 (gray). The values are the means of two independent experiments that produced similar results. **(C)** Growth phenotypes of strain L5487 (complemented with pRS415) carrying an empty vector pYES2, pYES2-CTR1, pYES2-CTR3, and pYES2-FRE1. After overnight cultivation in SC-Leu-Ura liquid medium, approximately 10^6 cells of each strain, and serial dilutions of 10^{-1} to 10^{-4} (from left to right) were spotted and incubated onto SCGal-Leu-Ura-Lys-Arg-His agar medium in the absence (Control) and presence of excess levels of basic amino acids (25 mM lysine (+ Lys) or 5 mM histidine (+ His)). **(D)** Growth phenotypes of strain L5487 (complemented with pRS415 and pRS416 (*URA3*)) under various concentrations of copper. After overnight cultivation in SD liquid medium, which contains 0.25 µM $CuSO_4$, approximately 10^6 cells of each strain, and serial dilutions of 10^{-1} to 10^{-4} (from left to right) were spotted and incubated onto SD-Cu agar medium in the presence of 0.0025, 0.025, 0.25, 2.5, or 25 µM $CuSO_4$ and in the absence (Control) and presence of 5 mM histidine (+ His). **(E)** Intracellular levels of copper ions of strain L5487 (complemented with pRS415 and pRS416) in the absence (0 mM His) and presence of excess histidine (5 or 10 mM His). The values are the means and standard deviations of three independent experiments. Asterisks indicate a significant increase in copper levels compared to the control sample (0 mM His) ($p < 0.05$). **(F)** Growth phenotypes of strain L5487 (complemented with pRS415 and pRS416) (shown as ρ^+) and its spontaneous ρ^- mutant. After overnight cultivation in SD liquid medium, approximately 10^6 cells of each strain, and serial dilutions of 10^{-1} to 10^{-4} (from left to right) were spotted and incubated onto SD agar medium in the absence (Control) and presence of 5 mM histidine (+ His) under aerobic or anaerobic conditions.

Cox11 and Sco1, both of which are essential for assembly of cytochrome *c* oxidase, which is the last enzyme in the respiratory electron transport chain [44]. In this study, we tested whether histidine cytotoxicity might be mediated by defective mitochondrial functions due to reduced copper availability. As shown in Figure 1F, L5487 cells were clearly sensitive to 5 mM histidine on SD plates, although histidine cytotoxicity was completely abrogated by a spontaneous cytoplasmic *petite* (ρ^-) mutation, which inactivates mitochondrial respiratory functions. It is also worth noting that yeast cells showed similar growth phenotypes in the absence or presence of excess histidine under anaerobic conditions (Figure 1F). Thus, histidine cytotoxicity was observed only when mitochondrial aerobic respiration should be functional. We hypothesize that the deficiency of the Cox17-bound copper ions due to excess of intracellular histidine might cause the abnormal assembly of cytochrome *c* oxidase complex in aerobically growing cells, leading to the observed toxic effect. This might be supported by the fact that the disturbance of cytochrome *c* oxidase induces apoptosis-like cell death in *S. cerevisiae* [45]. However, we cannot rule out the possibility that defective mitochondrial respiration reduces incorporation of histidine by some unknown mechanism(s), and hence, the excess level of histidine did not elicit the toxicity in ρ^- cells under aerobic conditions or in ρ^+ cells under anaerobic conditions.

In this study, we discovered the cytotoxicity of excess histidine in *S. cerevisiae*, which is tightly associated with the reduced availability of intracellular copper ions. Similarly, Pearce and Sherman [37] revealed that intracellular histidine synthesis is required for the detoxification of excess copper. Both studies commonly suggest that intracellular histidine has a novel and important role in copper homeostasis.

MATERIALS AND METHODS

Strains and Plasmids

The *S. cerevisiae* strain used in this study was Σ1278b wild-type strain L5487 (*MATα ura3-52 leu2::hisG*), which was generously provided by Gerald Fink (Whitehead Institute). A spontaneous cytoplasmic *petite* (ρ^-) mutant of L5487 was isolated, according to Fox *et al.* [46]. *Escherichia coli* strain DH5α (F⁻λ⁻ *Φ80lacZΔM15 Δ(lacZYA argF)U169 deoR recA1 endA1 hsdR17*($r_K^- m_K^+$) *supE44 thi-1 gyrA96*) was used to subclone the yeast gene and construct plasmids. Low-copy plasmids pRS415 and pRS416 [24] were used to complement auxotrophic mutations *ura3* and *leu2*, respectively. A galactose-inducible plasmid pYES2 (Life Technologies) was used for overexpression of the *CAN1*, *CTR1*, *CTR3*, and *FRE1* gene. A 2 µm-based yeast genomic library YEp51B [12] was used to identify multicopy suppressor genes of *CAN1*-overexpressing L5487 strain.

Culture Media

The media used for growth of *S. cerevisiae* were a synthetic complete medium, SC (2% glucose, 0.67% yeast nitrogen base without amino acids (Difco), supplemented with synthetic drop-out amino acid and nucleotide mixture as required), and a synthetic defined medium, SD (2% glucose, 0.67% yeast nitrogen base without amino acids (Difco)). For overexpression of *CAN1*, *CTR1*, *CTR3*, and *FRE1*, SC with galactose as a carbon source (SCGal) (2% galactose, 0.67% yeast nitrogen base without amino acids (Difco), supplemented with synthetic drop-out amino acid and nucleotide mixture as required) was used. To evaluate the effect of copper sulfate ($CuSO_4$) addition, SD-Cu media (2% glucose, 0.67% yeast nitrogen base without amino acids and copper (ForMedium)) containing different concentrations of $CuSO_4$ were used. All experiments were performed at 30°C, and all growth media were adjusted to pH 6.5 with HEPES buffer (pH 7.0) and sodium hydroxide. When necessary, 2% agar was added to solidify the medium. For anaerobic cultivation, the inoculated plates were incubated with O_2-absorber/CO_2-generator AnaeroPouch-Anaero (Mitsubishi Gas Chemical Company) and O_2 indicators. The *E. coli* recombinant strains were grown in Luria-Bertani complete medium containing 50 µg/ml ampicillin or M9 minimal medium plus 2% Casamino acids containing 50 µg/ml ampicillin. If necessary, 2% agar was added to solidify the medium.

Measurements of Intracellular Amino Acids Levels

According to a method described previously [3], intracellular amino acids were extracted by boiling from log-phase cells cultivated in SCGal-Leu-Ura liquid medium, and were subsequently quantified with an amino acid analyzer AminoTac JLC-500/V (JEOL).

Determination of Intracellular Copper Ions

Cell lysates were prepared from log-phase cells cultivated in SC-Leu-Ura-Lys-Arg-His liquid medium, and copper ions concentrations were determined by the CUPRAC-BCS assay [25].

ACKNOWLEDGMENTS

We are grateful to Dr. Gerald R. Fink and the NBRP of the MEXT, Japan, for providing the *S. cerevisiae* Σ1278b wild-type strain L5487 and a high-copy yeast genomic DNA library YEp51B, respectively.

CONFLICT OF INTEREST

The authors declare no conflict of interest.

REFERENCES

1. Takagi H, Iwamoto F, and Nakamori S (**1997**). Isolation of freeze-tolerant laboratory strains of *Saccharomyces cerevisiae* from proline-analogue-resistant mutants. **Appl Microbiol Biotechnol** 47(4): 405-411.

2. Morita Y, Nakamori S, and Takagi H (**2002**). Effect of proline and arginine metabolism on freezing stress of *Saccharomyces cerevisiae*. **J Biosci Bioeng** 94(5): 390-394.

3. Morita Y, Nakamori S, and Takagi H (**2003**). L-Proline accumulation and freeze tolerance in *Saccharomyces cerevisiae* are caused by a mutation in the *PRO1* gene encoding ⍰-glutamyl kinase. **Appl Environ Microbiol** 69(1): 212-219.

4. Nishimura A, Kotani T, Sasano Y, and Takagi H (**2010**). An antioxidative mechanism mediated by the yeast N-acetyltransferase Mpr1: oxidative stress-induced arginine synthesis and its physiological role. **FEMS Yeast Res** 10(6): 687-698.

5. Sasano Y, Haitani Y, Hashida K, Ohtsu I, Shima J, and Takagi H (**2012**). Enhancement of the proline and nitric oxide synthetic pathway improves fermentation ability under multiple baking-associated stress conditions in industrial baker's yeast. **Microb Cell Fact** 11: 40.

6. Nishimura A, Kawahara N, and Takagi H (**2013**). The flavoprotein Tah18-dependent NO synthesis confers high-temperature stress tolerance on yeast cells. **Biochem Biophys Res Commun** 430(1): 137-143.

7. Solomon JK and Geison RL (**1978**). Effect of excess dietary L-histidine on plasma cholesterol levels in weanling rats. **J Nutr** 108(6): 936-943.

8. Mizunuma T, Kawamura S, and Kishino Y (**1984**). Effects of injecting excess arginine on rat pancreas. **J Nutr** 114(3): 467-471.

9. Biczó G, Hegyi P, Dósa S, Shalbuyeva N, Berczi S, Sinervirta R, Hracskó Z, Siska A, Kukor Z, Jármay K, Venglovecz V, Varga IS, Iványi B, Alhonen L, Wittmann T, Gukovskaya A, Takács T, and Rakonczay Z Jr (**2011**). The crucial role of early mitochondrial injury in L-lysine-induced acute pancreatitis. **Antioxid Redox Signal** 15(10): 2669-2681.

10. Schmidt KH, Viebranz E, Doerfler L, Lester C, and Rubenstein A (**2010**). Formation of complex and unstable chromosomal translocations in yeast. **PLoS One** 5(8): e12007.

11. Grenson M, Mousset M, Wiame JM, and Bechet J (**1966**). Multiplicity of the amino acid permeases in *Saccharomyces cerevisiae*. I. Evidence for a specific arginine-transporting system. **Biochim Biophys Acta** 127(2): 325-338.

12. Cooper TG (**1982**). Nitrogen metabolism and *Saccharomyces cerevisiae*. In: Strathern JN, Jones EW, and Broach JR, editors. The molecular biology of the yeast *Saccharomyces*: metabolism and gene expression. Cold Spring Harbor Laboratory, New York; pp 339-461.

13. Sychrová H and Souciet JL (**1994**). *CAN1*, a gene encoding a permease for basic amino acids in *Candida albicans*. **Yeast** 10(12): 1647-1651.

14. Sychrova H and Chevallier MR (**1993**). Cloning and sequencing of the *Saccharomyces cerevisiae* gene *LYP1* coding for a lysine-specific permease. **Yeast** 9(7): 771-782.

15. Regenberg B, Düring-Olsen L, Kielland-Brandt MC, and Holmberg S (**1999**). Substrate specificity and gene expression of the amino-acid permeases in *Saccharomyces cerevisiae*. **Curr Genet** 36(6): 317-328.

16. Grenson M and Acheroy B (**1982**). Mutations affecting the activity and the regulation of the general amino-acid permease of *Saccharomyces cerevisiae*. Localisation of the *cis*-acting dominant *pgr* regulatory mutation in the structural gene of this permease. **Mol Gen Genet** 188(2): 261-265.

17. Jauniaux JC and Grenson M. (**1990**). *GAP1*, the general amino acid permease gene of *Saccharomyces cerevisiae*. Nucleotide sequence, protein similarity with the other bakers yeast amino acid permeases, and nitrogen catabolite repression. **Eur J Biochem** 190(1): 39-44.

18. Correa García S, Bermúdez Moretti M, Ramos E, and Batlle A. (**1997**). Carbon and nitrogen sources regulate delta-aminolevulinic acid and ⍰-aminobutyric acid transport in *Saccharomyces cerevisiae*. **Int J Biochem Cell Biol** 29(8-9): 1097-1101.

19. Uemura T, Kashiwagi K, and Igarashi K (**2005**). Uptake of putrescine and spermidine by Gap1p on the plasma membrane in *Saccharomyces cerevisiae*. **Biochem Biophys Res Commun** 328(4): 1028-1033.

20. Crabeel M and Grenson M (**1970**). Regulation of histidine uptake by specific feedback inhibition of two histidine permeases in *Saccharomyces cerevisiae*. **Eur J Biochem** 14(1): 197-204.

21. Tanaka J and Fink GR (**1985**). The histidine permease gene (*HIP1*) of *Saccharomyces cerevisiae*. **Gene** 38(1-3): 205-214.

22. Sopko R, Huang D, Preston N, Chua G, Papp B, Kafadar K, Snyder M, Oliver SG, Cyert M, Hughes TR, and Andrews B (**2006**). Mapping pathways and phenotypes by systematic gene overexpression. **Mol Cell** 21(3): 319-30.

23. Yoshikawa K, Tanaka T, Ida Y, Furusawa C, Hirasawa T, and Shimizu H (**2011**). Comprehensive phenotypic analysis of single-gene deletion and overexpression strains of *Saccharomyces cerevisiae*. **Yeast** 28(5): 349-361.

24. Deschamps P, Kulkarnip P, Gautam-Basak M, and Sarkar B (**2005**). The saga of copper(II)-L-histidine. **Coord Chem Rev** 249(9-10): 895-909.

25. Sarkar B and Kruck TPA (**1966**). Copper-amino acid complexes in human serum. In: Peisach J, Aisen P, and Blumberg WE, editors. Biochemistry of copper. Academic Press, New York; pp 183-196.

26. Darwish HM, Cheney JC, Schmitt RC, and Ettinger MJ (**1984**). Mobilization of copper(II) from plasma components and mechanisms of hepatic copper transport. **Am J Physiol** 246(1 Pt 1): G72-G79.

27. Hartter DE and Barnea A (**1988**). Brain tissue accumulates ^{67}copper by two ligand-dependent saturable processes. A high affinity, low capacity and a low affinity, high capacity process. **J Biol Chem** 263(2): 799-805.

28. Mas A and Sarkar B (**1992**). Uptake of ^{67}Cu by isolated human trophoblast cells. **Biochim Biophys Acta** 1135(2): 123-128.

29. McArdle HJ, Guthrie JR, Ackland ML, and Danks DM (**1987**). Albumin has no role in the uptake of copper by human fibroblasts. **J Inorg Biochem** 31(2): 123-131.

30. Sarkar B (**1999**). Treatment of Wilson and Menkes diseases. **Chem Rev** 99(9): 2535-2544.

31. Papadopoulou LC, Sue CM, Davidson MM, Tanji K, Nishino I, Sadlock JE, Krishna S, Walker W, Selby J, Glerum DM, Coster RV, Lyon G, Scalais E, Lebel R, Kaplan P, Shanske S, De Vivo DC, Bonilla E, Hirano M, DiMauro S, and Schon EA (**1999**). Fatal infantile cardioencephalomyopathy with COX deficiency and mutations in SCO2, a COX assembly gene. **Nat Genet** 23(3): 333-337.

32. Dancis A, Yuan DS, Haile D, Askwith C, Eide D, Moehle C, Kaplan J, and Klausner RD (**1994**). Molecular characterization of a copper transport protein in *S. cerevisiae*: an unexpected role for copper in iron transport. **Cell** 76(2): 393-402.

33. Knight SA, Labbé S, Kwon LF, Kosman DJ, and Thiele DJ (**1996**). A widespread transposable element masks expression of a yeast copper transport gene. **Genes Dev** 10(15): 1917-1929.

34. Hassett R and Kosman DJ (**1995**). Evidence for Cu(II) reduction as a component of copper uptake by *Saccharomyces cerevisiae*. **J Biol Chem** 270(1): 128-134.

35. Labbé S, Zhu Z, and Thiele DJ (**1997**). Copper-specific transcriptional repression of yeast genes encoding critical components in the copper transport pathway. **J Biol Chem** 272(25): 15951-15958.

36. Yamaguchi-Iwai Y, Serpe M, Haile D, Yang W, Kosman DJ, Klausner RD, and Dancis A (**1997**). Homeostatic regulation of copper uptake in yeast via direct binding of MAC1 protein to upstream regulatory sequences of FRE1 and CTR1. **J Biol Chem** 272(28): 17711-17718.

37. Pearce DA and Sherman F (**1999**). Toxicity of copper, cobalt, and nickel salts is dependent on histidine metabolism in the yeast *Saccharomyces cerevisiae* **J Bacteriol** 181(16): 4774-4779.

38. Malínská K, Malínský J, Opekarová M, and Tanner W (**2003**). Visu-

alization of protein compartmentation within the plasma membrane of living yeast cells. **Mol Biol Cell** 14(11): 4427-4436.

39. Akada R, Yamamoto J, and Yamashita I (**1997**). Screening and identification of yeast sequences that cause growth inhibition when overexpressed. **Mol Gen Genet** 254(3): 267-274.

40. Li C, Wang J, and Zhou B (**2010**). The metal chelating and chaperoning effects of clioquinol: insights from yeast studies. **J Alzheimers Dis** 21(4): 1249-1262.

41. Lin SJ, Pufahl RA, Dancis A, O'Halloran TV, and Culotta VC (**1997**). A role for the *Saccharomyces cerevisiae ATX1* gene in copper trafficking and iron transport. **J Biol Chem** 272(14): 9215-9220.

42. Culotta VC, Klomp LW, Strain J, Casareno RL, Krems B, and Gitlin JD (**1997**). The copper chaperone for superoxide dismutase. **J Biol Chem** 272(38): 23469-23472.

43. Sturtz LA, Diekert K, Jensen LT, Lill R, and Culotta VC (**2001**). A fraction of yeast Cu,Zn-superoxide dismutase and its metallochaperone, CCS, localize to the intermembrane space of mitochondria. A physiological role for SOD1 in guarding against mitochondrial oxidative damage. **J Biol Chem** 276(41): 38084-38089.

44. Horng YC, Cobine PA, Maxfield AB, Carr HS, and Winge DR (**2004**). Specific copper transfer from the Cox17 metallochaperone to both Sco1 and Cox11 in the assembly of yeast cytochrome *c* oxidase. **J Biol Chem** 279(34): 35334-35340.

45. Zuo X, Djordjevic JT, Bijosono Oei J, Desmarini D, Schibeci SD, Jolliffe KA, and Sorrell TC (**2011**). Miltefosine induces apoptosis-like cell death in yeast via Cox9p in cytochrome *c* oxidase. **Mol Pharmacol** 80(3): 476-485.

46. Fox TD, Folley LS, Mulero JJ, McMullin TW, Thorsness PE, Hedin LO, and Costanzo MC (**1991**). Analysis and manipulation of yeast mitochondrial genes. **Methods Enzymol** 194: 149-165.

4

Oxygen availability strongly affects chronological lifespan and thermotolerance in batch cultures of *Saccharomyces cerevisiae*

Markus M.M. Bisschops[1,3,#], Tim Vos[1,#], Rubén Martínez-Moreno[2,4], Pilar de la Torre Cortés[1], Jack T. Pronk[1], Pascale Daran-Lapujade[1,*]

[1] Department of Biotechnology, Delft University of Technology, Delft, The Netherlands.
[2] Instituto de Ciencias de la Vid y del Vino, CSIC, Universidad de La Rioja, Gobierno de La Rioja, Logroño, Spain.
[3] Current address: Division of Systems and Synthetic Biology, Department of Biology and Biological Engineering & The Novo Nordisk Foundation Center for Biosustainability, Chalmers University of Technology, Gothenburg, Sweden.
[4] Current address: Quercus Europe S.L., L'Hospitalet de Llobregat, Catalonia, Spain.
[#] Authors contributed equally to this work.
* Corresponding Author: Pascale Daran-Lapujade, Julianalaan 67, 2628 BC; Delft, The Netherlands;
E-mail: p.a.s.daran-lapujade@tudelft.nl

ABSTRACT Stationary-phase (SP) batch cultures of *Saccharomyces cerevisiae*, in which growth has been arrested by carbon-source depletion, are widely applied to study chronological lifespan, quiescence and SP-associated robustness. Based on this type of experiments, typically performed under aerobic conditions, several roles of oxygen in aging have been proposed. However, SP in anaerobic yeast cultures has not been investigated in detail. Here, we use the unique capability of *S. cerevisiae* to grow in the complete absence of oxygen to directly compare SP in aerobic and anaerobic bioreactor cultures. This comparison revealed strong positive effects of oxygen availability on adenylate energy charge, longevity and thermotolerance during SP. A low thermotolerance of anaerobic batch cultures was already evident during the exponential growth phase and, in contrast to the situation in aerobic cultures, was not substantially increased during transition into SP. A combination of physiological and transcriptome analysis showed that the slow post-diauxic growth phase on ethanol, which precedes SP in aerobic, but not in anaerobic cultures, endowed cells with the time and resources needed for inducing longevity and thermotolerance. When combined with literature data on acquisition of longevity and thermotolerance in retentostat cultures, the present study indicates that the fast transition from glucose excess to SP in anaerobic cultures precludes acquisition of longevity and thermotolerance. Moreover, this study demonstrates the importance of a preceding, calorie-restricted conditioning phase in the acquisition of longevity and stress tolerance in SP yeast cultures, irrespective of oxygen availability.

***Keywords**: chronological lifespan, thermotolerance, stationary phase, anaerobiosis, energetics, transcriptional response, conditioning.*

Abbreviations:
CLS - chronological lifespan,
ExP - exponential growth phase,
PD - post-diauxic phase,
ROS - reactive oxygen species,
SP - stationary-phase.

INTRODUCTION

Just like other living organisms, *Saccharomyces cerevisiae* cells age and have a finite chronological lifespan. The similarity of cellular processes in *S. cerevisiae* to those in higher eukaryotes and its accessibility to a wide range of experimental techniques have made this yeast a popular model for studying chronological aging of metazoan cells [1,2,3,4]. Chronological aging of *S. cerevisiae* is typically studied in aerobic batch cultures, in which growth arrest and quiescence are triggered by exhaustion of the available carbon sources in the growth medium [5,6]. Survival of individual yeast cells in such non-growing, stationary-phase (SP) cultures is then taken as a measure for their chronological lifespan (CLS). Over the past decade, studies on SP yeast cultures have contributed to our understanding of cellular mechanisms involved in aging, and several underlying cellular mechanisms were also found in higher eukaryotes [7].

Calorie restriction has been shown to extend lifespan in organisms ranging from yeast to man, with studies on many organisms pointing at an important role of nutrient-signaling cascades [8]. Turn-over of damaged macromolecules, and in particular proteins, has similarly been identified as a key process in aging in many organisms [9]. A third universal factor implicated in aging is respiration and, in particular, the associated formation of reactive oxygen species (ROS), which has been shown to enhance aging-related cellular deterioration in many organisms [10]. However, ROS have also been implicated in beneficial effects. In particular, mild ROS stress has been proposed to contribute to CLS extension by inducing stress-resistance genes, a phenomenon known as hormesis [11,12]. Similarly, increased mitochondrial respiration and ROS production rates in calorie-restricted yeast cultures have been linked to CLS extension [13,14,15].

ROS generation is not necessarily the only mechanism by which respiration and oxygen can affect CLS. In aerobic, glucose-grown batch cultures of *S. cerevisiae*, a fast and predominantly fermentative growth phase on glucose is followed by a second, respiratory growth phase in which the fermentation products ethanol and acetate are consumed [16]. This second growth phase, known as post-diauxic phase, is characterized by slow growth. During the post-diauxic phase, genes involved in SP are already expressed at an elevated level, as well as some features associated with SP cultures, such as increased stress resistance [6]. In anaerobic cultures of *S. cerevisiae*, the absence of oxygen prevents a respiratory post-diauxic growth phase. Instead, a phase of fast, fermentative exponential growth on glucose is immediately followed by SP, in which maintenance of viability and cellular integrity depends on metabolism of storage compounds. *S. cerevisiae* cells can contain two types of storage polymers: the storage carbohydrates trehalose and glycogen, and fatty acids, which are mostly stored in the form of di- and triacylglycerol esters [17,18,19]. In the absence of oxygen, yeast cells cannot catabolize fatty acids by β-oxidation and, moreover, conversion of storage carbohydrates via alcoholic fermentation yields 5-8 fold less ATP than their respiratory dissimilation [20].

Previous studies on the role of respiration in aging were predominantly based on the use of respiration-deficient *S. cerevisiae* mutants (e.g. ρ^0 strains and other mutants) [21,22,23,24,25] and respiratory inhibitors [14]. These approaches, however, have several drawbacks. Firstly, mitochondria are not only involved in respiration, but also in essential anabolic reactions (e.g., assembly of iron-sulfur complexes, amino acid biosynthesis and long-chain lipid biosynthesis [26]). Studies on petite or ρ^0 mutants may therefore cause unwanted 'side-effects' resulting from the absence or inefficiency of mitochondrial processes, rather than from direct effects of oxygen or respiration on aging. For example, the absence of mitochondrial DNA influences crosstalk between these organelles and the nucleus [27]. Furthermore, inhibition of respiration may result in reduced ROS levels [14], but can also result in ROS accumulation [28], depending on the intervention chosen. In addition, ROS may still be produced by other oxygen-consuming processes in yeast, such as disulfide-bond formation during oxidative protein folding [29].

Surprisingly, while *S. cerevisiae* is unique among yeasts and eukaryotes for its ability to grow fast under fully aero-

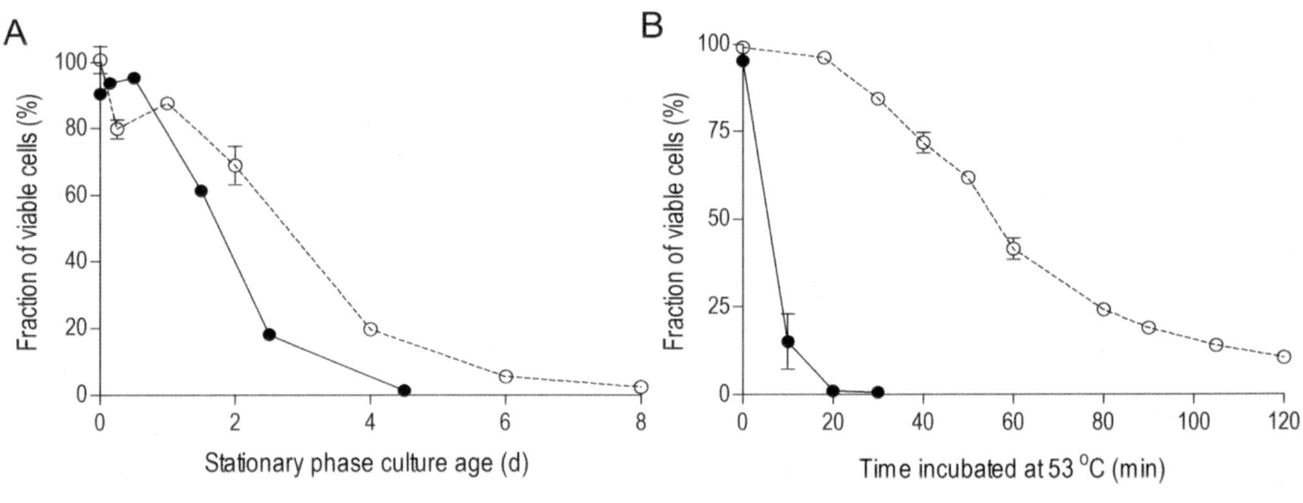

FIGURE 1: Chronological life span and thermotolerance of stationary-phase cultures is much lower under anaerobic than under aerobic conditions. Chronological life span **(A)**: survival expressed as ratio of colony forming units divided by the number of cells plated, during aerobic and anaerobic SP cultures. Time point zero indicates the onset of SP, which corresponds to exogenous glucose exhaustion in anaerobic cultures and exhaustion of all exogenous carbon sources including ethanol and organic acids in aerobic cultures. Thermotolerance **(B)**: loss of viability after sudden exposure of cells from aerobic and anaerobic cultures during early SP to 53°C. Open symbols (o) represent aerobic cultures, closed symbols (•) anaerobic cultures. Data represent the average and SEM of measurements on independent duplicate cultures.

bic as well as strictly anaerobic conditions [30], this ability has not been used to systematically investigate the impact of oxygen availability on entry into SP, on longevity and on robustness. The goal of the present study was therefore to investigate the impact of oxygen availability on yeast physiology in SP cultures. More specifically, we investigated whether the post-diauxic phase and respiratory mobilization of storage compounds in aerobic cultures affects CLS and thermotolerance during SP. To this end, aerobic and anaerobic bioreactor batch cultures of *S. cerevisiae* were grown into SP and subjected to detailed physiological and transcriptome analyses.

RESULTS

Anaerobicity reduces chronological lifespan and stress resistance in stationary phase cultures

To investigate the impact of oxygen availability on chronological lifespan in SP cultures of *S. cerevisiae,* survival kinetics were analyzed during SP in aerobic and anaerobic, glucose-grown bioreactor cultures. In aerobic cultures, the percentage of cells capable of colony formation on complex-medium agar plates typically decreased to ca. 2% in the 8 days following onset of SP, i.e. after exhaustion of all exogenous carbon sources including ethanol and organic acids (Figure 1A). Viability of anaerobic cultures decreased much faster, reaching values below 1% within 4.5 days after the onset of SP, that is after all exogenous glucose was consumed (Figure 1A).

Increased thermotolerance is a well-documented characteristic of SP cultures of *S. cerevisiae* [31,32]. Indeed, half of the cells in samples from aerobic, early-SP cultures survived a 60-min exposure to 53°C. Notably, up to 20 minutes incubation at 53°C hardly affected viability, suggesting that cells were well capable of repairing heat-induced damage during this period. In contrast, fewer than 20% of the cells from anaerobic early-SP cultures survived a 10-min incubation at this temperature (Figure 1B). Implementation of anaerobic conditions during sampling and heat-shock assays did not significantly affect this difference, indicating that heat-induced loss of viability was not influenced by exposure of anaerobically grown cells to oxygen during the assays (data not shown). Furthermore, washing of cells prior to the heat-shock experiments did not influence heat-shock resistance, indicating that the presence of low (< 1 g/L) ethanol concentrations in the assays did not cause the low thermotolerance of cells from anaerobic SP cultures.

Oxygen availability strongly affects the transcriptome of SP cultures

In aerobic yeast cultures, entry into SP is accompanied by a range of physiological changes that enhance survival in harsh, nutrient-poor environments [31]. This adaptation

TABLE 1. Functional categories overrepresented among genes with different expression levels in aerobic and anaerobic stationary phase cultures.

Category description	# of genes in dataset	# of genes in category	P-value[a]
574 genes with higher expression in aerobic stationary phase cultures			
Genes induced in stationary phase [34]	53	122	$1.2 \cdot 10^{-22}$
Genes induced by environmental stress response [39]	54	281	$3.2 \cdot 10^{-6}$
Fatty acid metabolic process	13	29	$5.4 \cdot 10^{-4}$
Fatty acid beta-oxidation	7	9	$2.0 \cdot 10^{-3}$
Transmembrane transport	52	303	$3.8 \cdot 10^{-3}$
Glyoxylate cycle	6	8	$1.7 \cdot 10^{-2}$
878 genes with higher expression in anaerobic stationary phase cultures			
Translation	131	345	$4.1 \cdot 10^{-28}$
Genes induced by environmental stress response [39]	78	281	$1.7 \cdot 10^{-8}$
Mitochondrial translation	36	81	$1.9 \cdot 10^{-8}$
Oxidation reduction	74	270	$1.6 \cdot 10^{-6}$
Metabolic process	93	389	$2.6 \cdot 10^{-5}$
Response to stress	49	161	$3.2 \cdot 10^{-5}$
Heme biosynthetic process	10	12	$1.6 \cdot 10^{-4}$
Methionine metabolic process	11	15	$3.5 \cdot 10^{-4}$
Sulfate assimilation	9	11	$1.0 \cdot 10^{-3}$
Porphyrin biosynthetic process	8	9	$1.4 \cdot 10^{-3}$
Carbohydrate metabolic process	28	93	$4.2 \cdot 10^{-2}$
Glycolysis	14	32	$4.6 \cdot 10^{-2}$
Methionine biosynthetic process	14	32	$4.6 \cdot 10^{-2}$

[a]Bonferroni-corrected P-value cut-off of 0.05 was used and P-values indicate the probability of finding the same numbers of genes in a random set of genes. Functional categories are obtained from the Gene Ontology set or, in italic font, directly from literature references. Details can be found in supplemental table S2.

coincides with a vast transcriptional reprogramming [33,34,35] that includes up-regulation of genes involved in resistance mechanisms to a wide array of stresses. Currently, no transcriptome data are available in the literature on anaerobic SP cultures of *S. cerevisiae*.

A transcriptome analysis, performed on culture samples taken 4 h after the onset of SP, revealed that a quarter of the yeast genome (1452 genes, Supplemental Table S1) was differentially expressed (fold-change cut-off of 2.0 and adjusted P-value below 0.05) in aerobic and anaerobic SP cultures. Among these genes were several genes known to be regulated by the heme and oxygen dependent transcription factors Hap1 and Rox1 [36,37,38]. Approximately 40% of the differentially expressed genes (574 genes, Supplemental Table S1) were transcribed at higher levels in the aerobic SP cultures. This gene set showed a strong overrepresentation of genes involved in fatty acid metabolism and, in particular, in β-oxidation (Table 1). This set of genes was also strongly enriched for genes that were up-regulated during SP in previous studies performed in shake flasks [34] (Table 1). Examples included the SP-genes *SPG1*, *SPG3*, *SPG4*, *SPG5*, and *SSA3*, which encodes a stress-induced ATPase. Furthermore, a significant number of genes (54) induced by the environmental stress response [39] was expressed at higher levels in aerobic SP cultures than in their anaerobic counterparts (Table 1).

Most of the genes that were differentially expressed in aerobic and anaerobic SP cultures (60%, 878 genes, Supplemental Table S1) showed a higher transcript level under anaerobic conditions. This gene set showed an overrepresentation of genes that were previously shown to be expressed at high levels in anaerobic yeast cultures and which were therefore not necessarily related to SP. This subgroup included genes involved in heme synthesis [38] and members of the multi-gene seripauperin family [40] (Table 1 and Supplemental Table S2). Interestingly, a strong overrepresentation (131 out of 345 genes) was found for the GO category 'translation' (Table 1). This subset included many genes encoding cytosolic and mitochondrial ribosomal proteins (65 and 24 genes respectively). Furthermore, several genes involved in carbohydrate metabolism, including glycogen metabolism, were expressed at higher levels in anaerobic SP cultures (Table 1). Finally, the set of 878 genes with higher expression in anaerobic SP cultures showed a very strong overrepresentation of genes induced by the environmental stress response [39].

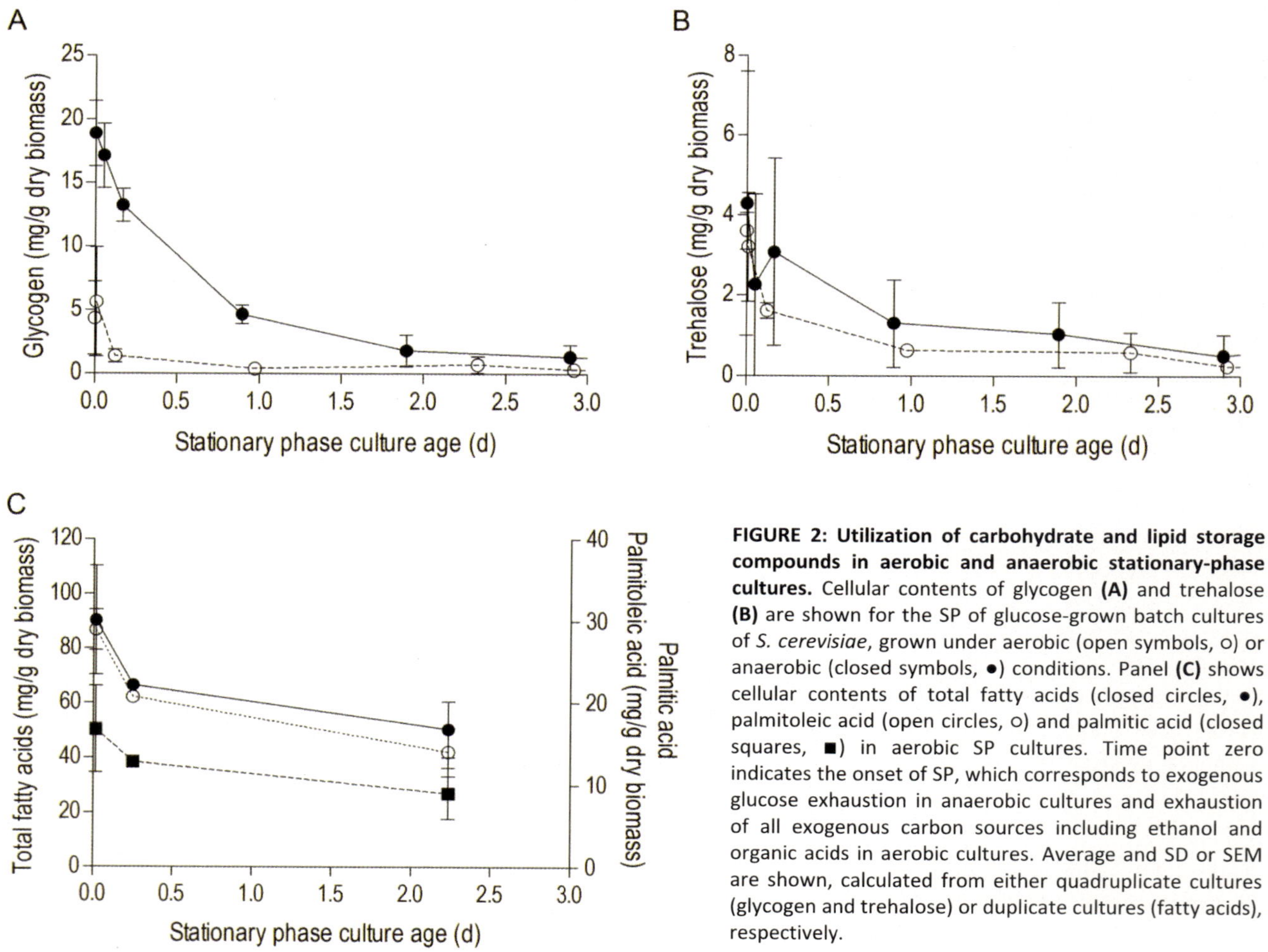

FIGURE 2: Utilization of carbohydrate and lipid storage compounds in aerobic and anaerobic stationary-phase cultures. Cellular contents of glycogen **(A)** and trehalose **(B)** are shown for the SP of glucose-grown batch cultures of *S. cerevisiae*, grown under aerobic (open symbols, ○) or anaerobic (closed symbols, ●) conditions. Panel **(C)** shows cellular contents of total fatty acids (closed circles, ●), palmitoleic acid (open circles, ○) and palmitic acid (closed squares, ■) in aerobic SP cultures. Time point zero indicates the onset of SP, which corresponds to exogenous glucose exhaustion in anaerobic cultures and exhaustion of all exogenous carbon sources including ethanol and organic acids in aerobic cultures. Average and SD or SEM are shown, calculated from either quadruplicate cultures (glycogen and trehalose) or duplicate cultures (fatty acids), respectively.

Anaerobicity negatively affects the energy status of SP cultures

Use of oxygen as an electron acceptor for respiration enables oxidative phosphorylation. As a consequence, ATP yields on glucose, glycogen and trehalose in respiratory cultures can be up to ca. 8-fold higher than in anaerobic, fermentative cultures [20,41]. Furthermore, since anaerobic yeast cultures cannot catabolize fatty acids, their use as energy reserves is restricted to aerobic cultures. To investigate the impact of storage metabolism and energy status on the short CLS of anaerobic SP cultures, cellular contents of storage materials and adenylate energy charge (a measure for the energetic status of living cells [42]) were analysed in aerobic and anaerobic SP cultures.

In aerobic cultures, intracellular pools of trehalose and glycogen were depleted within 1 day after entry into SP (Figure 2A and 2B). Cellular contents of the fatty acids palmitic and palmitoleic acid, also decreased during aerobic SP, but at a much slower rate than the storage carbohydrates and approached 6% of the total fatty acids content. This level is close to the membrane-associated fatty acid content previously reported in *S. cerevisiae* [43], indicating that yeast cells had exhausted most of their reserve lipids after 2 days in SP (Figure 2C). Together with the increased expression of genes involved in β-oxidation in aerobic SP cultures (Table 1), this observation indicates that aerobic SP cells use part of their fatty acids as an endogenous carbon and energy source. In aerobic cultures, the adenylate energy charge was 0.70 (± 0.08) and the intracellular ATP concentration was 5.45 (± 0.76) mM at the onset of SP. These results are in good agreement with published data [44,45]. In the days after the onset of SP, both parameters gradually decreased. Two full days after the onset of SP, the adenylate energy charge was still above 0.25 (Figure 3).

Trehalose and glycogen are the only known carbon and energy reserves in anaerobic *S. cerevisiae* cultures. The initial trehalose content and utilization profile in anaerobic SP cultures strongly resembled those observed in aerobic cultures. The initial glycogen concentration in anaerobic SP cultures was ca. four-fold higher than in aerobic SP cultures. Nevertheless, intracellular glycogen was exhausted after 2 days in SP (Figure 2A and 2B). At the onset of SP, intracellular ATP concentration and adenylate energy charge of anaerobic SP cultures were already lower than in aerobic cultures. Moreover, they decreased to very low levels within half a day after the onset of SP (Figure 3).

Different entry trajectories into SP in aerobic and anaerobic cultures

The results described above reveal clear differences in transcriptome, energy status, thermotolerance and CLS of aerobic and anaerobic SP cultures. Some of these parameters already differed at the onset of SP, indicating the importance of different 'entry trajectories' of aerobic and anaerobic cultures into SP. A major difference between aerobic and anaerobic batch cultures is the absence, in the latter, of a respiratory post-diauxic phase, in which ethanol and minor fermentation products acetate and glycerol are consumed. Growth in the post-diauxic phase, in which metabolism is completely respiratory, is slower than during the preceding glucose phase [46]. In this study, the maximum specific growth rate of anaerobic cultures (0.31 ± 0.01 h^{-1}) during the glucose phase was lower than that of aerobic cultures (0.39 ± 0.02 h^{-1}, Figure 4). In aerobic cultures, the specific growth rate during the post-diauxic phase (0.10 ± 0.01 h^{-1}) was ca. four-fold lower than during the fast growth phase on glucose. As a consequence, the specific growth rate in the hours preceding the onset of SP was ca. three-fold lower in aerobic cultures than in anaerobic cultures. To investigate whether the post-diauxic

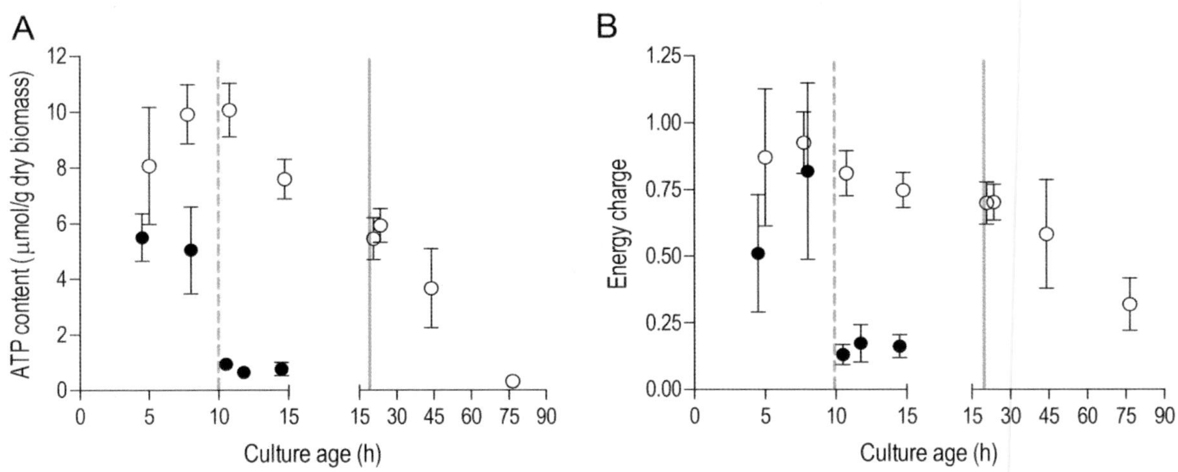

FIGURE 3: Intracellular ATP concentration and adenylate energy charge in aerobic and anaerobic stationary-phase cultures. Cellular ATP content **(A)** and adenylate energy charge **(B)** in aerobic (open symbols, ○) and anaerobic (closed symbols, ●) cultures. The dashed vertical line represents glucose exhaustion and the onset of anaerobic SP, the solid vertical line represents ethanol exhaustion and the onset of aerobic SP. Values are shown as averages of duplicate cultures (+/- SEM).

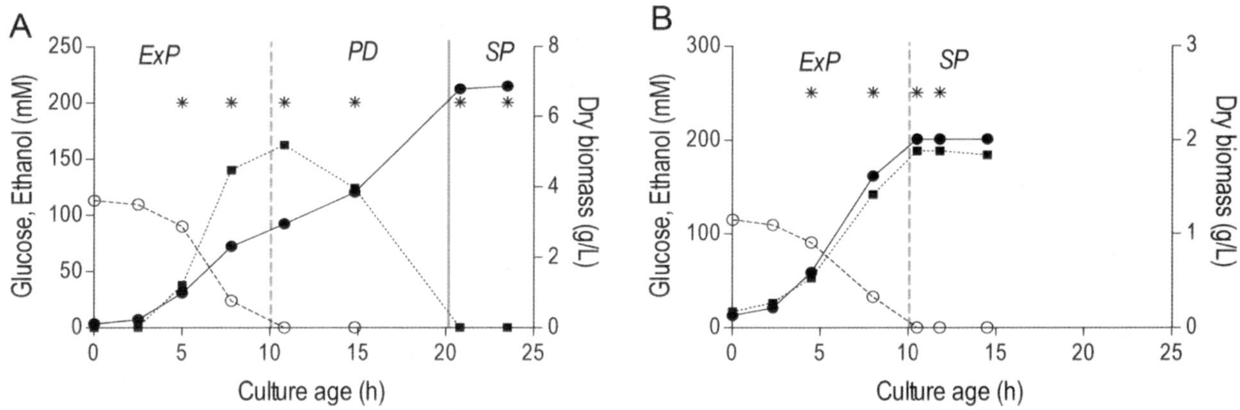

FIGURE 4: Growth phases in aerobic and anaerobic batch cultures. Biomass (closed circles, •), glucose (open circles, o) and ethanol (closed squares, ■) concentration during the different growth phases of aerobic **(A)** and anaerobic **(B)** batch cultures of *S. cerevisiae*. The initial phase of exponential growth on glucose (**ExP**), the following post-diauxic phase of slower growth on non-fermentable carbon sources (**PD**) and final stationary phase (**SP**) are indicated. Values shown are from single representative batch cultures, independent replicate cultures yielded essentially the same results. Vertical lines indicate depletion of glucose (dashed line) and of fermentation products (solid line). Asterisks (*) indicate time points at which samples were taken for transcriptome analysis.

phase may have 'conditioned' aerobic cultures for entry into SP, analysis of aerobic and anaerobic batch cultures was extended to include the growth phases that precede SP.

A much higher thermotolerance in aerobic cultures was already evident in the exponential growth phase and further increased during the post-diauxic phase, to reach a maximum upon entry into SP (Figure 5). Conversely, thermotolerance of anaerobic cultures did not increase during entry into SP and, consequently, remained much lower than that of aerobic cultures (Figure 5).

To further compare the different 'entry trajectories' into SP of aerobic and anaerobic batch cultures, transcriptome analyses were performed at different time points

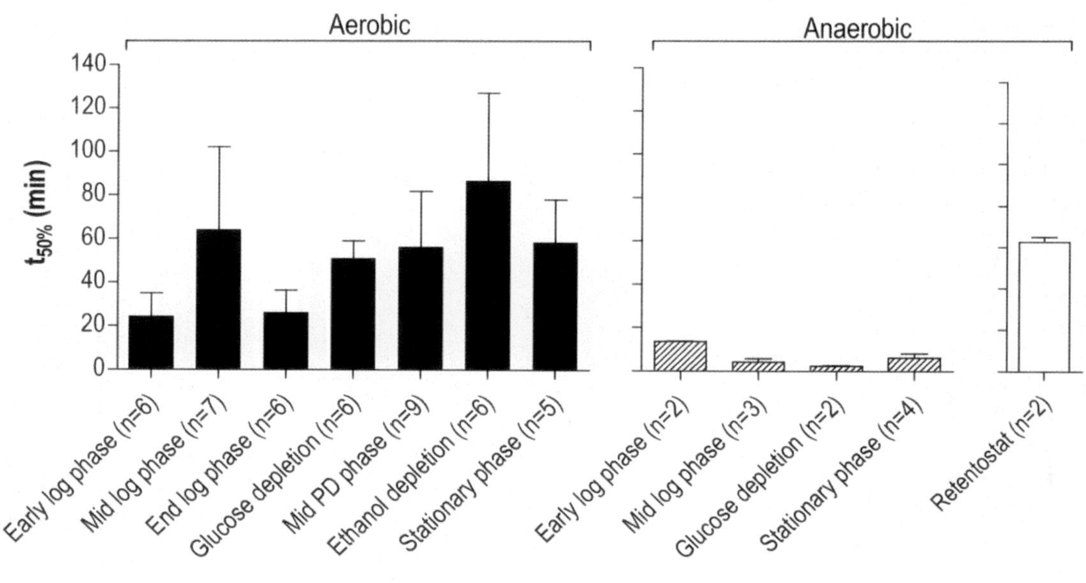

FIGURE 5: Thermotolerance of aerobic and anaerobic cultures during different growth phases. Thermotolerance of cells during different growth phases of aerobic (**black bars**) and anaerobic (**hatched bars**) batch cultures of *S. cerevisiae* (Figure 4). The **white bar** depicts the thermotolerance of *S. cerevisiae* grown for 8 days in anaerobic retentostats [20]. Thermotolerance was assayed by monitoring viability during incubation at 53°C and is shown as the incubation time resulting in a 50% decrease in viability (t_{50}) (see Materials and Methods for more details). The number of independent culture replicates for each of the growth phases is denoted on the x-axis labels.

during exponential phase, post-diauxic phase (aerobic cultures only) and SP. Genes were grouped in 9 clusters, based on their time-dependent expression profiles in aerobic and anaerobic cultures (Figure 6). A full dataset is available in Supplemental Data (Tables S3 and S4).

Less than one tenth (126 of 1452) of the genes that were differentially expressed in aerobic and anaerobic SP cultures (Table 1) already showed corresponding differences during the mid-exponential growth phase in aerobic and anaerobic cultures. For over half (834 of 1452) of the differentially expressed genes in SP cultures, the differences rose after glucose exhaustion, i.e. during the post-diauxic phase in aerobic cultures (Figure 6A, clusters 1-3). Cluster 1 comprises genes whose expression increased during the aerobic and anaerobic exponential growth phases, with a further increase during the aerobic post-diauxic phase (Figure 6A). Genes involved in fatty acid catabolism were overrepresented in this cluster, as well as genes that were previously shown to be induced in aerobic SP (including *SPG1*, *SPG3*, *SPG4*, *SPG5* and Hsp70-family-member *SSA3* [34]) (Figures 6A and 7). Genes in cluster 2 showed similar transcript levels during the aerobic and anaerobic exponential growth phases on glucose. However, due to a pronounced decrease in expression during the post-diauxic phase in aerobic cultures, expression levels were higher in anaerobic SP cultures than in aerobic SP cultures. This cluster was markedly enriched for genes involved in amino acid synthesis and translation (Figure 6A), suggesting that a down-regulation of protein synthesis occurred during the post-diauxic phase. Cluster 3 comprised of genes whose transcript levels increased during the exponential phase of both aerobic and anaerobic cultures but, subsequently, only decreased in aerobic SP cultures. Cluster 3 showed an overrepresentation of genes involved in mitochondrial translation and respiration (Figure 6A).

For 595 of the 'oxygen-responsive in SP' genes listed in Table 1, differences in expression occurred already upon glucose exhaustion (Figure 6B, cluster 4-8). Genes in Cluster 5, characterized by a specific up-regulation upon entry into anaerobic SP, showed an overrepresentation of stress-responsive genes (Figure 6B, cluster 5). Several of these are known to be specifically expressed under anaerobic conditions (e.g., the cell-wall mannoprotein-encoding gene *DAN4* and members of the seripauperin family [40]), but

A

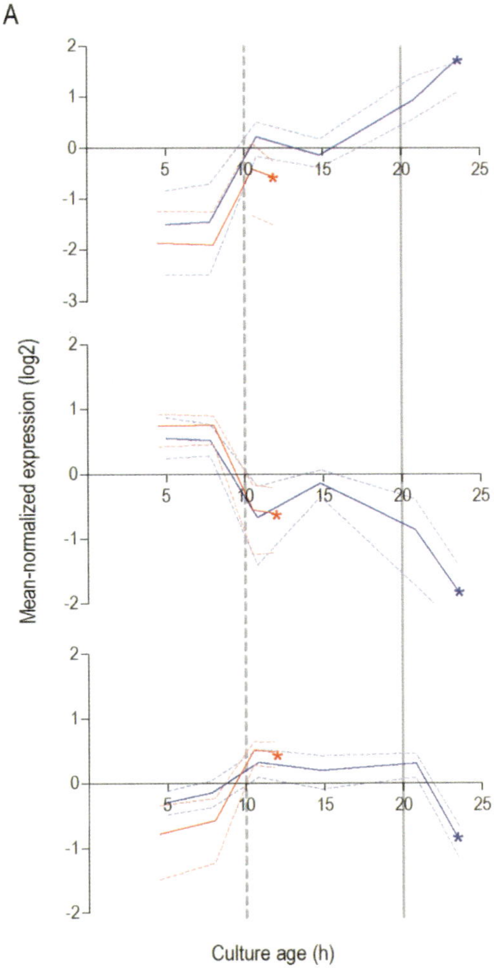

Cluster 1 (316 genes)

Functional category	*# in cluster*	*# in category*	*P-value*[a]
Fatty acid metabolism	10	29	$1.8 \cdot 10^{-3}$
SP-genes [34]	46	122	$2.1 \cdot 10^{-27}$
ESR-induced genes [39]	51	281	$1.7 \cdot 10^{-14}$

Cluster 2 (306 genes)

Functional category	*# in cluster*	*# in category*	*P-value*[a]
Translation	89	345	$3.2 \cdot 10^{-40}$
rRNA export from nucleus	17	48	$9.2 \cdot 10^{-8}$
Cellular amino acid biosynthetic process	22	101	$4.7 \cdot 10^{-6}$
ESR-repressed genes	86	563	$9.6 \cdot 10^{-22}$

Cluster 3 (212 genes)

Functional category	*# in cluster*	*# in category*	*P-value*[a]
Mitochondrial translation	34	81	$2.5 \cdot 10^{-26}$
Translation	40	345	$4.9 \cdot 10^{-9}$
tRNA aminoacylation for protein translation	10	35	$4.1 \cdot 10^{-4}$
Aerobic respiration	12	71	$1.0 \cdot 10^{-2}$

FIGURE 6A. See next page for figure 6B and figure legend.

B

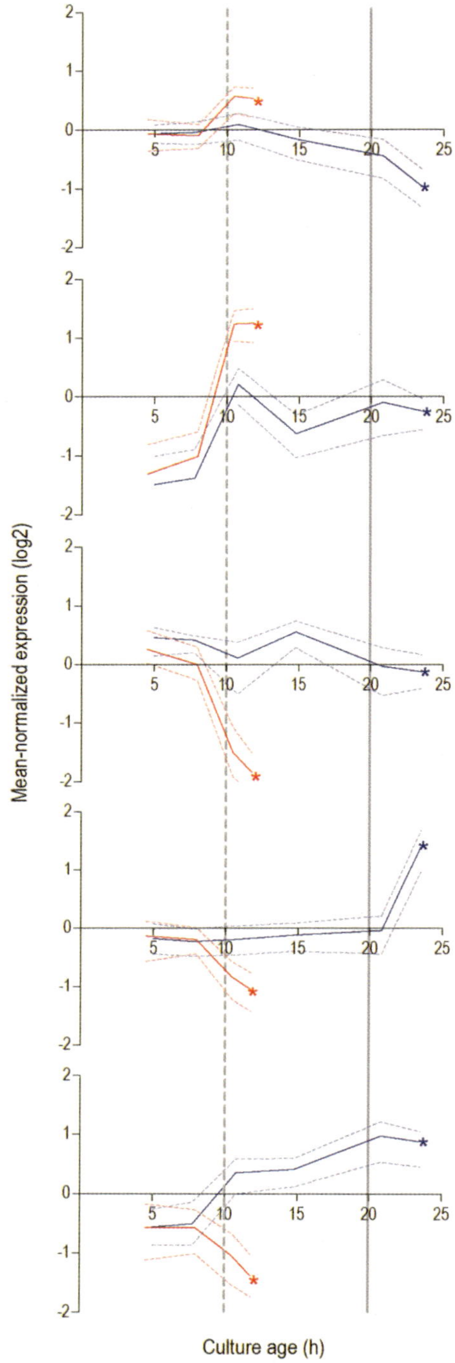

Cluster 4 (83 genes)

Functional category	*# in cluster*	*# in category*	*P-value[a]*
No significant terms			

Cluster 5 (254 genes)

Functional category	*# in cluster*	*# in category*	*P-value[a]*
Response to stress	20	161	$1.7 \cdot 10^{-2}$
ESR-induced genes [39]	69	281	$2.9 \cdot 10^{-35}$

Cluster 6 (53 genes)

Functional category	*# in cluster*	*# in category*	*P-value[a]*
tRNA transcription from RNA polymerase III promoter	4	15	$1.5 \cdot 10^{-2}$

Cluster 7 (84 genes)

Functional category	*# in cluster*	*# in category*	*P-value[a]*
Regulation of transcription	20	505	$1.6 \cdot 10^{-2}$

Cluster 8 (121 genes)

Functional category	*# in cluster*	*# in category*	*P-value[a]*
Carboxylic acid metabolic process	4	9	$4.0 \cdot 10^{-2}$

FIGURE 6: Clustering of genes differentially expressed between aerobic and anaerobic SP cultures according to their expression profiles during the growth phases preceding SP. Genes whose differential expression between aerobic and anaerobic SP cultures originated from changes after glucose depletion **(A)**. Clusters of genes whose differential expression between aerobic and anaerobic stationary phase cultures from changes upon glucose exhaustion **(B).** Each graph presents the expression profiles of genes from aerobic culture (**blue lines**) and anaerobic cultures (**red lines**) in a particular gene cluster. The solid lines represent the average of the mean-normalized expression of all genes in the cluster. The dashed lines represent the first and third quartile of these mean-normalized expression values, giving information on the scatter in the expression of genes in the cluster. Asterisks (*) indicate the SP samples from anaerobic and aerobic batches. Vertical lines indicate glucose exhaustion (**dashed line**) and carbon exhaustion (**solid line**, for aerobic cultures only). For each cluster a table reports the overrepresentation of functional categories, including the number of genes in the cluster belonging to a specific functional category (# in cluster), the total number of genes in this functional category (# in category), and the [a]Bonferroni-corrected P-values that indicate the likelihood of obtaining such enrichment in a random set of genes. Only categories with [a]Bonferroni-corrected P-value below 0.05 were deemed significant and presented in the tables. More details can be found in Supplemental Table S3.

cluster 5 also included heat-shock genes whose expression is not specifically linked to anoxic conditions (e.g. *HSP30* and *SSA4*). Genes that showed a specific downregulation during anaerobic SP, but a constant (Figure 6B, cluster 6) or increased expression in aerobic SP (Figure 6B, clusters 7 and 8) showed an overrepresentation of genes involved in transcription-related processes and carboxylic acid metabolism. The latter of which plays a role in the respiration of exogenous carboxylic acids during the post-diauxic phase (Figure 6B, cluster 8). All 23 genes whose transcript levels were higher under anaerobic conditions, irrespective of the growth phase (Supplemental Table S4), were previously described to be upregulated under anaerobic conditions. The majority (14) of these genes belonged to the seripauperin family [40].

Two clusters (Figure 6, cluster 1 and 5) comprised genes whose transcript levels increased in aerobic as well as in anaerobic batch cultivation, but to different final levels. These clusters were enriched for genes induced by the environmental stress response [39]. The extreme differences in thermotolerance of aerobic and anaerobic SP cultures (Figure 5) were therefore only partially mirrored at the transcript level, indicating that factors other than transcriptional reprogramming contribute to these differences.

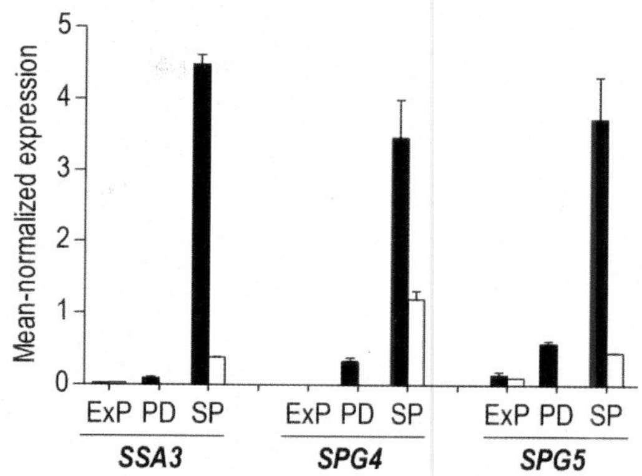

FIGURE 7: Expression levels of the SP-associated genes *SSA3*, *SPG4* and *SPG5* during different growth phases in aerobic and anaerobic batch cultures. The mean-normalized expression values during exponential growth on glucose phase (**ExP**), post-diauxic growth phase (**PD**, only aerobic cultures) and stationary phase (**SP**, i.e. 4 hours after exhaustion of exogenous consumable carbon-sources) in aerobic (**black bars**) and anaerobic (**white bars**) cultures of *S. cerevisiae* of the genes *SSA3, SPG4* and *SPG5*. Average values of duplicate cultures are shown (± SEM).

DISCUSSION

This study demonstrates a strong impact of oxygen availability on chronological lifespan and stress tolerance in SP batch cultures of *S. cerevisiae* and, thereby, confirms and extends earlier observations on its physiology in aerobic and anaerobic cultures [13,19,47]. The CLS of anaerobically grown SP cultures was much shorter than that of their aerobic counterparts and an even more dramatic difference was observed for thermotolerance. As will be discussed below, these differences involved a different conditioning of aerobic and aerobic cultures during the growth phases preceding SP, as well as energetic constraints imposed on yeast cells in anaerobic SP cultures.

The post-diauxic growth phase enables transcriptional conditioning of aerobic yeast cultures for stationary phase

Our transcriptome data revealed that 57% of the transcriptional differences between aerobic and anaerobic SP cultures originated from transcriptional reprogramming during the aerobic post-diauxic growth phase. Although several well-known hallmark transcripts of SP cultures, previously identified in (semi-) aerobic shake-flask cultures, such as *SPG4*, *SPG5* and *SSA3* [34], showed increased levels in anaerobic SP cultures, their levels did not reach those observed in aerobic cultures (Figure 7). Moreover, many genes involved in biosynthesis were strongly downregulated during the aerobic post-diauxic phase and SP, but retained expression levels close to those in the exponential growth phase in anaerobic cultures (Figure 6). These transcriptome data are consistent with the hypothesis that the post-diauxic phase in aerobic cultures conditions cells for entry into SP and that, conversely, absence of a post-diauxic phase prevents anaerobic batch cultures from adequately adapting to SP and starvation.

Hormesis could potentially explain the difference in robustness between aerobic and anaerobic cultures. Indeed respiration can generate low levels of ROS and thereby induce stress tolerance via increased expression of stress tolerance genes [13,21]. However, among a set of 22 genes encoding enzymes involved in ROS-protective mechanisms [48], including the superoxide dismutase genes *SOD1* and *SOD2*, whose expression is strongly upregulated during exposure to ROS [13,39], only the peroxisomal catalase *CTA1* was higher expressed in aerobic SP cultures. These findings argue against a dominant role of ROS-based hormesis in the acquisition of increased robustness by aerobic SP cultures.

Caloric restriction: a key factor in conditioning yeast cells for stationary phase and starvation

Thermotolerance is negatively correlated with specific growth rate in *S. cerevisiae* as Lu *et al.* (2009) demonstrated in nutrient-limited chemostat cultures [49]. Although these authors did not evaluate the impact of specific growth rate on CLS nor investigate anaerobic growth, they showed that the negative correlation between thermotolerance and growth rate also held in a respiratory-deficient *S. cerevisiae* strain [49]. Our data are fully consistent with the hypothesis that the strong reduction of specific growth rate (from 0.39 to 0.10 h^{-1}) during transition from fast exponential growth on glucose to the post-diauxic phase in aerobic batch cultures could similarly trigger increased thermotolerance and extended CLS during the starvation phase. We have also recently shown that a gradual decrease of the specific growth rate to near-zero values in glucose-limited retentostats [50] yielded yeast cells with a thermotolerance that is as high as that of aerobic SP cultures (Figure 5), and with an even longer CLS during subsequent starvation [20]. The transcriptional reprogramming observed in these anaerobic severely calorie-restricted cultures [51] strongly resembled the transcriptome changes observed in the present study for aerobic cultures entering SP and proteome analysis showed increased levels of proteins involved in stress resistance [52]. Deletion of Rim15, a kinase under control of several nutrient signaling pathways [53], strongly reduced the acquisition of robustness in both anaerobic and aerobic calorie-restricted cultures [54,55], suggesting a strong role for nutrient signaling independent of oxygen availability.

The present study, combined with our previous retentostat studies, therefore clearly demonstrates that prior conditioning by a period of caloric restriction (e.g. by slow growth during the aerobic post-diauxic phase or in extremely glucose-limited cultures) is a prerequisite for acquisition of a prolonged CLS by non-growing, starving cultures of *S. cerevisiae*. This conclusion, which supports earlier proposals based on starvation experiments by Thomsson *et al.* [47,56], has important implications for the design and interpretation of yeast studies on chronological aging, for example when such studies involve mutants that are impeded in energy metabolism.

The low thermotolerance of exponentially growing anaerobic cultures does not correlate with expression of heat shock genes

Although both aerobic and anaerobic, non-growing yeast cultures can acquire a similar thermotolerance by an appropriate preceding conditioning phase, a drastic difference was observed in the thermotolerance of exponentially growing aerobic and anaerobic cultures (Figure 5). The similar adenylate energy charge and intracellular ATP concentrations of aerobic and anaerobic cultures during exponential growth on glucose appear to rule out cellular energy status as a major cause of this difference. Since the high temperature (53°C) during the thermotolerance assays precludes *de novo* synthesis of mRNA or protein synthesis [57], this difference must already be expressed in the batch cultures themselves. Transcript levels of genes that were previously implicated in heat-shock resistance (including *HSP* genes [58] such as *HSP104* [59], *HSP26* [60], *HSP12* [61], *SSA3* [62]) were similar during the exponential growth phase on glucose in aerobic and anaerobic cultures. Moreover, of 59 genes identified as essential for heat-shock survival by Gibney and coworkers [57], only one gene was differentially expressed, *LIA1*, and showed a higher transcript level in anaerobic cultures.

Oxygen availability strongly influences sterol and unsaturated fatty acid composition of yeast cells [19], especially because these compounds have to be added to growth media as anaerobic growth factors [43]. These differences in membrane composition might partially explain the observed differences in thermotolerance between aerobically and anaerobically grown *S. cerevisiae* cells. The hypothesis that membrane composition is a key determinant in thermotolerance of *S. cerevisiae* [63,64] is consistent with a recent study, in which the acquisition of increased thermotolerance by laboratory-evolved strains was shown to be caused by changes in their sterol composition [65].

A low energy status of anaerobic SP cultures limits metabolic flexibility

Consistent with earlier reports [19,66], anaerobic batch cultures of *S. cerevisiae* displayed a substantially higher glycogen content than aerobic cultures. However, after the onset of SP, anaerobic cultures showed a much faster decrease of the adenylate energy charge. This difference can be attributed to several factors. Firstly, since *S. cerevisiae* cannot derive metabolic energy from lipids and amino acids in the absence of oxygen [67], anaerobic cultures are entirely dependent on glycogen and trehalose as energy storage compounds and anaerobic catabolism of these storage carbohydrates yields less ATP than respiration. The estimated ATP synthesis rate from anaerobic glycogen dissimilation of ca. 5 µmol per g biomass dry weight per hour during the first day in SP (based on a maximum ATP yield of 3 ATP per glucose residue [20]), was two orders of magnitude lower than the cellular ATP demand for maintenance estimated from chemostat and retentostat cultures (m_{ATP} = 1 mmol ATP per g biomass dry weight per hour [50]). A similar extreme reduction of ATP turnover rates was observed when anaerobic retentostat cultures were switched to carbon starvation [20]. Together, these observations indicate that an extremely low ATP turnover is an intrinsic feature of anaerobic, starving yeast cultures. In addition to this extreme low ATP-turnover, it is even conceivable that the apparent inability of anaerobic batch cultures to efficiently down-regulate energy-consuming processes, including protein synthesis, the single most expensive biosynthetic process in living cells [68,69], may have exacerbated the fast decline of their energy status after entry into SP (Figure 3).

The low energy status of cells may at the same time have put strong constraints on these energy consuming processes. Proteome analyses should reveal whether the increased transcription of *HSP* genes, implicated in thermotolerance, which took place late in the exponential growth phase (Figure 6), was too late to enable synthesis of the corresponding proteins before the decline in cellular energy status in anaerobic SP cultures. Such a scenario would explain the discrepancy between the oxygen-independent upregulation of these genes (with notable exception of *SSA3*) and the absence of increased thermotolerance in anaerobic SP cultures.

Taken together, the results from the present study indicate that, in the short time lapse between the moment at which anaerobic cultures sense that glucose reaches critically low levels and the actual exhaustion of glucose, they lack the time and resources to perform the energy-intensive remodeling of their transcriptome and proteome required to robustly face starvation. Our data are therefore entirely consistent with the notion that the low CLS and thermotolerance of anaerobic SP cultures, in comparison with aerobic cultures, is due to the absence of a proper conditioning phase and a limited metabolic flexibility due to a lower cellular energy status.

Outlook

Many studies in which SP yeast cultures are used as a model system to investigate aspects of aging, still rely on shake-flask cultures. Due to their low and poorly controlled oxygen-transfer capacity, the aeration status of shake-flask cultures is generally unclear. The strong impact of oxygen availability on aging-related characteristics [14] underlines the value of controlled cultivation techniques, e.g. in bioreactors, including batch, chemostat and retentostat cultures [50,70,71,72] or flow-through cells [73], in yeast-based aging studies. In particular, the use of anaerobic cultures as a model offers interesting possibilities to clarify the role of respiration and ROS in aging, apoptosis and longevity.

The short life span and low robustness of anaerobic SP cultures of *S. cerevisiae* is directly relevant for industrial applications. Robustness of SP cultures is especially important for processes in which biomass from anaerobic batch cultures is recycled, e.g. in industrial bioethanol production and beer brewing [74,75]. Clearly, results from (semi-)aerobic shake-flask cultures cannot be used to predict the performance of such anaerobic processes and improvement of robustness in these industrial processes will have to be based on studies in anaerobic systems.

Saccharomyces yeasts have the capability, rare among eukaryotes, to grow fast in the complete absence of oxygen and it is often assumed that they are well adapted to anaerobic environments [30,76,77]. While the natural habitat of *S. cerevisiae* is still a matter of debate [77], lower biomass concentrations frequently encountered in natural environments combined with the low affinity of yeast glucose transporters [78] may lead to a transition into SP that is sufficiently slow to enable acquisition of longevity and robustness under anaerobic conditions. Further research is therefore needed to investigate the ecological relevance of this laboratory study.

MATERIALS AND METHODS

Strains and cultivation

The prototrophic *Saccharomyces cerevisiae* strain CEN.PK113-7D (*MAT*a *MAL*2-8c *SUC*2) used in this study is a congenic member of the CEN.PK family [79,80]. Stock cultures were grown at 30°C in shake flasks containing yeast extract (1% w/v), peptone (2% w/v) and dextrose (2% w/v) (YPD) medium. Glycerol, final concentration 20% (v/v), was added to overnight cultures and 1 mL aliquots were stored at -80°C.

Previously described synthetic medium [81] was used with 20 g/L glucose as sole carbon-source and 0.2 g/L antifoam Emulsion C (Sigma, St. Louise, USA). In case of anaerobic cultivations, the medium was supplemented with anaerobic growth factors ergosterol (10 mg/L) and Tween 80 (420 mg/L) dissolved in ethanol. Inocula for batch fermentations consisted of 100 mL yeast culture grown overnight to an OD_{660} of 4 in synthetic medium with 20 g/L glucose. Aerobic and anaerobic batch fermentations were carried out at 30°C in 2 L bioreactors (Applikon, Schiedam, The Netherlands), with a working volume of 1.4 L. Cultures were stirred at 800 rpm and sparged at a flow-rate of 700 mL/min with either dried air or nitrogen gas (< 10 ppm oxygen, Linde Gas Benelux, The Netherlands). The bioreactors were equipped with Norprene tubing (Saint-Gobain Performance Plastics, Courbevoie, France) and Viton O-rings (Eriks, Alkmaar, The Netherlands) to minimize diffusion of oxygen. During aerobic cultivations, dissolved oxygen levels remained above 40% of the initial saturation level as measured by Clark electrodes (Mettler Toledo, Greifensee, Switzerland). The culture pH was maintained at 5.0 by automatic addition of 2 M KOH and 2 M H_2SO_4.

Analysis of biomass, metabolites, substrate and exhaust gas

Biomass concentration as culture dry weight was determined as described previously [82].

For substrate and extracellular metabolite concentration determination, culture supernatants were obtained by centrifugation of culture samples (3 min at 20.000 g) and analysed by high-performance liquid chromatograph (HPLC) analysis on a Waters Alliance 2690 HPLC (Waters, Milford, MA) equipped with a Bio-Rad HPX 87H ion exchange column (BioRad, Veenendaal, The Netherlands), operated at 60°C with 5 mM H_2SO_4 as the mobile phase at a flow rate of 0.6 ml/min. Detection was by means of a dual-wavelength absorbance detector (Waters 2487) and a refractive index detector (Waters 2410).

The exhaust gas from batch cultures was cooled with a condenser (2°C) and dried with a PermaPure Dryer (model MD 110-8P-4; Inacom Instruments, Veenendaal, the Netherlands) prior to online analysis of carbon dioxide and oxygen with a Rosemount NGA 2000 Analyser (Baar, Switzerland).

Colony forming units

To determine culture viability, small aliquots of culture broth were taken from the reactor and cells were counted on a Z2 Coulter Counter (Beckman Coulter, Woerden, Netherlands) equipped with a 50 µm orifice (Multisizer II, Beckman Coulter, Woerden, Netherlands). Cells were diluted in 0.1% peptone and 100 µL suspensions containing approximately 30, 300 and 3000 cells were plated on yeast extract (1% w/v), peptone (2% w/v) and dextrose (0.5% w/v)) (YPD) agar plates and incubated at 30°C for at least 3 days before counting the colonies. CFU was calculated as the number of colonies formed divided by the number of plated cells.

Thermotolerance assay

Cells from culture broth were counted with a Z2 Coulter Counter and diluted in pre-warmed (53°C) isotone diluent II (Beckman Coulter, Woerden, Netherlands) to yield 50 mL cell suspensions with a density of $1{\cdot}10^7$ cells/mL. Cell suspensions were incubated in a waterbath at 53°C and 4 mL aliquots were sampled in 10 min intervals. Samples were cooled on ice and assayed for viability using the *Funga*Light 5-CFDA, AM (acetoxymethyl ester 5-carboxyfluorescein diacetate)/propidum iodide (PI) yeast viability kit (Invitrogen, Carlsbrad, CA) by counting 10,000 cells on a Cell Lab Quanta SC MPL flow cytometer (Beckman Coulter, Woerden, Netherlands) as described previously [20]. 5-CFDA, AM is a cell-permeant substrate for intracellular non-specific esterase activity. Hydrolytic cleavage of the lipophilic blocking acetoxymethyl and diacetate groups of 5-CFDA, AM results in a green fluorescent signal in metabolically active cells. Propidium iodide intercalates with DNA in cells with a compromised cell membrane, which results in red fluorescence. Cells stained with PI were considered dead cells. For each independent sample, the t_{50} value (the time after which 50% of the initial viable population was dead) was estimated by fitting the viability data with a sigmoidal dose-response curve in Graphpad Prism 4.03. Both measurements of viability, i.e. metabolic activity based on 5-CFDA, AM and membrane integrity based on PI gave similar t_{50}-values, therefore only results based on PI are shown.

Storage carbohydrate measurements

1.2 mL culture broth was quenched in 5 mL of cold methanol (-40°C) using a rapid sampling setup described previously [83], mixed and subsequently pelleted (4,400 *g*, 5 min) at -19°C. Cells were washed with 5 mL of cold methanol (-40°C) and pellets stored at -80°C. Pellets were resuspended in Na_2CO_3 (0.25 M) and further processed according to a previously described procedure [84].

Fatty acids measurements

Culture volumes corresponding to 50 mg biomass were sampled on ice, centrifuged (10,000 *g*, 10 min at 4°C), washed twice, resuspended in 5 ml ice-cold water and stored at -20°C. Lipid extraction was performed as described previously [85]. Aliquots of 0.15 mL were added to 15 mL tubes and 1.5 mL of a mixture of concentrated HCl and 1-propanol (1:4) and 1.5 mL of dichlorethane were added. 400 µg of myristic acid (a 15:0 fatty acid) was included as internal standard. Samples were incubated at 100°C for 2 h. Subsequently, 3 mL of water was added to cooled samples. 1 mL of the organic phase was filtered over water-free sodium sulfate into GC vials. The fatty acid propyl esters in the organic phase were analyzed by gas chromatography (model 6890N, Agilent, U.S.A.) using a DB-wax column (length, 30 m; inside diameter, 0.25 mm; film thickness, 0.25 µm; J&W Scientific, Folsom, CA) and helium as the carrier gas. The sample volume was 1 µL, and the split was set to 1:20. The injection temperature was 230°C, and the following temperature gradient was used: 120°C at the start, increasing at a rate of 10°C/min up to 240°C, and then 240°C for 8 min. The fatty acid propyl esters were detected using a flame ionization detector at 250°C.

Analysis of intracellular adenosine-phosphate concentrations

Samples for internal metabolite analysis were obtained by rapid sampling [83]. 1.2 mL of culture broth was rapidly quenched into 5 mL of 100% methanol, pre-cooled to -40°C. Samples were washed with cold methanol and extracted with boiling ethanol. Intracellular AMP and ADP were determined enzymatically, using a previously described assay based on myokinase, pyruvate kinase and lactate dehydrogenase reactions [86]. Assays were performed in white, flat bottom 96-well microtiter plates (Corning Inc., USA). NADH fluorescence was measured in a TECAN GENios Pro microtiterplate reader (Tecan, Männedorf, Switzerland) as previously described [83]. Intracellular ATP was also assessed enzymatically. The assay contained 115 mM triethanolamine (pH 7.6), 11.5 mM $MgSO_4x7H_2O$, 1.15 mM $NADP^+$ per well including sample extract, total volume was 205 µL per well. The reaction to measure ATP was initiated by adding 12 mM glucose and 30 U hexokinase (Sigma-Aldrich Chemie B.V, Zwijndrecht, The Netherlands). Assays were performed in black, flat bottom 96-well microtiter plates (Corning Inc., USA). NADPH fluorescence was measured in a TECAN GENios Pro microtiterplate reader. The adenylate energy charge was calculated according to the previously described [42] equation:

$$\mathrm{EC} = \frac{([\mathrm{ATP}] + 0.5[\mathrm{ADP}])}{([\mathrm{ATP}] + [\mathrm{ADP}] + [\mathrm{AMP}])}$$

Transcriptome analysis

Independent duplicate aerobic and anaerobic batch cultures were sampled at six and four different time points respectively (see Figure 4A and 4B) for microarray analysis, resulting in a total dataset of 20 microarrays. Sampling from batch cultures for transcriptome analysis was performed using liquid nitrogen for rapid quenching of mRNA turnover [87]. Prior to RNA extraction, samples were stored in a mixture of phenol/chloroform and TEA buffer at -80°C. Total RNA extraction, isolation of mRNA, cDNA synthesis, cRNA synthesis, labelling and array hybridization was performed as previously described [88]. The quality of total RNA, cDNA, aRNA and fragmented aRNA was checked using an Agilent Bioanalyzer 2100 (Agilent Technologies, Santa Clara, CA). Hybridization of labelled fragmented aRNA to the microarrays and staining, washing and scanning of the microarrays was performed according to Affymetrix instructions (EukGE_WS2v5).

The 6383 yeast open reading frames were extracted from the 9335 transcript features on the YG-S98 microarrays. To allow comparison, all expression data were normalized to a target value of 240 using the average signal from all gene features [89]. The microarray data used in this study are available via GEO series accession number GSE69485. To eliminate variation in genes that are not expressed, genes with expression values below 12 were set to 12 and the gene features for which the maximum expression was below 20 for all 20 arrays were discarded [51]. The average deviation of the mean transcript data of replicate batches was ca. 11%, similar to the reproducibility usually observed in replicate steady state chemostat cultures [90]. The expression of housekeeping genes *ACT1, PDA1, TFC1, ALG9, TAF10* and *UBC6* [91] remained stable for both conditions and all sample points (average coefficient of variation 17% ± 5%).

To identify genes that were differentially expressed between aerobic and anaerobic SP cultures, a pairwise comparison was performed between aerobic samples taken at time point 6 (Figure 4A) and anaerobic samples taken at time point 4 (Figure 4B) as previously described [89]. Similarly, genes differently expressed during growth on glucose under aerobic or anaerobic conditions were identified through a pairwise comparison of aerobic and anaerobic samples taken at time point 1 (Figure 4A and 4B). Differences with adjusted P-values lower than 0.05 and a fold difference of 2 or higher were considered statistically significant. Time-dependent expression profiles of selected genes were clustered according to optimal k-means clustering using positive correlation as distance metric (Expressionist Pro version 3.1, Genedata, Basel, Switzerland) resulting in an optimal number of 9 clusters. For display of time-dependent transcript levels, expression values were normalized per gene by dividing single expression values by the average expression value of both conditions and all time points. Mean values of these average-normalized values for all genes in each cluster are shown, as well as the first and third quartile of average-normalized values.

Gene expression clusters were analysed for overrepresentation of functional annotation categories of the Gene Ontology (GO) database (http://www.geneontology.org/), based on a hypergeometric distribution analysis tool [92]. Additional categories describing genes expressed in SP cultures [34], genes commonly induced by several environmental stresses [39] or essential for heat-shock survival [57] were extracted from the respective references.

ACKNOWLEDGMENTS

We thank for Pauline Folch, Nuria Barrajón-Simanca, Xavier Hakkaart for assistance in performing the experiments. Martínez-Moreno, R is recipient of a CSIC JAE-Predoc grant co-funded by the European Social Fund of the EU.

CONFLICT OF INTEREST

The authors declare no conflict of interest.

REFERENCES

1. Denoth Lippuner A, Julou T, and Barral Y (**2014**). Budding yeast as a model organism to study the effects of age. **FEMS Microbiol Rev** 38(2): 300-325.

2. Longo VD, Shadel GS, Kaeberlein M, and Kennedy B (**2012**). Replicative and chronological aging in *Saccharomyces cerevisiae*. **Cell Metab** 16(1): 18-31.

3. Breitenbach M, Jazwinski SM, and Laun P, editors (**2012**). Aging research in yeast. **Springer Netherlands, Dordrecht**.

4. Mirisola MG, Braun RJ, and Petranovic D (**2014**). Approaches to study yeast cell aging and death. **FEMS Yeast Res** 14(1): 109-118.

5. Longo VD and Fabrizio P (**2012**). Chronological aging in *Saccharomyces cerevisiae*. In: Breitenbach M, Jazwinski SM, Laun P, editors. Aging Research in Yeast. **Springer Netherlands, Dordrecht;** pp 101-121.

6. Gray JV, Petsko GA, Johnston GC, Ringe D, Singer RA, and Werner-Washburne M (**2004**). "Sleeping beauty": quiescence in *Saccharomyces cerevisiae*. **Microbiol Mol Biol Rev** 68(2): 187-206.

7. Fabrizio P and Longo VD (**2003**). The chronological life span of *Saccharomyces cerevisiae*. **Aging Cell** 2(2): 73-81.

8. Sinclair DA (**2005**). Toward a unified theory of caloric restriction and longevity regulation. **Mech Ageing Dev** 126(9): 987-1002.

9. Ryazanov AG and Nefsky BS (**2002**). Protein turnover plays a key role in aging. **Mech Ageing Dev** 123(2-3): 207-213.

10. Balaban RS, Nemoto S, and Finkel T (**2005**). Mitochondria, Oxidants, and Aging. **Cell** 120(4): 483-495.

11. Ludovico P and Burhans WC (**2014**). Reactive oxygen species, ageing and the hormesis police. **FEMS Yeast Res** 14(1): 33-39.

12. Sharma PK, Agrawal V, and Roy N (**2011**). Mitochondria-mediated hormetic response in life span extension of calorie-restricted *Saccharomyces cerevisiae*. **AGE** 33(2): 143-154.

13. Mesquita A, Weinberger M, Silva A, Sampaio-Marques B, Almeida B, Leão C, Costa V, Rodrigues F, Burhans WC, and Ludovico P (**2010**). Caloric restriction or catalase inactivation extends yeast chronological lifespan by inducing H2O2 and superoxide dismutase activity. **Proc Natl Acad Sci U S A** 107(34): 15123-15128.

14. Ocampo A, Liu J, Schroeder EA, Shadel GS, and Barrientos A (**2012**). Mitochondrial respiratory thresholds regulate yeast chronological life span and its extension by caloric restriction. **Cell Metab** 16(1): 55-67.

15. Smith J, McClure JM, Matecic M, and Smith JS (**2007**). Calorie restriction extends the chronological lifespan of *Saccharomyces cerevisiae* independently of the Sirtuins. **Aging Cell** 6(5): 649-662.

16. van Dijken JP, Weusthuis RA, and Pronk JT (**1993**). Kinetics of growth and sugar consumption in yeasts. **Antonie van Leeuwenhoek** 63(3-4): 343-352.

17. François J and Parrou JL (**2001**). Reserve carbohydrates metabolism in the yeast *Saccharomyces cerevisiae*. **FEMS Microbiol Rev** 25(1): 125-145.

18. Kohei H and Satoshi Y (**1984**). Regulatory role of phosphatidate phosphatase in triacylglycerol synthesis of *Saccharomyces cerevisiae*. **BBA-Lipid Lipid Met** 796(1): 110-117.

19. Wilson K and McLeod BJ (**1976**). The influence of conditions of growth on the endogenous metabolism of *Saccharomyces cerevisiae*: effect on protein, carbohydrate, sterol and fatty acid content and on viability. **Antonie van Leeuwenhoek** 42(4): 397-410.

20. Boender LGM, Almering MJH, Dijk M, van Maris AJA, de Winde JH, Pronk JT, and Daran-Lapujade P (**2011**). Extreme calorie restriction and energy source starvation in *Saccharomyces cerevisiae* represent distinct physiological states. **BBA-Mol Cell Res** 1813(12): 2133-2144.

21. Tahara EB, Cunha FM, Basso TO, Della Bianca BE, Gombert AK, and Kowaltowski AJ (**2013**). Calorie restriction hysteretically primes aging *Saccharomyces cerevisiae* toward more effective oxidative metabolism. **PLoS ONE** 8(2): e56388.

22. Lin SJ, Kaeberlein M, Andalis AA, Sturtz LA, Defossez PA, Culotta VC, Fink GR, and Guarente L (**2002**). Calorie restriction extends *Saccharomyces cerevisiae* lifespan by increasing respiration. **Nature** 418(6895): 344-348.

23. Bonawitz ND, Chatenay-Lapointe M, Pan Y, and Shadel GS (**2007**). Reduced TOR signaling extends chronological life span via increased respiration and upregulation of mitochondrial gene expression. **Cell Metab** 5(4): 265-277.

24. Aerts AM, Zabrocki P, Govaert G, Mathys J, Carmona-Gutierrez D, Madeo F, Winderickx J, Cammue BPA, and Thevissen K (**2009**). Mitochondrial dysfunction leads to reduced chronological lifespan and increased apoptosis in yeast. **FEBS Lett** 583(1): 113-117.

25. Barros MH, da Cunha FM, Oliveira GA, Tahara EB, and Kowaltowski AJ (**2010**). Yeast as a model to study mitochondrial mechanisms in ageing. **Mech Ageing Dev** 131(7-8): 494-502.

26. Breitenbach M, Rinnerthaler M, Hartl J, Stincone A, Vowinckel J, Breitenbach-Koller H, and Ralser M (**2014**). Mitochondria in ageing: there is metabolism beyond the ROS. **FEMS Yeast Res** 14(1): 198-212.

27. Woo DK and Poyton RO (**2009**). The absence of a mitochondrial genome in rho0 yeast cells extends lifespan independently of retrograde regulation. **Exp Gerontol** 44(6-7): 390-397.

28. Leadsham JE, Sanders G, Giannaki S, Bastow EL, Hutton R, Naeimi WR, Breitenbach M, and Gourlay CW (**2013**). Loss of cytochrome c oxidase promotes RAS-dependent ROS production from the ER resident NADPH oxidase, Yno1p, in yeast. **Cell Metab** 18(2): 279-286.

29. Rosenfeld E and Beauvoit B (**2003**). Role of the non-respiratory pathways in the utilization of molecular oxygen by *Saccharomyces cerevisiae*. **Yeast** 20(13): 1115-1144.

30. Visser W, Scheffers WA, Batenburg-van der Vegte WH, and van Dijken JP (**1990**). Oxygen requirements of yeasts. **Appl Environ Microbiol** 56(12): 3785-3792.

31. Werner-Washburne M, Braun E, Johnston GC, and Singer RA (**1993**). Stationary phase in the yeast *Saccharomyces cerevisiae*. **Microbiol Rev** 57(2): 383-401.

32. Allen C, Buttner S, Aragon AD, Thomas JA, Meirelles O, Jaetao JE, Benn D, Ruby SW, Veenhuis M, Madeo F, and Werner-Washburne M (**2006**). Isolation of quiescent and nonquiescent cells from yeast stationary-phase cultures. **J Cell Biol** 174(1): 89-100.

33. Wanichthanarak K, Wongtosrad N, and Petranovic D (**2015**). Genome-wide expression analyses of the stationary phase model of ageing in yeast. **Mech Ageing Dev** 149:65-74.

34. Martinez MJ, Roy S, Archuletta AB, Wentzell PD, Anna-Arriola SS, Rodriguez AL, Aragon AD, Quiñones GA, Allen C, and Werner-Washburne M (**2004**). Genomic analysis of stationary-phase and exit in *Saccharomyces cerevisiae*: Gene expression and identification of novel essential genes. **Mol Biol Cell** 15(12): 5295-5305.

35. Galdieri L, Mehrotra S, Yu S, and Vancura A (**2010**). Transcriptional regulation in yeast during diauxic shift and stationary phase. **OMICS** 14(6): 629-638.

36. Ter Linde JJM and Steensma HY (**2002**). A microarray-assisted screen for potential Hap1 and Rox1 target genes in *Saccharomyces cerevisiae*. **Yeast** 19(10): 825-840.

37. Harbison CT, Gordon DB, Lee TI, Rinaldi NJ, Macisaac KD, Danford TW, Hannett NM, Tagne JB, Reynolds DB, Yoo J, Jennings EG, Zeitlinger J, Pokholok DK, Kellis M, Rolfe PA, Takusagawa KT, Lander ES, Gifford DK, Fraenkel E, and Young RA (**2004**). Transcriptional regulatory code of a eukaryotic genome. **Nature** 431(7004): 99-104.

38. Keng T (**1992**). HAP1 and ROX1 form a regulatory pathway in the repression of *HEM13* transcription in *Saccharomyces cerevisiae*. **Mol Cell Biol** 12(6): 2616-2623.

39. Gasch AP, Spellman PT, Kao CM, Carmel-Harel O, Eisen MB, Storz G, Botstein D, and Brown PO (**2000**). Genomic expression programs in the response of yeast cells to environmental changes. **Mol Biol Cell** 11(12): 4241-4257.

40. Rachidi N, Martinez MJ, Barre P, and Blondin B (**2000**). *Saccharomyces cerevisiae* PAU genes are induced by anaerobiosis. **Mol Microbiol** 35(6): 1421-1430.

41. Verduyn C, Postma E, Scheffers WA, and van Dijken JP (**1990**). Energetics of *Saccharomyces cerevisiae* in anaerobic glucose-limited chemostat cultures. **J Gen Microbiol** 136(3): 405-412.

42. Atkinson DE (**1968**). Energy charge of the adenylate pool as a regulatory parameter. Interaction with feedback modifiers. **Biochemistry** 7(11): 4030-4034.

43. Verduyn C, Postma E, Scheffers WA, and van Dijken JP (**1990**). Physiology of *Saccharomyces cerevisiae* in anaerobic glucose-limited chemostat cultures. **J Gen Microbiol** 136(3): 395-403.

44. Choi JS and Lee CK (**2013**). Maintenance of cellular ATP level by caloric restriction correlates chronological survival of budding yeast. **Biochem Biophys Res Commun** 439(1): 126-131.

45. Ball WJ and Atkinson DE (**1975**). Adenylate energy charge in *Saccharomyces cerevisiae* during starvation. **J Bacteriol** 121(3): 975-982.

46. Sonnleitner B and Käppeli O (**1986**). Growth of *Saccharomyces cerevisiae* is controlled by its limited respiratory capacity: Formulation and verification of a hypothesis. **Biotechnol Bioeng** 28(6): 927-937.

47. Thomsson E, Larsson C, Albers E, Nilsson A, Franzén CJ, and Gustafsson L (**2003**). Carbon starvation can induce energy deprivation and loss of fermentative capacity in *Saccharomyces cerevisiae*. **Appl Environ Microbiol** 69(6): 3251-3257.

48. Perrone GG, Tan SX, and Dawes IW (**2008**). Reactive oxygen species and yeast apoptosis. **BBA-Mol Cell Res** 1783(7): 1354-1368.

49. Lu C, Brauer MJ, and Botstein D (**2009**). Slow growth induces heat-shock resistance in normal and respiratory-deficient yeast. **Mol Biol Cell** 20(3): 891-903.

50. Boender LGM, de Hulster E, van Maris AJA, Daran-Lapujade P, and Pronk JT (**2009**). Quantitative physiology of *Saccharomyces cerevisiae* at near-zero specific growth rates. **Appl Environ Microbiol** 75(17): 5607-5614.

51. Boender LGM, van Maris AJA, de Hulster EAF, Almering MJH, van der Klei IJ, Veenhuis M, de Winde JH, Pronk JT, and Daran-Lapujade P (**2011**). Cellular responses of *Saccharomyces cerevisiae* at near-zero growth rates: transcriptome analysis of anaerobic retentostat cultures. **FEMS Yeast Res** 11(8): 603-620.

52. Binai NA, Bisschops MMM, van Breukelen B, Mohammed S, Loeff L, Pronk JT, Heck AJR, Daran-Lapujade P, and Slijper M (**2014**). Proteome adaptation of *Saccharomyces cerevisiae* to severe calorie restriction in retentostat cultures. **J Proteome Res** 13(8): 3542-3553.

53. Swinnen E, Wanke V, Roosen J, Smets B, Dubouloz F, Pedruzzi I, Cameroni E, De Virgilio C, and Winderickx J (**2006**). Rim15 and the crossroads of nutrient signalling pathways in *Saccharomyces cerevisiae*. **Cell Div** 1(1): 3.

54. Wei M, Fabrizio P, Hu J, Ge H, Cheng C, Li L, and Longo VD (**2008**). Life span extension by calorie restriction depends on Rim15 and transcription factors downstream of Ras/PKA, Tor, and Sch9. **PLoS Genet** 4(1):e13.

55. Bisschops MMM, Zwartjens P, Keuter SGF, Pronk JT, and Daran-Lapujade P (**2014**). To divide or not to divide: A key role of Rim15 in calorie-restricted yeast cultures. **BBA-Mol Cell Res** 1843(5): 1020-1030.

56. Thomsson E, Svensson M, and Larsson C (**2005**). Rapamycin pre-treatment preserves viability, ATP level and catabolic capacity during carbon starvation of *Saccharomyces cerevisiae*. **Yeast** 22(8): 615-623.

57. Gibney PA, Lu C, Caudy AA, Hess DC, and Botstein D (**2013**). Yeast metabolic and signaling genes are required for heat-shock survival and have little overlap with the heat-induced genes. **Proc Natl Acad Sci U S A** 110(46): E4393-E4402.

58. Westerheide SD and Morimoto RI (**2005**). Heat shock response modulators as therapeutic tools for diseases of protein conformation. **J Biol Chem** 280(39): 33097-33100.

59. Lindquist S and Kim G (**1996**). Heat-shock protein 104 expression is sufficient for thermotolerance in yeast. **Proc Natl Acad Sci U S A** 93(11): 5301-5306.

60. Haslbeck M, Walke S, Stromer T, Ehrnsperger M, White HE, Chen S, Saibil HR, and Buchner J (**1999**). Hsp26: a temperature-regulated chaperone. **EMBO J** 18(23): 6744-6751.

61. Praekelt UM and Meacock PA (**1990**). *HSP12*, a new small heat shock gene of *Saccharomyces cerevisiae*: Analysis of structure, regulation and function. **Mol Gen Genet** 223(1): 97-106.

62. Werner-Washburne M, Stone DE, and Craig EA (**1987**). Complex interactions among members of an essential subfamily of hsp70 genes in *Saccharomyces cerevisiae*. **Mol Cell Biol** 7(7): 2568-2577.

63. Dufourc EJ (**2008**). Sterols and membrane dynamics. **J Chem Biol** 1(1-4): 63-77.

64. Guyot S, Gervais P, Young M, Winckler P, Dumont J, and Davey HM (**2015**). Surviving the heat: heterogeneity of response in *Saccharomyces cerevisiae* provides insight into thermal damage to the membrane. **Environ Microbiol** 17(8): 2982-2992.

65. Caspeta L, Chen Y, Ghiaci P, Feizi A, Buskov S, Hallström BM, Petranovic D, and Nielsen J (**2014**). Altered sterol composition renders yeast thermotolerant. **Science** 346(6205): 75-78.

66. Chester VE (**1963**). The dissimilation of the carbohydrate reserves of a strain of *Saccharomyces cerevisiae*. **Biochem J** 86:153-160.

67. van Roermund CWT, Waterham HR, Ijlst L, and Wanders RJA (**2003**). Fatty acid metabolism in *Saccharomyces cerevisiae*. **Cell Mol Life Sci** 60(9): 1838-1851.

68. Verduyn C, Stouthamer AH, Scheffers WA, and van Dijken JP (**1991**). A theoretical evaluation of growth yields of yeasts. **Antonie van Leeuwenhoek** 59(1): 49-63.

69. Stouthamer AH (**1973**). A theoretical study on the amount of ATP required for synthesis of microbial cell material. **Antonie van Leeuwenhoek** 39(1): 545-565.

70. Ziv N, Brandt NJ, and Gresham D (**2013**). The use of chemostats in microbial systems biology. **J Vis Exp** 80): e50168.

71. Büchs J (**2001**). Introduction to advantages and problems of shaken cultures. **Biochem Eng J** 7(2): 91-98.

72. Nagarajan S, Kruckeberg AL, Schmidt KH, Kroll E, Hamilton M, McInnerney K, Summers R, Taylor T, and Rosenzweig F (**2014**). Uncoupling reproduction from metabolism extends chronological lifespan in yeast. **Proc Natl Acad Sci U S A** 111(15): E1538-E1547.

73. Lee SS, Vizcarra IA, Huberts DHEW, Lee LP, and Heinemann M (**2012**). Whole lifespan microscopic observation of budding yeast aging through a microfluidic dissection platform. **Proc Natl Acad Sci U S A** 109(13): 4916-4920.

74. Basso LC, Basso TO, and Rocha SN (**2011**). Ethanol production in Brazil: the industrial process and its impact on yeast fermentation. In: dos Santos Bernardes MA, editor. Biofuel production - Recent developments and prospects. **Intech, Rijeka;** pp 85-100.

75. Gibson BR, Lawrence SJ, Leclaire JPR, Powell CD, and Smart KA (**2007**). Yeast responses to stresses associated with industrial brewery handling. **FEMS Microbiol Rev** 31(5): 535-569.

76. Hagman A and Piškur J (**2015**). A study on the fundamental mechanism and the evolutionary driving forces behind aerobic fermentation in yeast. **PLoS ONE** 10(1): e0116942.

77. Goddard MR and Greig D (**2015**). *Saccharomyces cerevisiae*: a nomadic yeast with no niche? **FEMS Yeast Res** 15(3): 1-6.

78. Reifenberger E, Boles E, and Ciriacy M (**1997**). Kinetic characterization of individual hexose transporters of *Saccharomyces cerevisiae* and their relation to the triggering mechanisms of glucose repression. **Eur J Biochem** 245(2): 324-333.

79. Entian KD and Kötter P (**2007**). Yeast genetic strain and plasmid collections. **Methods Microbiol** 36(06): 629-666.

80. Nijkamp JF, van den Broek M, Datema E, de Kok S, Bosman L, Luttik M, Daran-Lapujade P, Vongsangnak W, Nielsen J, Heijne WHM, Klaassen P, Paddon CJ, Platt D, Kötter P, van Ham RC, Reinders MJT, Pronk JT, de Ridder D, and Daran JM (**2012**). *De novo* sequencing, assembly and analysis of the genome of the laboratory strain *Saccharomyces cerevisiae* CEN.PK113-7D, a model for modern industrial biotechnology. **Microb Cell Fact** 11:36.

81. Verduyn C, Postma E, Scheffers W, and van Dijken JP (**1992**). Effect of benzoic acid on metabolic fluxes in yeasts: a continuous-culture study on the regulation of respiration and alcoholic fermentation. **Yeast** 8(7): 501-517.

82. Postma E, Verduyn C, Scheffers WA, and van Dijken JP (**1989**). Enzymic analysis of the crabtree effect in glucose-limited chemostat cultures of *Saccharomyces cerevisiae*. **Appl Environ Microbiol** 55(2): 468-477.

83. Canelas AB, Ras C, ten Pierick A, van Dam JC, Heijnen JJ, and van Gulik WM (**2008**). Leakage-free rapid quenching technique for yeast metabolomics. **Metabolomics** 4(3): 226-239.

84. Parrou JL and François J (**1997**). A simplified procedure for a rapid and reliable assay of both glycogen and trehalose in whole yeast cells. **Anal Biochem** 248(1): 186-188.

85. Johnson K, Kleerebezem R, and van Loosdrecht MCM (**2009**). Model-based data evaluation of polyhydroxybutyrate producing mixed microbial cultures in aerobic sequencing batch and fed-batch reactors. **Biotechnol Bioeng** 104(1): 50-67.

86. Jaworek D and Welsch J (**1985**). Adenosine 5'-diphosphate and adenosine 5'-monophosphate. In: Bergmeyer HU, Bergmeyer J, Graßl M, editors. Methods of enzymatic analysis. **Verlagsgesellschaft mbH, Weinheim;** pp 365-370.

87. Piper MDW, Daran-Lapujade P, Bro C, Regenberg B, Knudsen S, Nielsen J, and Pronk JT (**2002**). Reproducibility of oligonucleotide microarray transcriptome analyses: an interlaboratory comparison using chemostat cultures of *Saccharomyces cerevisiae*. **J Biol Chem** 277(40): 37001-37008.

88. Mendes F, Sieuwerts S, de Hulster E, Almering MJH, Luttik MAH, Pronk JT, Smid EJ, Bron PA, and Daran-Lapujade P (**2013**). Transcriptome-based characterization of interactions between *Saccharomyces cerevisiae* and *Lactobacillus delbrueckii* subsp. *bulgaricus* in lactose-grown chemostat cocultures. **Appl Environ Microbiol** 79(19): 5949-5961.

89. Hebly M, de Ridder D, de Hulster EAF, de la Torre Cortes P, Pronk JT, and Daran-Lapujade P (**2014**). Physiological and transcriptional responses of anaerobic chemostat cultures of *Saccharomyces cerevisiae* subjected to diurnal temperature cycles. **Appl Environ Microbiol** 80(14): 4433-4449.

90. Daran-Lapujade P, Jansen MLA, Daran JM, van Gulik W, de Winde JH, and Pronk JT (**2004**). Role of transcriptional regulation in controlling fluxes in central carbon metabolism of *Saccharomyces cerevisiae*: a chemostat culture study. **J Biol Chem** 279(10): 9125-9138.

91. Teste MA, Duquenne M, François J, and Parrou JL (**2009**). Validation of reference genes for quantitative expression analysis by real-time RT-PCR in *Saccharomyces cerevisiae*. **BMC Mol Biol** 10:99.

92. Knijnenburg T, de Winde JH, Daran JM, Daran-Lapujade P, Pronk JT, Reinders MJT, and Wessels LF (**2007**). Exploiting combinatorial cultivation conditions to infer transcriptional regulation. **BMC genomics** 8:25.

5

A central role for TOR signalling in a yeast model for juvenile CLN3 disease

Michael E. Bond[1], Rachel Brown[1], Charalampos Rallis[3,4], Jürg Bähler[3,4] and Sara E. Mole[1,2,3,*]

[1] MRC Laboratory for Molecular Cell Biology, University College London, London WC1E 6BT, UK.

[2] UCL Institute of Child Health, 30 Guilford Street, London WC1N 1EH, UK.

[3] Department of Genetics, Evolution and Environment, University College London, London WC1E 6BT, UK.

[4] Institute of Healthy Ageing, University College London, London WC1E 6BT, UK.

* Corresponding Author: Sara E. Mole, MRC Laboratory for Molecular Cell Biology, University College London; London WC1E 6BT, UK; E-mail: s.mole@ucl.ac.uk

ABSTRACT Yeasts provide an excellent genetically tractable eukaryotic system for investigating the function of genes in their biological context, and are especially relevant for those conserved genes that cause disease. We study the role of *btn1*, the orthologue of a human gene that underlies an early onset neurodegenerative disease (juvenile CLN3 disease, neuronal ceroid lipofuscinosis (NCLs) or Batten disease) in the fission yeast *Schizosaccharomyces pombe*. A global screen for genetic interactions with *btn1* highlighted a conserved key signalling hub in which multiple components functionally relate to this conserved disease gene. This signalling hub includes two major mitogen-activated protein kinase (MAPK) cascades, and centers on the Tor kinase complexes TORC1 and TORC2. We confirmed that yeast cells modelling CLN3 disease exhibit features consistent with dysfunction in the TORC pathways, and showed that modulating TORC function leads to a comprehensive rescue of defects in this yeast disease model. The same pathways may be novel targets in the development of therapies for the NCLs and related diseases.

***Keywords**: Batten disease, NCL, CLN3, btn1, Tor, TORC, S. pombe, yeast.*

Abbreviations:

CWI - cell wall integrity,
NCLs - neuronal ceroid lipofuscinosis,
SAPK - stress-associated protein kinase,
SGA - synthetic genetic array,
WT - wild-type.

INTRODUCTION

Yeasts have long been used as model systems to shed light on basic eukaryotic cell biology, and can provide a rapid and comprehensive route of investigation for genes of unknown function. This is particularly relevant for those genes that are conserved across diverse eukaryotic species and which can be presumed to play a fundamental biological role. A significant health challenge facing current research and drug development is the increasing incidence of age-related neurodegenerative disorders. Elucidating the mechanisms that underlie neurodegeneration is complex, particularly so in common age-related dementias that may have numerous contributing factors. There are, however, monogenic inherited neurodegenerative diseases that present a simpler alternative for investigation and real opportunities to determine these basic cellular changes, and some of the underlying genes are conserved even in yeasts. The genetic tractability of yeasts offers particular advantages to the challenges of understanding disease mechanisms in a relevant biological context and this knowledge can inform therapeutic development for such conserved genetic diseases.

The neuronal ceroid lipofuscinoses (NCLs) are such a group of monogenic neurodegenerative disorders that generally affect children [1]. The most common of these is juvenile CLN3 disease [2]. This disease is characterised by progressive neuronal atrophy that causes visual failure, seizures and a progressive decline in cognitive and motor function. This disease is accompanied by cellular features characteristic of many neurodegenerative conditions [3, 4] that include the dysfunction of core cellular processes, such as reduced lysosomal and autophagic clearance [5, 6] and mitochondrial abnormalities [7]. Moreover, CLN3 disease also leads to specific cellular pathologies characteristic of more common dementias. These include the accumulation of lipofuscin [1], as observed in aged neurons [4], aberrant amyloid-β processing [5], a feature of Alzheimer's disease [8] and α-synuclein accumulation [9], a feature of Parkinson's disease [10].

Juvenile CLN3 disease is caused by mutations in a single gene (*CLN3*) [2], whose function is unknown but which is

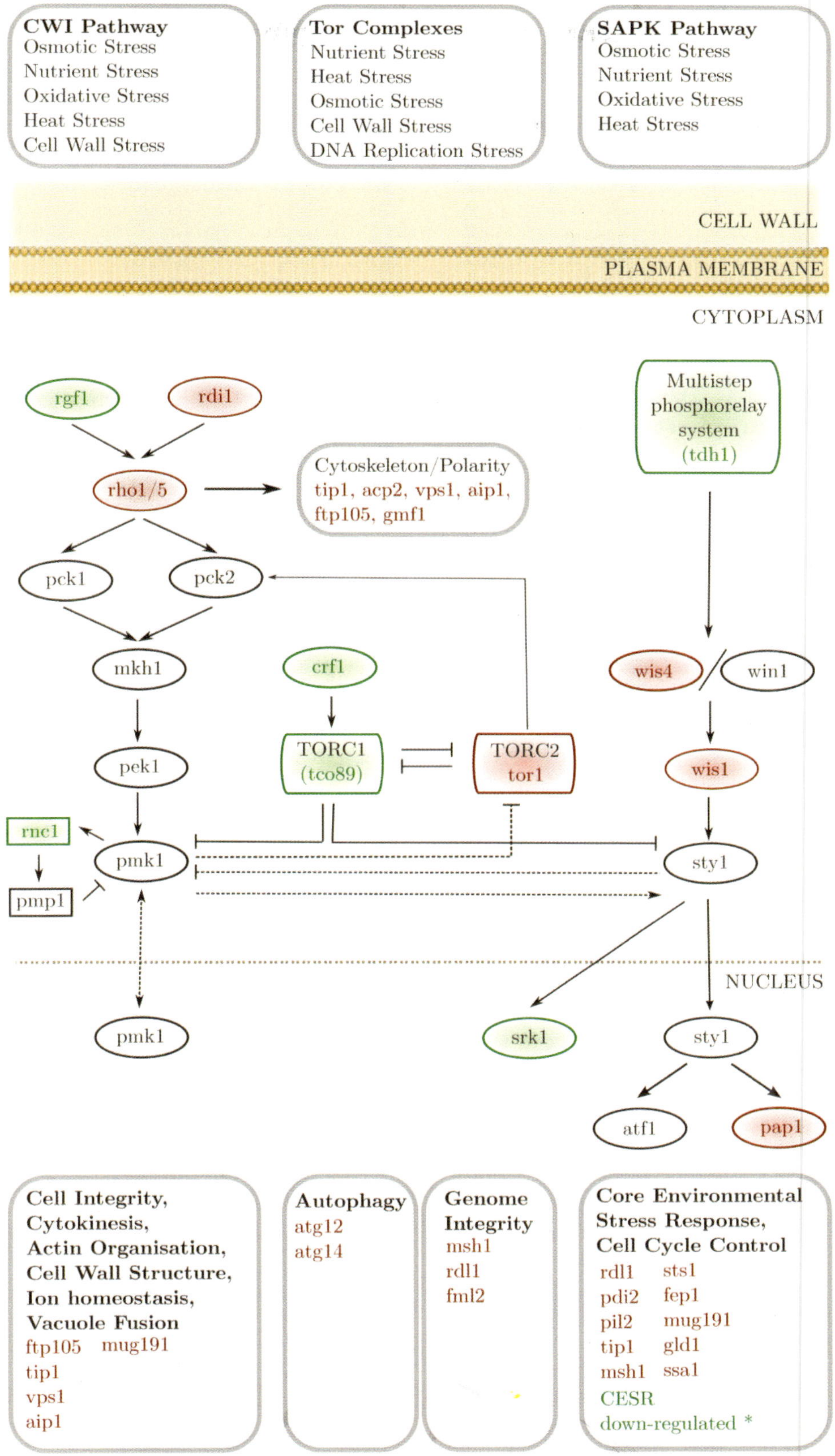

FIGURE 1: Synthetic genetic array (SGA) analysis of the genetic interactions of *btn1*. A schematic of the genes involved in stress responses that genetically interact with *btn1.* Genes in red were identified as negative interactors and those in green as positive. Grey genes are involved in these pathways but were not found to interact with *btn1.* * CESR refers to 'core environmental stress response' genes, a group of genes that respond to most environmental stressors as described by Chen *et al.* (2003) [76].

highly conserved across eukaryotic species [11]. As a consequence, this disease is relatively straightforward to reproduce in experimental systems and, given its similarity to more common dementias, such experimental models are ideal paradigms to study the basic cellular changes that occur in neurodegenerative diseases. The fission yeast *Schizosaccharomyces pombe* contains a single orthologue of *CLN3* (*btn1*) [11]. Work in this yeast has revealed a role for *btn1* in many cellular processes. Like patient cells, yeast lacking *btn1* (*btn1Δ*) have enlarged and less acidic vacuoles [12] and further work has supported roles for *btn1* in vacuolar homeostasis [13]. The ability of *btn1* modeling the most common CLN3 mutation (a 1 Kb deletion) to rescue these vacuolar defects found in the *btn1Δ* null strain has also revealed that this mutant allele retains some function, making juvenile CLN3 disease a mutation-specific disease rather than the consequence of a complete loss of function as originally assumed [14].

A host of further morphological defects exhibited in the *btn1Δ* strain support additional roles for *btn1* in cytokinesis [12] and the organisation of cell polarity [15]. Furthermore, *btn1Δ* cells also display cell wall defects [16]. Importantly, work in this model has expanded our understanding of the disease by revealing that *btn1* is involved in two independent pathways; one pH-dependent and one pH-independent, thus providing the first suggestion that Batten disease is more than a pH-related lysosome disorder [16]. Indeed, a comprehensive metabolomics approach has revealed that *btn1* is required for the regulation of glycolysis and amino acid homeostasis [17]. The involvement of *btn1* in numerous, apparently disparate, pathways may be a result of alterations at the Golgi apparatus, as the number, morphology, and location of thus organelle are affected by its deletion [13]. Lastly, this model has been used successfully to model disease mutations in Btn1p, an effort that has provided valuable insight into their consequences on protein trafficking and function. It was found that equivalent CLN3 disease mutations in *btn1* affect the yeast phenotype in a way that can accurately predict the severity of disease, further establishing yeast as an accurate disease model despite its simplicity. Importantly, the observations reported in the fission yeast model have consistently been confirmed in mammalian systems [5, 18, 19]. Unfortunately however, despite these insights, the molecular processes that underlie cell death in this disease are poorly understood, the function of *CLN3* is unknown, and there remains a significant need for protective therapeutic targets.

A particular advantage of yeast model systems is the availability of genome-wide techniques. Synthetic genetic arrays (SGAs) have proved a particularly powerful means of exploring genetic interactions in yeast species [20]. This approach highlights genes involved in pathways parallel to, or converging with, the query gene. This provides information about functional relationships among genes, as well as processes that suppress the defects associated with a particular mutation. As they are hypothesis-free, SGAs are particularly valuable in the investigation of complex biological problems and those where gene function is unclear. These advantages are particularly relevant for neurodegeneration in general, due to the complexity of the problem, and for juvenile CLN3 disease in particular, due to the lack of a clear gene function. SGAs have previously been employed in budding yeast to investigate mutant huntingtin and α-synuclein toxicity [21]. Such an approach can place the gene under investigation within its biological context and thereby uncover much-needed protective pathways for neurodegenerative disease.

We applied SGA analysis to identify pathways that are altered as a consequence of loss of function of *btn1* in an effort to better understand the molecular consequences of CLN3 disease, and to provide new candidate target pathways and processes for therapeutic development. A third of the genetic interactions that were identified centered on a set of conserved and connected signalling pathways. Manipulation of these pathways leads to a complete rescue of the pleiotropic array of *btn1Δ* phenotypes. This approach represents the most successful rescue of cellular dysfunction in any model for juvenile CLN3 disease to date.

RESULTS

Genome-wide analysis of genetic interactions with *btn1* reveals a central role for TOR kinases

We applied an SGA approach as an unbiased, genome-wide strategy to probe the interactions of a conserved neurodegenerative disease gene (*btn1*). We identified 331 positive interactors and 131 negative interactors of *btn1* (Table S1A and B). A large number of negative interactions (n = 39; 29.7%) mapped to a set of highly interconnected signalling processes, which center on the Tor kinase complexes TORC1 and TORC2 (Fig. 1). Core components of both complexes display genetic interactions with *btn1*; *tco89,* encoding a component of TORC1, was found to interact positively with *btn1,* while *tor1,* encoding the Tor kinase of TORC2, was found to interact negatively with *btn1*. Such a pattern suggests that TORC2 signalling is beneficial to the fitness of *btn1Δ* cells, whereas TORC1 signalling is detrimental, an observation that is in keeping with the idea that the two complexes play opposing roles and undergo mutual repression [22-24]. *btn1* was also found to negatively interact with genes encoding components of the connected cell wall integrity (CWI) and stress-associated protein kinase (SAPK) MAP kinase cascades, and a number of downstream stress regulated genes (Table S1A and B).

btn1Δ cells display features consistent with dysfunctional Tor signalling

The interaction of *btn1* with core TORC components, and the link to surrounding signalling processes, provides compelling evidence for the importance of Tor signalling in cells lacking *btn1*. To validate this observation, we investigated *btn1Δ* cells for features consistent with Tor dysfunction. Repression of TORC1 activity is required to mount a correct response to nitrogen limitation [24]. As loss of TORC1 activity was beneficial in cells lacking *btn1*, we hypothesised that these cells may display features consistent with dysregulation of TORC1 repression, and respond poorly to nitrogen limitation. To test this idea, we grew wild-type and *btn1Δ* cells in minimal media, and minimal media lacking a nitrogen source, over 72 h. Viability was monitored at 24 h intervals using propidium iodide to label dead cells and calcofluor white to label the total cell population (Fig. 2A). Cells lacking *btn1* displayed a consistently lower viability in media lacking nitrogen, falling to 81.5 ± 2.2% after 24 h compared to 98.7 ± 0.27% in minimal media containing nitrogen ($P < 0.0015$, unpaired t test), and remaining lower throughout the time course. Wild-type cells displayed no change in viability when cultured in media lacking nitrogen.

TORC1 repression and TORC2 signalling is also required to mount a response to glucose limitation [25, 26]. Cells that are defective in TORC2 function do not respond appropriately to glucose limitation by reducing their cell length [26]. This is a relatively uncommon phenotype, with only *tor1* mutants and mutants in the Ca^{2+}/calmodulin-dependent-like gene *ssp1* known to display this phenotype [26, 27]. To test the ability of *btn1Δ* cells to adapt to conditions of limited glucose, we assessed both cell growth and the morphological response of these cells when grown with an alternative carbon source (glycerol). Cells lacking *btn1* displayed a clear growth defect under these conditions (Fig. 2B). Wild-type and *btn1Δ* cells grown for 6 h in media containing either glucose or glycerol as a carbon source were stained with calcofluor white to visualise the cell wall. Cell elongation was determined using a measure of cell circularity and given a score of between 0 and 1, where 0 represents a perfectly round cell (Fig. 2C and D). Wild-type cells displayed a significant reduction in cell elongation upon growth in glycerol (0.25 ± 0.01 to 0.16 ± 0.01, $P \leq 0.01$). Cells lacking *btn1,* however, displayed no significant change (0.32 ± 0.02 to 0.31 ± 0.02), consistent with a defect in their response to glucose limitation that is most likely linked to Tor1 function.

Lastly, TORC2 and the connected CWI pathway are also required for resistance to high temperature [26]. This temperature sensitivity of TORC2 and CWI pathway components is osmoremedial, i.e. it can be rescued by hypertonic growth media [28]. We confirmed the temperature sensitivity of *btn1Δ* cells, and that it can be rescued with 2M

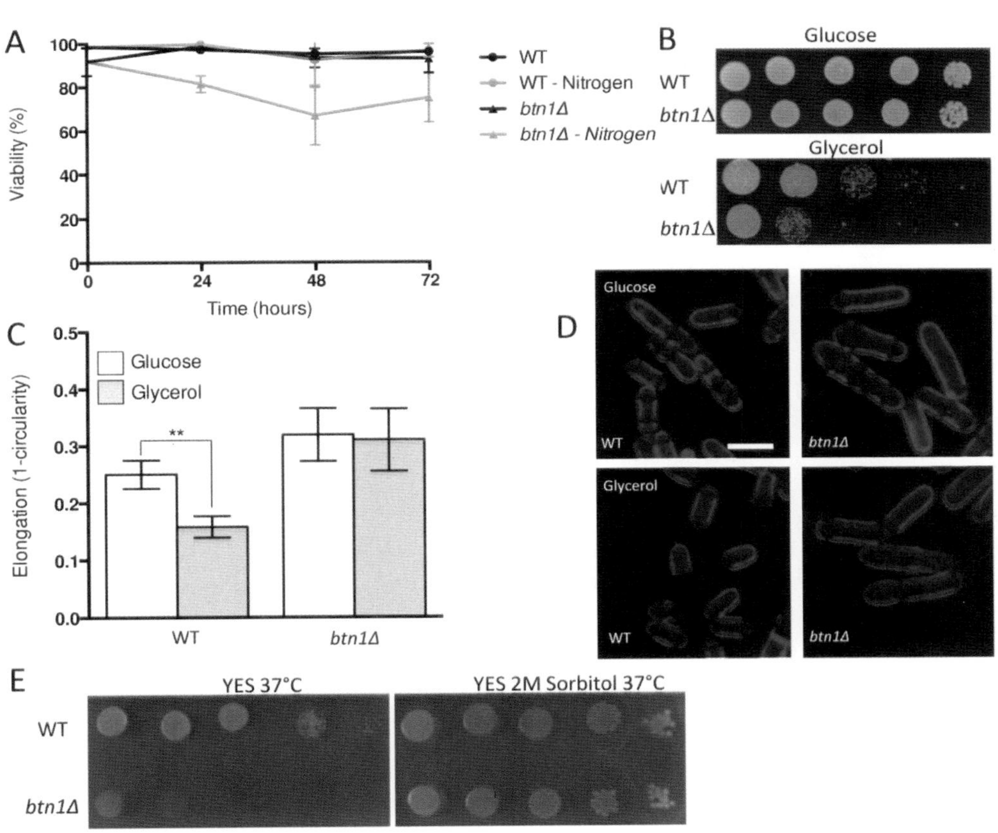

FIGURE 2: *btn1Δ* cells display features consistent with dysfunction in TOR signalling processes. (A) Cell viability upon nitrogen limitation was determined over periods of up to 72 hours in wild-type (WT) and *btn1Δ* cells, using the cell impermeable nucleotide stain propidium iodide to stain dead cells and calcofluor white to stain the total cell population. Cells were cultured in either MM or MM lacking a nitrogen source (NH_4Cl). 500 cells were scored for viability per data set, and data shown is a mean (± SEM) of 3 independent experiments. **(B)** WT cells and *btn1Δ* cells were serially diluted from a log-phase culture (1 x 10^6 cells/ml), and spotted onto plates containing either glucose or glycerol as a carbon source. Plates were then incubated at 30°C for 6 - 7 days to determine growth on fermentative and non-fermentative carbon sources. Images are representative of three independent experiments. **(C)** The morphological response of WT and *btn1Δ* cells to growth on glycerol was analysed following 6 hours in culture using a measure of cell elongation, on a scale of 0 to 1, where 0 represents a perfectly round cell $(1 - (\frac{4\,\pi\,\text{area}}{\text{perimeter}^2}))$. Data shown is a mean (± SEM) of 5 independent experiments. Statistical significance between each condition was determined using a one-way ANOVA with a Tukey's multiple comparison post-test (** = $P < 0.01$). **(D)** Representative images of experiments as performed in (C) are shown. Scale bar represents 10 μm. **(E)** WT and *btn1Δ* cells were also serially diluted from a log-phase culture (1 x 10^6 cells/ml) and spotted onto YES plates and YES plates containing 2M sorbitol. Plates were then incubated at 37°C for 3 - 4 days to determine the growth at high temperature and the influence of osmotic stabilisation. Images are representative of three independent experiments.

sorbitol (Fig. 2E), consistent with previous observations [16].

These data together indicate that cells lacking *btn1* display many features consistent with defects in Tor signalling and function. Further, ectopic expression of *btn1* was able to rescue all these aspects of the mutant phenotype (Fig. S1A - D). Combined with the SGA results, these results suggest that Tor signalling is critical to the defects observed in

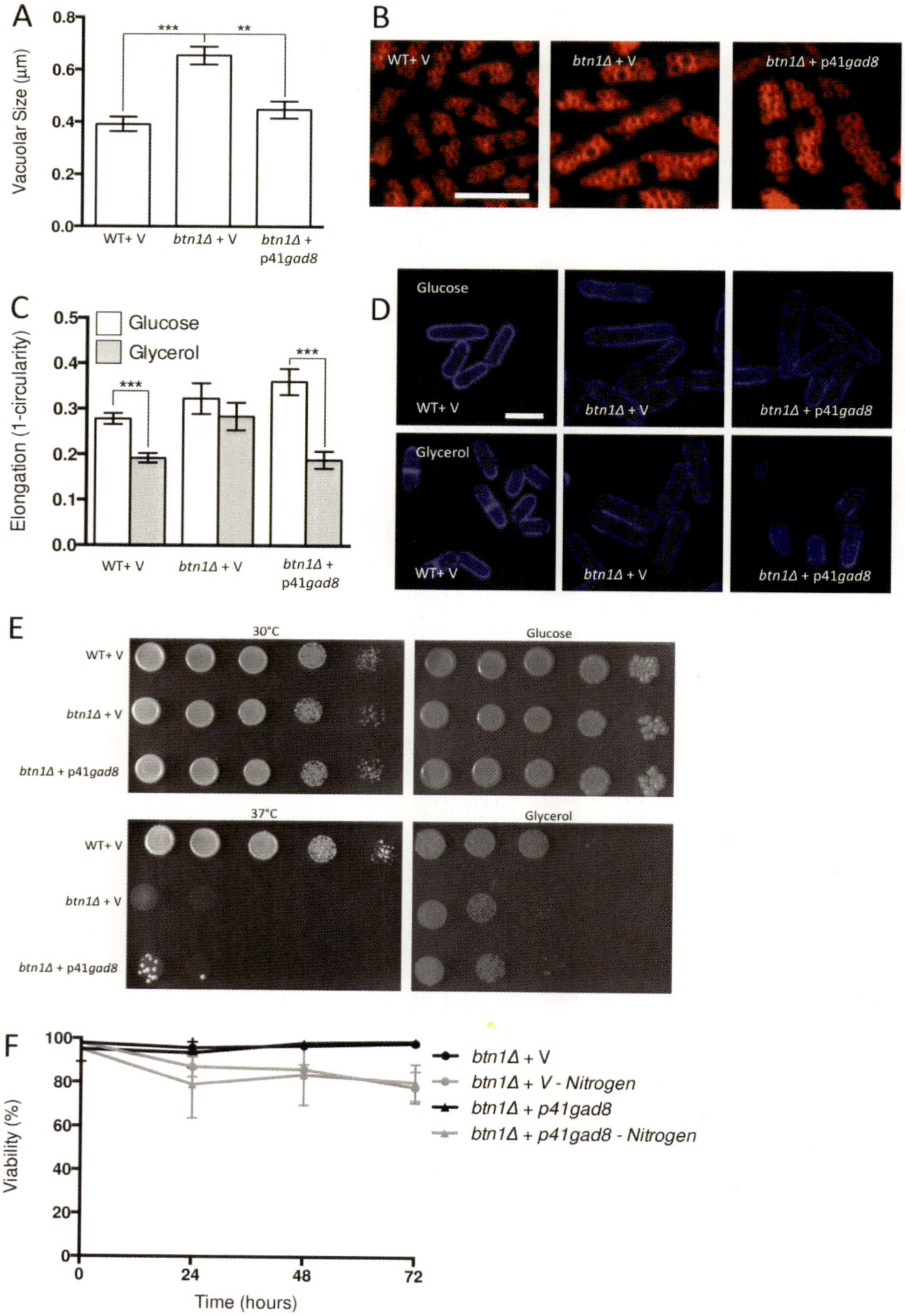

FIGURE 3: Increasing TORC2-dependent signalling rescues aspects of *btn1Δ* phenotype. (A) Cells lacking *btn1* were transformed with an expression vector containing *gad8*. Vacuole size was measured in these cells, in addition to WT and *btn1Δ* cells containing empty vector, following staining with the vital dye FM4-64. The diameter of 300 vacuoles was determined per data set, and data shown is a mean (± SEM) of 4 independent experiments. Statistical analysis was performed using a one-way ANOVA with a Tukey's multiple comparison post-test (** = $P < 0.01$, *** = $P < 0.001$). **(B)** Representative images of experiments as performed in (A) are shown. Scale bar represents 10 μm. **(C)** The morphological response of these populations to growth on glycerol was analysed following 6 hours in culture using a measure of cell elongation, on a scale of 0 to 1, where 0 represents a perfectly round cell $(1-(\frac{4\,\pi\,\text{area}}{\text{perimeter}^2}))$. Data shown is a mean (± SEM) of 5 independent experiments. Statistical significance between each condition was determined using a one-way ANOVA with a Tukey's multiple comparison post-test (*** = $P < 0.001$). **(D)** Representative images of experiments as performed in (C) are shown. Scale bar represents 10 μm. **(E)** These cells were serially diluted from a log-phase culture (1 x 10^6 cells/ml) and spotted onto YES plates. Plates were then incubated at 30°C or 37°C for 3 - 4 days to determine growth at high temperature. They were also spotted onto plates containing either glucose or glycerol as a carbon source. Plates were then incubated at 30°C for 6 - 7 days to determine growth under non-fermentative conditions. Images are representative of three independent experiments. **(F)** Viability upon nitrogen limitation was determined over periods of up to 72 hours in these cell populations, using propidium iodide to stain dead cells and calcofluor white to stain all cells. Cells were cultured in either MM or MM lacking a nitrogen source (NH_4Cl). 500 cells were scored for viability per data set, and data shown is a mean (± SEM) of 3 independent experiments.

btn1Δ cells.

TORC2 and CWI pathway: modulating distinct signalling nodes elicit different levels of correction in *btn1Δ* cells

Given the negative genetic interaction between *btn1* and *tor1*, increasing TORC2 activity could alleviate aspects of the *btn1Δ* phenotype. We initially explored this approach through overexpression of the downstream kinase Gad8, which is a multicopy suppressor of aspects of the loss of TORC2 [29]. A key phenotype in cells lacking *btn1*, and one that is strongly linked to the phenotype of CLN3 disease, is a change in vacuole homeostasis [12]. One of the clearest manifestations of this feature in *btn1Δ* cells is an increase in vacuole size [12]. We investigated vacuole size in wild-type and *btn1Δ* cells containing empty vector alone (as a control), as well as *btn1Δ* cells expressing *gad8* from the *nmt1* promoter of the pREP41 plasmid (Fig. 3A and B). Consistent with previous reports, *btn1Δ* cells exhibited significantly larger vacuoles than wild-type cells (0.65 ± 0.03 μm compared to 0.39 ± 0.01 μm, P ≤ 0.001). The over-expression of *gad8* in *btn1Δ* cells significantly rescued vacuole size (0.40 ± 0.03 μm) (P ≤ 0.01), restoring them to near that of wild-type. These data provide the first indication that increasing TORC2 function significantly rescues a disease-relevant phenotype of *btn1Δ* cells.

As discussed above, the ability of cells to respond to glucose-limitation with a change in cell size requires TORC2 function [26]. To investigate whether the inability of *btn1Δ* cells to undergo this morphological response can be rescued by increasing TORC2 pathway activity, we assessed the elongation of *btn1Δ* cells overexpressing *gad8* following growth for 6 hours in media containing either glucose or glycerol as a carbon source (Fig. 3C and D). Wild-type and *btn1Δ* cells containing empty vector alone displayed comparable responses to cells without vector (WT glucose - 0.28 ± 0.01, WT glycerol - 0.19 ± 0.01, P ≤ 0.001; *btn1Δ* glucose - 0.32 ± 0.02, *btn1Δ* glycerol - 0.28 ± 0.01, ns). The overexpression of *gad8* was able to significantly rescue this morphological response, despite these cells being elongated under growth in glucose (glucose - 0.36 ± 0.01, glycerol - 0.19 ± 0.01, P ≤ 0.001).

Despite this rescue, expression of *gad8* was unable to significantly rescue the growth of *btn1Δ* cells on glycerol or at high temperature, suggesting that the ability of these cells to adapt to heat and nutrient stress is still impaired (Fig. 3E). Similarly, expression of *gad8* was unable to rescue the viability of *btn1Δ* cells grown in the absence of a nitrogen source (Fig. 3F).

Given that the phenotypes not rescued by increasing TORC2 activity were linked to stress and nutrient adaptation, we reasoned that modulating processes more closely linked to the CWI pathway might be more effective in correcting these phenotypes. These processes seem particularly relevant in this instance, as this pathway is known to respond to both heat and glucose limitation. In particular, mutants of the *pmk1* MAP kinase, like *btn1Δ* mutants, display a partial growth defect on glycerol. Further, this pathway is required for proper activation of Sty1 (the MAP kinase of the SAPK pathway) under glucose-limitation, integrating these processes with further identified *btn1* interactors [30].

To increase CWI activity, we chose to increase Rho GTPase levels in these cells, which represent a key hub in CWI pathway regulation, by overexpressing the *btn1* interactor *rho5* and its essential paralogue *rho1*. Overexpression of *rho1* significantly decreased vacuole size in *btn1Δ* cells (P ≤ 0.001) to near that of wild-type cells containing empty vector alone (WT - 0.30 ± 0.02 μm, *btn1Δ* - 0.47 ± 0.04 μm, *btn1Δ* with p41*rho1* - 0.32 ± 0.01 μm) (Fig. 4A and B). The overexpression of *rho5* led to an over-correction of this phenotype (0.19 ± 0.01 μm), producing vacuoles significantly smaller than those of *btn1Δ* cells (P ≤ 0.001) and wild-type cells containing vector alone (P ≤ 0.05). Consistent with these observations, over-expression of *pmp1*, which inhibits the activity of *pmk1* [31], caused an increase in vacuole size in *btn1Δ* cells (*btn1Δ* with p41*pmp1* - 0.80 ± 0.08 μm vs *btn1Δ* - 0.52 ± 0.04 μm, P ≤ 0.01) but not in wild-type cells (WT with p41*pmp1* – 0.24 ± 0.01 μm vs WT – 0.23 ± 0.02, ns) (Fig. 4C). Such data indicate that increasing CWI pathway activity is able to correct the vacuole defect of *btn1Δ* cells, and directly inhibiting this pathway exacerbates the vacuole defect. The over-correction by *rho5* could indicate a strong dose-dependence in this particular rescue. Indeed, previous work has demonstrated such dose dependence in the relationship between *btn1* and vacuole size [12].

We next investigated whether the Rho GTPases were able to rescue the morphological response of these cells to glycerol, as seen with *gad8*. We looked just at *rho1* in this case, as *rho5* overexpression has a pronounced effect on cell shape and septation [32], making morphological comparisons difficult (Fig. S2). Overexpression of *rho1* elicited a rescue of this response, with cells displaying a significant reduction in cell elongation when grown in glycerol containing media (0.28 ± 0.02 to 0.16 ± 0.01, P ≤ 0.01) (Fig. 4D and E).

Further to a rescue of the morphological response to glycerol, both *rho1* and *rho5* were able to correct the growth defect of *btn1Δ* cells on glycerol as a carbon source. In addition, *rho1* was able to rescue the temperature sensitivity phenotype of these cells (Fig. 4F). The overexpression of *rho5* did not correct this phenotype, although, given the link between heat-sensitivity and septation defects in *btn1Δ* cells [15] and the enhanced septation defect of cells overexpressing *rho5* (Fig. S2), this is not wholly unexpected.

In addition to these phenotypes, overexpression of *rho1* and *rho5* rescued the viability of *btn1Δ* cells when grown in the absence of a nitrogen source, with viability remaining higher than 95% throughout the course of the experiment (Fig. 4G). These data provide evidence for a profound positive interaction between CWI pathway signalling and *btn1*, indicating that such processes are intimately linked with the defects that occur in *btn1Δ* cells.

Reducing TORC1 activity corrects defects in *btn1Δ* cells

TORC1 activity is a key signal of a favourable environment, promoting proliferation and suppressing stress-responsive

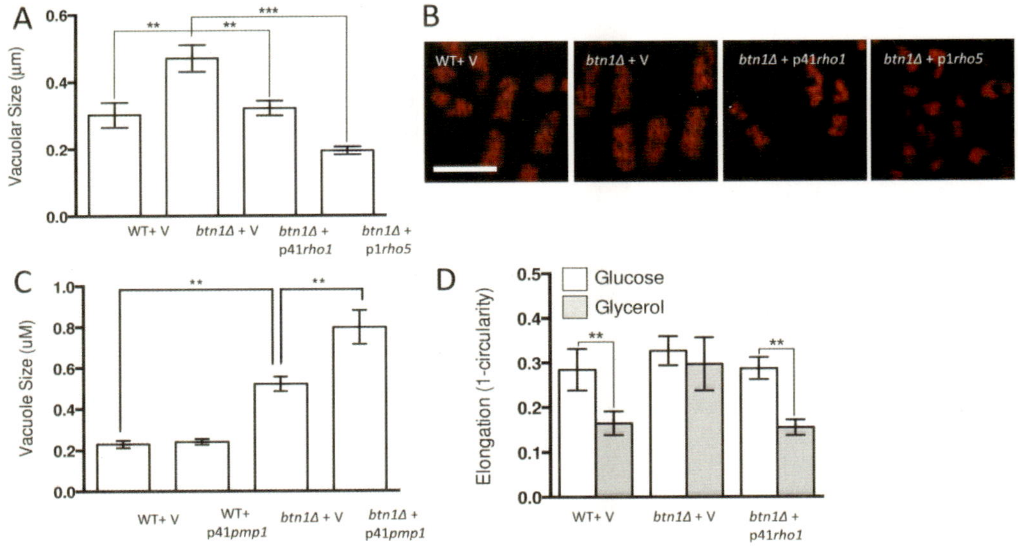

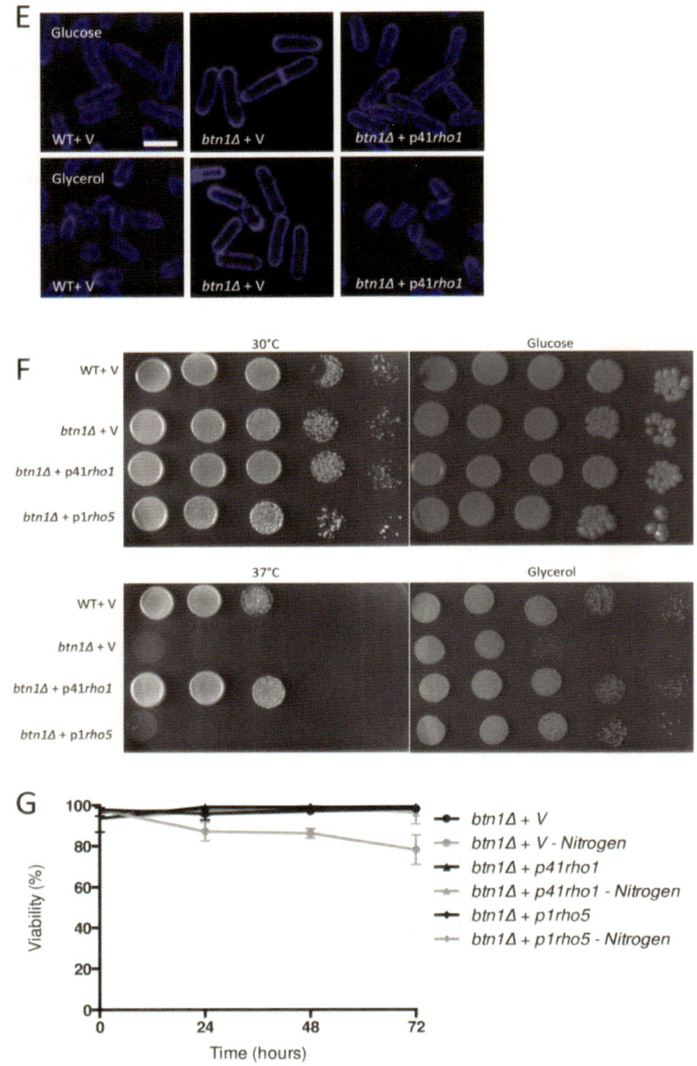

FIGURE 4: Overexpression of Rho rescues all aspects of *btn1Δ* phenotype. (A) Cells lacking *btn1* were transformed with an expression vector containing *rho1* or *rho5*. Vacuole size was measured in these cells, in addition to wild-type (WT) and *btn1Δ* cells containing empty vector, following staining with the vital dye FM4-64. The diameter of 300 vacuoles was measured per data set, and data shown is a mean (±SEM) of 4 independent experiments. Statistical analysis was performed using a one-way ANOVA with a Tukey's multiple comparison post-test (** = P < 0.01, *** = P < 0.001). **(B)** Representative images are shown in lower panels. Scale bar represents 10 µm. **(C)** WT and cells lacking *btn1* were transformed with an expression vector containing *pmp1*. Vacuoles were stained and measured as for (A). Statistical analysis was performed using a one-way ANOVA with a Tukey's multiple comparison post-test (** = P < 0.01). **(D)** The morphological response of WT and *btn1Δ* cells containing empty vector as well as *btn1Δ* cells overexpressing *rho1* to growth on glycerol was analysed following 6 hours in culture using a measure of cell elongation, on a scale of 0 to 1, where 0 represents a perfectly round cell $(1-(\frac{4\,\pi\,\text{area}}{\text{perimeter}^2}))$. Data shown is a mean (±SEM) of 5 independent experiments. Statistical significance between each condition was determined using a one-way ANOVA with a Tukey's multiple comparison post-test (** = P < 0.01). **(E)** Representative images of experiments as performed in (D) are shown. Scale bar represents 10 µm. **(F)** Wild-type and *btn1Δ* cells containing empty vector as well as *btn1Δ* cells overexpressing *rho1* and *rho5* were serially diluted from a log-phase culture (1 x 10^6 cells/ml) and spotted onto YES plates. Plates were then incubated at 30°C or 37°C for 3-4 days to determine growth at high temperature. They were also spotted onto plates containing either glucose or glycerol as a carbon source. Plates were then incubated at 30°C for 6-7 days to determine growth under non-fermentative conditions. Images are representative of three independent experiments. **(G)** Viability upon nitrogen limitation was determined over periods of up to 72 hours in these cell populations, using propidium iodide to stain dead cells and calcofluor white to stain all cells. Cells were cultured in either MM or MM lacking a nitrogen source (NH_4Cl). 500 cells were scored for viability per data set, and data shown is a mean (±SEM) of 3 independent experiments.

processes [24]. As a consequence, it is antagonistic to many of the processes explored in this study that improved the *btn1Δ* phenotype. TORC1 represses TORC2 function [23], and negatively regulates Sty1 [33], which displays extensive cross-talk with the CWI pathway [30]. This relationship is highlighted by the fact that the core TORC1 component gene *tco89* is a positive interactor of *btn1*.

In order to test whether repression of TORC1 activity could rescue the phenotypes of *btn1Δ* cells, we first overexpressed a dominant-negative form of the upstream activator of TORC1, Rhb1 (*rhb1*^{D60K}) [34]. Upon overexpression of dominant-negative *rhb1* (*dnrhb1*) in *btn1Δ* cells, we observed a significant reduction ($P \leq 0.001$) in vacuole size compared to *btn1Δ* cells containing vector alone (*btn1Δ* with p41*dnrhb1* - 0.40 ± 0.04 µm, *btn1Δ* with empty vector - 0.65 ± 0.03 µm) (Fig. 5A and B). This result was comparable to the vacuole size of wild-type cells (0.39 ± 0.01 µm), indicating a rescue of vacuole morphology upon repression of TORC1 activity.

We next investigated the ability of TORC1 repression to rescue the nutrient-sensing defects of cells lacking *btn1*. The overexpression of *dnrhb1* led to a significant rescue of the morphological response of these cells to glucose-limitation, following 6 hours growth in glycerol as a carbon source (Fig. 5C and D). Cell elongation fell from 0.29 ± 0.02 to 0.18 ± 0.01 when grown in glycerol as opposed to glucose, a response comparable to that observed in wild-type cells (glucose - 0.28 ± 0.01, glycerol - 0.19 ± 0.01). This also corresponded to a rescue of the growth defect of these cells on glycerol (Fig. 5E). Further to the glycerol growth

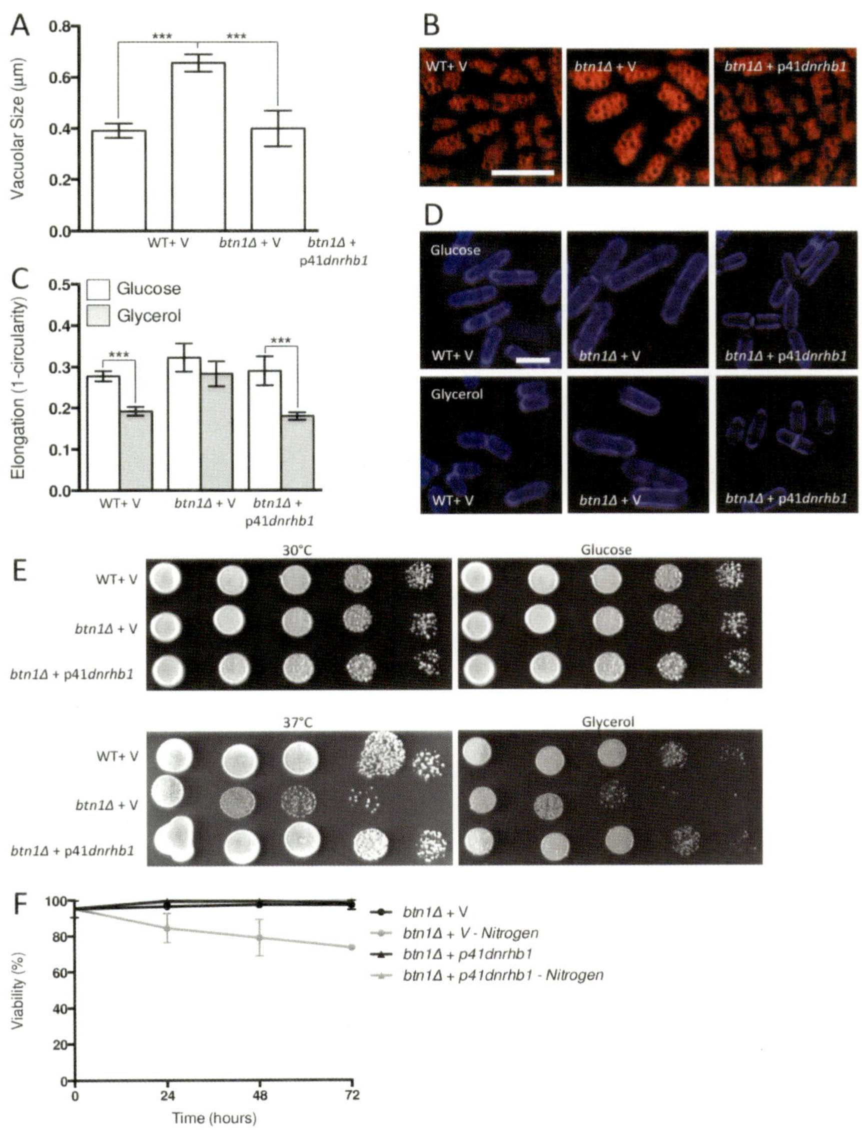

FIGURE 5: Inhibiting TORC1 function rescues the defects in *btn1Δ* cells. (A) Cells lacking *btn1* were transformed with an expression vector containing a dominant-negative form of *rhb1* (*dnrhb1*). Vacuole size was measured in these cells, in addition to wild-type (WT) and *btn1Δ* cells containing empty vector, following staining with the vital dye FM4-64. The diameter of 300 vacuoles was measured per data set, and data shown is a mean (± SEM) of 4 independent experiments. Statistical analysis was performed using a one-way ANOVA with a Tukey's multiple comparison post-test (*** = P < 0.001). **(B)** Representative images of experiments as performed in (A) shown. Scale bar represents 10 µm. **(C)** The morphological response of these populations to growth on glycerol was analysed following 6 hours in culture using a measure of cell elongation, on a scale of 0 to 1, where 0 represents a perfectly round cell $(1 - (\frac{4\,\pi\,\text{area}}{\text{perimeter}^2}))$. Data shown is a mean (± SEM) of 5 independent experiments. Statistical significance between each condition was determined using a one-way ANOVA with a Tukey's multiple comparison post-test (*** = P < 0.001). **(D)** Representative images of experiments as performed in (C) shown. Scale bar represents 10 µm. **(E)** These cells were serially diluted from a log-phase culture (1 x 10^6 cells/ml) and spotted onto YES plates. Plates were then incubated at 30°C or 37°C for 3-4 days to determine growth at high temperature. They were also spotted onto plates containing either glucose or glycerol as a carbon source. Plates were then incubated at 30°C for 6-7 days to determine growth under non-fermentative conditions. Images are representative of three independent experiments. **(F)** Viability upon nitrogen limitation was determined over periods of up to 72 hours in these cell populations, using propidium iodide to stain dead cells and calcofluor white to stain all cells. Cells were cultured in either MM or MM lacking a nitrogen source (NH_4Cl). 500 cells were scored for viability per data set, and data shown is a mean (± SEM) of 3 independent experiments. Scale bar represents 10 µm.

defect, *dnrhb1* overexpression also rescued the heat sensitivity of *btn1Δ* cells (Fig. 5E). Finally, this construct was also able to elicit a complete rescue of viability under nitrogen limiting conditions, with viability remaining above 95% throughout the course of the experiment (Fig. 5F).

Given the substantial rescue of *btn1Δ* cells by *dnrhb1*-mediated activation of TORC1, we wanted to confirm the interaction between *btn1* and TORC1 by another means. We chose to use two pharmacological inhibitors of TORC1, rapamycin (allosteric) and caffeine (competitive inhibitor of ATP binding) [35]. TORC1 antagonism leads to cell rounding, a response that mimics the morphological change ordinarily exhibited by wild-type cells in response to nitrogen and amino acid starvation. Therefore, we examined the effect of TORC1 antagonism on cell rounding [36] and on vacuole size, heat sensitivity, and growth in glycerol. As expected, wild-type cells displayed a significant reduction in cell elongation upon rapamycin treatment for 6 hours (0.26 ± 0.02 to 0.15 ± 0.01, P ≤ 0.01) (Fig. 6A and B). This reduction in cell elongation did not change further upon exposure to both rapamycin and caffeine [35]. Cells lacking *btn1,* however, did not respond to rapamycin, displaying only a slight morphological change (0.30 ± 0.01 to 0.25 ± 0.02) that was slightly enhanced by the addition of caffeine and rapamycin, but again not significantly (0.22 ± 0.03) (Fig. 6A and B). The addition of caffeine and rapamycin to wild-type cells did not significantly affect vacuole size (vehicle - 0.37 ± 0.05 μm, rapamycin - 0.31 ± 0.03, caffeine and rapamycin 0.24 ± 0.01) (Fig. 6C and D). Rapamycin alone had no effect on the vacuole size of cells lacking *btn1*, but, in contrast, a combination of caffeine and rapamycin significantly (P ≤ 0.001) reduced vacuole size (vehicle - 0.58 ± 0.03 μm, rapamycin - 0.55 ± 0.04, caffeine and rapamycin 0.30 ± 0.02) (Fig. 6C and D). Further, treatment with rapamycin alone elicited a rescue of both the heat sensitivity defect and glycerol growth defect of cells lacking *btn1* (Fig. 6E). The ability of these pharmacological treatments to rescue the defects in vacuole size, heat sensitivity and growth in glycerol of *btn1Δ* cells appears to be closely linked to their ability to target TORC1 signalling itself, as opposed to changes

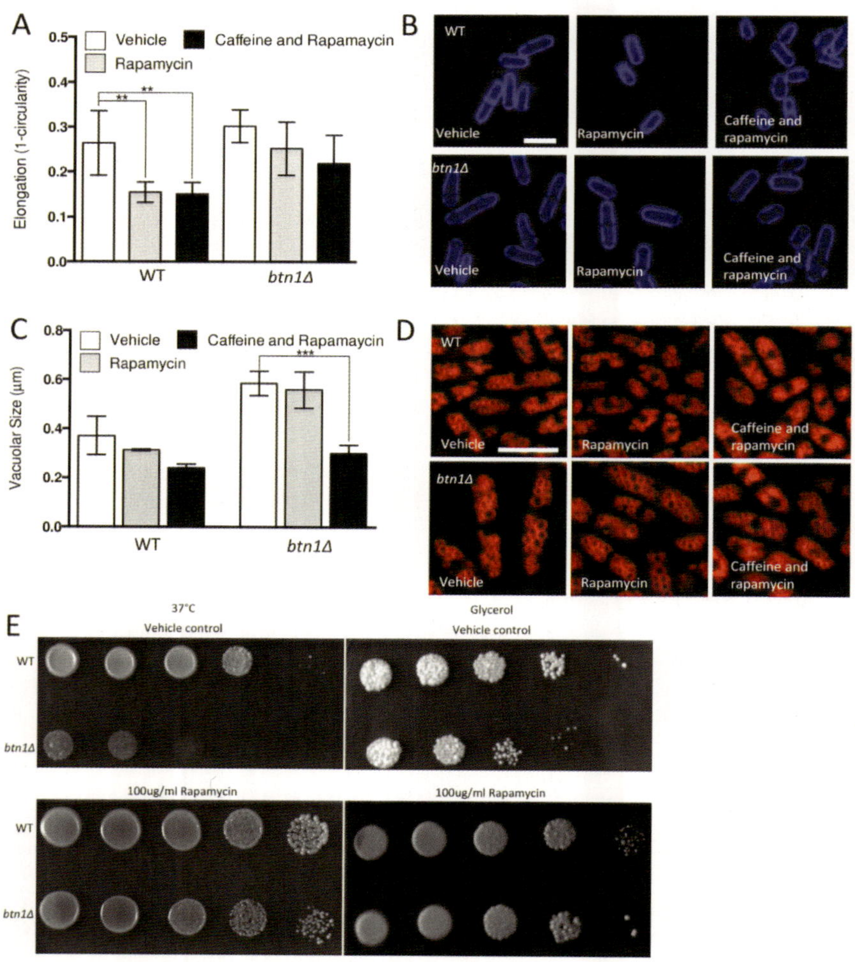

FIGURE 6: Pharmacological inhibition of TORC1 function replicates rescue by overexpression of dominant-negative *rhb1.* **(A)** The morphological response of wild-type (WT) and *btn1Δ* cells to the TORC1 antagonists rapamycin (100 μg/ml) and caffeine and rapamycin combined (10 mM and 100 μg/ml respectively) was analysed following 6 hours in culture using a measure of cell elongation, on a scale of 0 to 1, where 0 represents a perfectly round cell $(1-(\frac{4\,\pi\,\text{area}}{\text{perimeter}^2}))$. Data shown is a mean (± SEM) of 6 independent experiments. Statistical significance between each condition was determined using a one-way ANOVA with a Tukey's multiple comparison post-test (** = P < 0.01). **(B)** Representative images of experiments as performed in (A) are shown. Scale bar represents 10 μm. **(C)** Wild-type and *btn1Δ* cells were grown for 6 hours in MM alone or with the TORC1 antagonists rapamycin (100 μg/ml) and caffeine and rapamycin (10 mM and 100 μg/ml respectively). Vacuole size was then measured in these cells following staining with the vital dye FM4-64. The diameter of 300 vacuoles was measured per data set, and data shown is a mean (± SEM) of 4 independent experiments. Statistical analysis was performed using a one-way ANOVA with a Tukey's multiple comparison post-test (*** = P < 0.001). **(D)** Representative images of experiments as performed in (C) are shown. Scale bar represents 10 μm. **(E)** WT and *btn1Δ* cells were serially diluted from a log-phase culture (1 x 10^6 cells/ml) and spotted onto YES plates either containing or lacking rapamycin (100 μg/ml). Plates were then incubated at 37°C for 3-4 days to determine growth at high temperature. They were also spotted onto plates containing glycerol as a carbon source either containing or lacking rapamycin (100 μg/ml). Plates were then incubated at 30°C for 6-7 days to determine growth under non-fermentative conditions. Images are representative of three independent experiments.

in the activity of the connected SAPK pathway, as expression of a constitutively active *wis1* mutant (*wis1DD*) [37] did not rescue these phenotypes (Fig S3A - C). These data confirm that activation of the TORC1 pathway pharmacologically in *btn1*Δ cells rescues most of the phenotypes arising from loss of *btn1* function. However, the pharmacological inhibition of the TORC1 pathway is not completely equivalent to the inhibition of the TORC1 pathway using *dnrhb1* in *btn1*Δ cells.

DISCUSSION

In this study, SGA analysis was used as an unbiased approach to identify the genetic interactors of *btn1,* and place *btn1* into its biological context within the whole cell. This approach highlighted a key role for the Tor kinase complexes, TORC1 and TORC2, in the dysfunctional changes that occur in fission yeast lacking *btn1*, as one third of the interactions connected directly with pathways in which these complexes are active. Further, we have shown that *btn1*Δ cells display defects in their stress response to nitrogen and glucose limitation, in addition to osmoremedial heat sensitivity, consistent with TORC1 and TORC2 dysfunction in these cells (Fig. 7).

Specifically, the SGA identified negative interactions in *tor1*, the core kinase component of TORC2, multiple components of the stress-activated protein kinase (SAPK) pathway, and a component of the linked CWI pathway. In addition, *tco89*, encoding a core component of TORC1, which antagonises both the SAPK cascade and TORC2 signalling, was identified as a positive genetic interactor of *btn1*. *btn1* was also found to interact negatively with a number of genes that display stress-responsive expression.

TORC1 negatively regulates a number of processes associated with cellular catabolism and adaptation to stress [24]. One of these roles is in the repression of autophagy. Autophagy similarly appears important to the fitness of *btn1*Δ cells, as *atg14* and *atg12* were identified as negative interactors of *btn1*, and both are required for autophagosome formation [38]. In support of this work, and the use of yeast as a model, defective autophagy has previously been linked to CLN3 disease [39]. In repressing stress-responsive processes, TORC1 is known to negatively regulate *sty1*, the mitogen-activated protein (MAP) kinase and component of the stress-activated protein kinase (SAPK) cascade [33]. Both *wis4* and *wis1*, encoding the MAP kinase kinase kinase and MAP kinase kinase of the SAPK pathway,

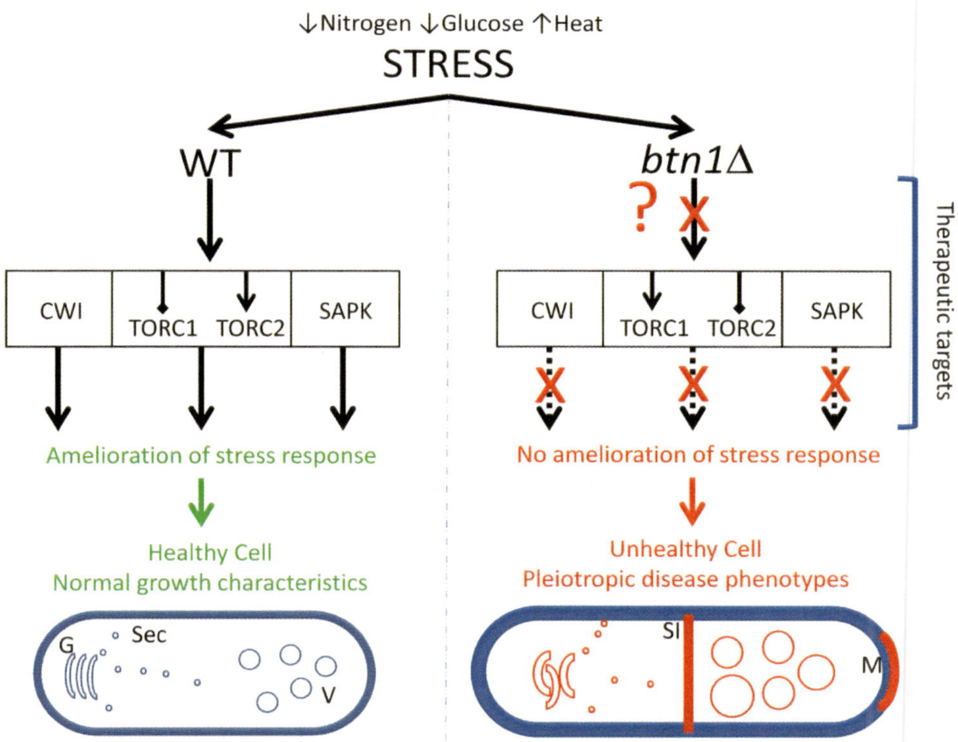

FIGURE 7: Diagrammatic summary of the effect of loss of *btn1* on the response to stress. The left panel shows the response of healthy cells to stresses applied in this study, such as low nitrogen or glucose or raised temperature. These cells are able to respond via interconnected signaling pathways such as CWI, TORC and SAPK. Activation of these pathways allows the cell to respond to the effects of stress and restore cell homeostasis and integrity. In contrast, cells lacking *btn1* are unable to respond to the applied stresses (represented by '?' and 'X') and undergo a variety of previously reported responses that include an increase in vacuole size, swollen and disorganized Golgi and mistrafficking of vacuole enzymes such as carboxypeptidase Y, increased septation index and longer cell cycle and cell length, monopolar growth, defective cell wall, absence of cell rounding (in response to low glucose). Genetic or pharmacological activation of the three signalling pathway provides partial or full rescue of these pleiotropic phenotypes of *btn1*Δ cells. G, Golgi; Sec, secretory vesicles; V, vacuole; SI, division septum; M, monopolar growth.

respectively, interact negatively with *btn1*, as does the transcription factor gene *pap1*, which is a downstream target of *sty1* [40]. Furthermore, *btn1* interacts negatively with a number of genes that display stress-responsive expression.

TORC2 is a positive regulator of stress-responsive processes [29], and is also involved in cytoskeletal organization, vacuolar morphology and glucose sensing [22]. In addition to a negative interaction with the gene encoding the Tor kinase component of TORC2 (*tor1*), *btn1* also negatively interacts with genes encoding components of the connected CWI MAP kinase cascade. The relationship between TORC2 and the CWI pathway is complex. TORC2 is required for the response to cell wall stressors [41], and is a direct regulator of the CWI pathway through the activation of the Rho guanine-nucleotide exchange factor (GEF) ROM2 in budding yeast [42]. In fission yeast, it has recently been shown that Tor1, acting as part of TORC2, enhances Pck2 synthesis to activate the CWI under both cell wall stress and glucose limitation conditions [43].

Despite this positive relationship, recent work in fission yeast has indicated a negative regulation of TORC2 by *pmk1*, the MAP kinase component of the fission yeast CWI pathway, at least in certain conditions [25]. In fission yeast both Rho1 and Rho2 are well established as regulators of the CWI pathway [44, 45]. Rho5 acts as a functional homologue of *rho1,* sharing the role of regulating cell wall homeostasis [46]. More recently, *rho5* has also been shown to be an upstream regulator of the CWI [47]. The regulation of this pathway is complex as Rho1 both positively [45] and negatively [48] regulates the downstream MAP kinase cascade. This potentially intricate relationship is reflected in the interactors of *btn1* that fall within this pathway, that exhibit both positive and negative effects upon signalling. *btn1* exhibits a negative interaction with the *rho1* homologue gene *rho5*, and a number of genes involved in downstream cytoskeletal and polarity components. Despite this, the Rho GEF gene *rgf1*, which is a positive regulator of the CWI pathway [49], is a positive interactor of *btn1* whereas *rdi1* (encoding a GDP dissociation factor) is a negative interactor. This multifaceted pattern of interactions is consistent with current understanding of this pathway, given the complex interactions between these signalling nodes.

Although a process that maintains CWI might seem fungal-specific, and not relevant to disease in a higher eukaryote, previous work in budding yeast revealed similar dysfunctional processes in cells overexpressing α-synuclein as a model for Parkinson's disease, suggesting that there is a mammalian counterpart to the CWI pathway in yeast. Such observations are of specific relevance to CLN3 disease, as α-synuclein is also upregulated in this condition [9]. RHO1 signalling and the CWI pathway are disrupted by overexpression of α-synuclein [50]. Further, it was demonstrated that similar changes in downstream kinase cascades could be observed in mammalian cells overexpressing α-synuclein. Affected kinases included c-Jun, which has previously been linked to CLN3 disease, and is part of a pathway closely related to the SAPK pathway in yeast [51]. This is not the only instance in which Rho signalling has been linked to neurodegeneration. Leucine-rich repeat kinase 2 (LRRK2), mutations in which represent the most common cause of familial late-onset Parkinson's disease, has been shown to bind Rac1, and to a lesser extent RhoA and Cdc42 [52]. All of these binding partners are members of the Rho GTPase family. The binding of LRRK2 to Rac1 led to its activation and relocalisation, and was lost when disease-causing mutations were introduced into LRRK2. Further, the overexpression of Rac1 was shown to be protective in these cells. Other studies have also demonstrated a protective effect of Rho signalling, focusing both on Rho GEFs [53] and the Rho GTPases themselves [54]. Such studies highlight how processes as seemingly specialized as CWI in yeast can be highly informative of processes relevant to disease in higher order eukaryotes. Further, they demonstrate that Rho-dependent signalling could be a vital neuroprotective process in more common neurodegenerative disease, as well as being a key pathway in the neurodegenerative disease model presented in this study.

Importantly, the validation of these interactions during this work revealed interventions that produced a complete rescue of all phenotypes investigated in the *btn1Δ* strain, including phenotypes linked to TORC1 function, TORC2 function, CWI and vacuole homeostasis. Among the interactions investigated, increasing Rho1 levels, and thereby upregulating the main signalling node in the CWI pathway, or repressing TORC1 function (pharmacologically or genetically) led to complete rescues.

More notably, a neuroprotective role of TORC1 repression is already well established in a number of different systems [55]. A protective effect of the TORC1 antagonist rapamycin has been demonstrated in models for Parkinson's disease [57], Huntington's disease [56], spinoceribellar ataxia type 3 [58] and retinal degeneration mediated by mitochondrial dysfunction [59]. In addition, Tor signalling is also elevated in Alzheimer's disease models [60].

It is important to note, however, that targeting TORC1 in disease presents some technical challenges. Although in some studies an increase in lysosome function has been observed upon rapamycin treatment [61], other work indicates that rapamycin is an incomplete antagonist of TORC1 and a poor inducer of lysosomal functions [62], in concert with our observations. Another complication of targeting TORC1 in disease is the antagonistic role of TORC2. TORC2 signalling and the connected CWI pathway were both protective in our model for CLN3 disease. Further to the protective role of Rho signalling in a Parkinson's disease model [50], TORC2/AKT signalling has also been shown to be neuroprotective in a separate study of Parkinson's disease [63]. As a result, general Tor antagonists, such as Torin1, which target both TORC1 and TORC2, do not produce the same positive outcomes that TORC1 specific antagonists are able to elicit [57]. In addition, chronic rapamycin exposure also leads to a reduction in TORC2 activity. This fact could explain why rapamycin has been shown to be detrimental in certain studies, for example in one study using a fruit fly model for Alzheimer's disease [64].

Conclusion

The data presented here highlight a set of interconnected pathways that, when modulated correctly, produce a profound rescue of the dysfunctional changes observed in a yeast model for neurodegeneration caused by mutations in a single gene (Fig. 7). Links between CLN3 and the response to stress have previously been reported in other model organisms [51, 65], supporting the relevance of this yeast work to human CLN3 disease. Importantly however, a link between CLN3 and Tor signalling specifically, as reported here, is a novel mechanistic and therapeutic insight into CLN3 disease, the most common paediatric neurodegenerative disease. Furthermore, the pathways identified display strong parallels with known protective pathways in more common dementias, providing not only strong biological plausibility for the importance of these interactions, but also an indication that common processes may be exploited in therapeutic development for seemingly disparate neurodegenerative diseases.

MATERIALS AND METHODS

Yeast strains and cell growth

The strains used in this study were the wild-type strains SP23 (*h–, ura4-D18, leu1-32, ade6-M210*) and SP38 (*h–, ura4-D18, leu1-32, ade6-210, his-*) and *btn1Δ* strains SP29 (*h-, btn1::NatMX, ura4-D18, leu1-32, ade6-210,*) and SP35 (*h-, btn1::NatMX, ura4-D18, leu1-32, ade6-210, his-*). SP29 and SP35 were generated using the long primer method described in [66], using pFA6a-*NatMX* as a template [67]. Both strain backgrounds were used for SGA analysis (two isolates of each) to account for strain background differences, and SP23 and SP29 were used in all other analyses. Cell growth and manipulations were performed as described previously [68]. Minimal media (MM) was used in all experiments, unless otherwise stated. In nitrogen starvation experiments NH_4Cl was omitted. Glucose and glycerol media were made as described previously [69]. Appropriate supplements were added to the media as required. All assays were performed using log-phase cultures (1 x 10^6 cells/ml). The *dnrhb1* construct (pREP4X-*rhb1*^{D60K}) was obtained from Dr. Elizabeth Henske [34], and the *gad8* construct (pREP41-*gad8*) from Dr. Ronit Weisman [70]. pREP1-*rho5* was obtained from Dr. Kentaro Nakano [46], pREP41-*rho1* from Dr. Pilar Pérez [48] and pREP1-*wis1DD-3HA6His* from Dr. Kaz Shiozaki [71]. The *GFP-btn1* construct used in this study was described previously (pREP41-*GFP-btn1*) [12]. SP23 containing empty vector (pREP41) was used as a control in all experiments using these constructs.

Synthetic genetic arrays (SGAs)

SGA analysis was performed as described in [20], using a Singer RoToR robot (*S. pombe* settings) and the Bioneer haploid library (v2.0) and analysed as described previously [72]. A control SGA using *ade6::NatMX4* as a query was performed simultaneously. The *ade6Δ* mutant does not alter the fitness of the deletion collection, and arising double mutants are used to determine Bioneer strain fitness and for colony size normalisation. After germination, haploid double mutants were pinned to YES + G418 + Nat in a 1536 format. After three days growth at 30°C, agar plates were imaged using a MultiDoc-It Imaging System (UVP) and processed for analysis using the Gitter image analysis R package [73]. Using Excel, the colony sizes produced by Gitter were normalized to the median colony size of each plate to account for within-plate positional growth effects and differential plate growth. These normalized values were used to establish the ratio of query *btn1Δ* mutant size to control (*ade6:NatMX4*) colony size. Ratios from the four *btn1Δ* isolates used for the screen were used to calculate a mean ratio. Using this mean value, colony-size ratio scores of < 0.8 were defined as negative interactions, and ratios of > 1.2 as positive interactions. Genetic interactions were scored from two independent biological experiments. Only interactions common to both replicates were considered robust enough for follow-up. Negative and positive interaction sets were tested using PANTHER [74].

Microscopy

All images were obtained using a Leica SPE2 true scanning confocal microscope with 63 x 1.4 NA oil immersion objective. Images were recorded using Leica AF software. Calcofluor-white (Sigma-Aldrich) staining was performed as described previously [68]. Calcofluor-white was used to allow cell segmentation for assays of cell morphology. For assays of cell viability, propidium iodide (Sigma-Aldrich) was added to the cells (15 µg/ml final concentration) at the same point as the calcofluor-white. Staining was performed in a volume of 1 ml, allowed to proceed at room temperature for 10 min, and the cells were washed twice in MM before visualisation. No cell fixation was applied. The vital dye FM4-64 (Molecular Probes) was used to visualise vacuolar morphology. The staining procedure has been described previously [12].

Image analysis

All microscopy image analysis was performed using ImageJ software. For analysis of cell morphology, cells were segmented using the BOA plugin of the quantitative imaging of membrane proteins (Quimp) package [75]. Cell elongation was then determined using the equation: $1-(\frac{4\,\pi\,\text{area}}{\text{perimeter}^2})$. Thirty cells were counted per dataset.

Statistical analysis

All statistical analysis was performed using Prism software, version 6.0C (Graphpad). In instances where only two samples were compared, significance was determined using an unpaired t test. In instances where multiple columns were compared, an ordinary one-way ANOVA was used with a Tukey's multiple-comparison post-test. All error bars represent mean ± standard error of the mean (SEM). In all cases, the statistical test used, p-values and the number of samples analysed are highlighted in the figure legend for clarity.

ACKNOWLEDGMENTS

We would like to thank Dr. Elizabeth Henske, Dr. Ronit Weismann, Prof. Kentaro Nakano, Prof. Pilar Pérez, and Prof. Kaz Shiozaki for generously sharing resources. The research leading to these results (MB, SM) received funding from the European Community's Seventh Framework Programme (FP7/2007- 2013) under Grant Agreement N° 281234 (DEM-CHILD), with additional funding from the Batten Disease Family Association, UK, and the Children's Batten Disease Foundation, USA. RB is funded by the University College London (UCL) MRC 4yr Doctoral Training Account in Life and Biomedical Sciences. Research in the Bähler laboratory (CR, JB) is supported by a BBSRC Research Grant [grant number BB/I012451/1] and a Wellcome Trust Senior Investigator Award [grant number 095598/Z/11/Z].

CONFLICT OF INTEREST

The authors declare no conflict of interest.

REFERENCES

1. Jalanko A, Braulke T (**2009**). Neuronal ceroid lipofuscinoses. **Biochim Biophys Acta** 1793(4): 697-709.

2. Williams RE, Mole SE (**2012**). New nomenclature and classification scheme for the neuronal ceroid lipofuscinoses. **Neurology** 79(2): 183-191.

3. Nixon RA, Yang DS, Lee JH (**2008**). Neurodegenerative lysosomal disorders: a continuum from development to late age. **Autophagy** 4(5): 590-599.

4. Brunk UT, Terman A (**2002**). The mitochondrial-lysosomal axis theory of aging: accumulation of damaged mitochondria as a result of imperfect autophagocytosis. **Eur J Biochem** 269(8): 1996-2002.

5. Golabek AA, Kida E, Walus M, Kaczmarski W, Michalewski M, Wisniewski KE (**2000**). CLN3 protein regulates lysosomal pH and alters intracellular processing of Alzheimer's amyloid-beta protein precursor and cathepsin D in human cells. **Mol Genet Metab** 70(3): 203-213.

6. Holopainen JM, Saarikoski J, Kinnunen PK, Jarvela I (**2001**). Elevated lysosomal pH in neuronal ceroid lipofuscinoses (NCLs). **Eur J Biochem** 268(22): 5851-5856.

7. Luiro K, Kopra O, Blom T, Gentile M, Mitchison HM, Hovatta I, Tornquist K, Jalanko A (**2006**). Batten disease (JNCL) is linked to disturbances in mitochondrial, cytoskeletal, and synaptic compartments. **J Neurosci Res** 84(5): 1124-1138.

8. Glenner GG (**1988**). Alzheimer's disease: its proteins and genes. **Cell** 52(3): 307-308.

9. Kang S, Heo TH, Kim SJ (**2014**). Altered levels of alpha-synuclein and sphingolipids in Batten disease lymphoblast cells. **Gene** 539(2): 181-185.

10. Nussbaum RL, Polymeropoulos MH (**1997**). Genetics of Parkinson's disease. **Hum Mol Genet** 6(10): 1687-1691.

11. Bond M, Holthaus SM, Tammen I, Tear G, Russell C (**2013**). Use of model organisms for the study of neuronal ceroid lipofuscinosis. **Biochim Biophys Acta** 1832(11): 1842-1865.

12. Gachet Y, Codlin S, Hyams JS, Mole SE (**2005**). *btn1*, the fission yeast homologue of the human Batten disease gene, *CLN3*, regulates vacuole homeostasis. **J Cell Sci** 118(23): 5525-5536.

13. Codlin S, Mole SE (**2009**). btn1, the fission yeast orthologue of CLN3, is required for Golgi exit of Vps10 and vacuole protein sorting. **J Cell Sci** 122(8): 1163-1173.

14. Kitzmüller C, Haines R, Codlin S, Cutler DF, Mole SE (**2008**). A function retained by the common mutant CLN3 protein is responsible for the late onset of juvenile neuronal ceroid lipofuscinosis (JNCL). **Hum Mol Genet** 17(2): 303-312.

15. Codlin S, Haines RL, Mole SE (**2008**). btn1 affects endocytosis, polarisation of sterol-rich membrane domains and polarised growth in Schizosaccharomyces pombe. **Traffic** 9(6): 936-950.

16. Codlin S, Haines RL, Burden JJE, Mole SE (**2008**). btn1 affects cytokinesis and cell-wall deposition by independent mechanisms, one of which is linked to dysregulation of vacuole pH. **J Cell Sci** 121(17): 2860-2870.

17. Pears MR, Codlin S, Haines RL, White IJ, Mortishire-Smith RJ, Mole SE, Griffin JL (**2010**). Deletion of btn1, an orthologue of CLN3, increases glycolysis and perturbs amino acid metabolism in the fission yeast model of Batten disease. **Molecular bioSystems** 6(6): 1093-1102.

18. Luiro K, Yliannala K, Ahtiainen L, Maunu H, Jarvela I, Kyttala A, Jalanko A (**2004**). Interconnections of CLN3, Hook1 and Rab proteins link Batten disease to defects in the endocytic pathway. **Hum Mol Genet** 13(23): 3017-3027.

19. Metcalfe DJ, Calvi AA, Seamann MNJ, Mitchison HM, Cutler DF (**2008**). Loss of the Batten disease gene CLN3 prevents exit from the TGN of the mannose 6-phosphate receptor. **Traffic** 11(11): 1905-1914.

20. Baryshnikova A, Costanzo M, Dixon S, Vizeacoumar FJ, Myers CL, Andrews B, Boone C (**2010**). Synthetic genetic array (SGA) analysis in Saccharomyces cerevisiae and Schizosaccharomyces pombe. **Methods Enzymol** 470:145-179.

21. Willingham S, Outeiro TF, DeVit MJ, Lindquist SL, Muchowski PJ (**2003**). Yeast genes that enhance the toxicity of a mutant huntingtin fragment or alpha-synuclein. **Science** 302(5651): 1769-1772.

22. Ikai N, Nakazawa N, Hayashi T, Yanagida M (**2011**). The reverse, but coordinated, roles of Tor2 (TORC1) and Tor1 (TORC2) kinases for growth, cell cycle and separase-mediated mitosis in Schizosaccharomyces pombe. **Biol Open** 1(3): 110007.

23. Weisman R, Roitburg I, Schonbrun M, Harari R, Kupiec M (**2007**). Opposite effects of tor1 and tor2 on nitrogen starvation responses in fission yeast. **Genetics** 175(3): 1153-1162.

24. Matsuo T, Otsubo Y, Urano J, Tamanoi F, Yamamoto M (**2007**). Loss of the TOR kinase Tor2 mimics nitrogen starvation and activates the sexual development pathway in fission yeast. **Molecular and cellular biology** 27(8): 3154-3164.

25. Cohen A, Kupiec M, Weisman R (**2014**). Glucose activates TORC2-Gad8 protein via positive regulation of the cAMP/cAMP-dependent protein kinase A (PKA) pathway and negative regulation of the Pmk1 protein-mitogen-activated protein kinase pathway. **J Biol Chem** 289(31): 21727-21737.

26. Ikeda K, Morigasaki S, Tatebe H, Tamanoi F, Shiozaki K (**2008**). Fission yeast TOR complex 2 activates the AGC-family Gad8 kinase essential for stress resistance and cell cycle control. **Cell cycle** 7(3): 358-364.

27. Hanyu Y, Imai KK, Kawasaki Y, Nakamura T, Nakaseko Y, Nagao K, Kokubu A, Ebe M, Fujisawa A, Hayashi T, Obuse C, Yanagida M (**2009**). Schizosaccharomyces pombe cell division cycle under limited glucose requires Ssp1 kinase, the putative CaMKK, and Sds23, a PP2A-related phosphatase inhibitor. **Genes Cells** 14(5): 539-554.

28. Levin DE (**2005**). Cell wall integrity signaling in Saccharomyces cerevisiae. **Microbiol Mol Biol Rev** 69(2): 262-291.

29. Matsuo T, Kubo Y, Watanabe Y, Yamamoto M (**2003**). Schizosaccharomyces pombe AGC family kinase Gad8p forms a conserved signaling module with TOR and PDK1-like kinases. **EMBO** 22(12): 3073-3083.

30. Madrid M, Fernandez-Zapata J, Sanchez-Mir L, Soto T, Franco A, Vicente-Soler J, Gacto M, Cansado J (**2013**). Role of the fission yeast cell integrity MAPK pathway in response to glucose limitation. **BMC Microbiol** 13:34.

31. Sugiura R, Toda T, Shuntoh H, Yanagida M, Kuno T (**1998**). pmp1+, a suppressor of calcineurin deficiency, encodes a novel MAP kinase phosphatase in fission yeast. **EMBO J** 17(1): 140-148.

32. Mutoh T, Nakano K, Mabuchi I (**2005**). Rho1-GEFs Rgf1 and Rgf2 are involved in formation of cell wall and septum, while Rgf3 is involved in cytokinesis in fission yeast. **Genes Cells** 10(12): 1189-1202.

33. Hartmuth S, Petersen J (**2009**). Fission yeast Tor1 functions as part

of TORC1 to control mitotic entry through the stress MAPK pathway following nutrient stress. **J Cell Sci** 122(Pt 11): 1737-1746.

34. van Slegtenhorst M, Carr E, Stoyanova R, Kruger WD, Henske EP (**2004**). Tsc1+ and tsc2+ regulate arginine uptake and metabolism in Schizosaccharomyces pombe. **J Biol Chem** 279(13): 12706-12713.

35. Rallis C, Codlin S, Bahler J (**2013**). TORC1 signaling inhibition by rapamycin and caffeine affect lifespan, global gene expression, and cell proliferation of fission yeast. **Aging Cell** 12(4): 563-573.

36. Nakashima A, Sato T, Tamanoi F (**2010**). Fission yeast TORC1 regulates phosphorylation of ribosomal S6 proteins in response to nutrients and its activity is inhibited by rapamycin. **J Cell Sci** 123(5): 777-786.

37. Nguyen AN, Shiozaki K (**1999**). Heat-shock-induced activation of stress MAP kinase is regulated by threonine- and tyrosine-specific phosphatases. **Genes & development** 13(13): 1653-1663.

38. Mukaiyama H, Nakase M, Nakamura T, Kakinuma Y, Takegawa K (**2010**). Autophagy in the fission yeast Schizosaccharomyces pombe. **FEBS Lett** 584(7): 1327-1334.

39. Chandrachud U, Walker MW, Simas AM, Heetveld S, Petcherski A, Klein M, Oh H, Wolf P, Zhao WN, Norton S, Haggarty SJ, Lloyd-Evans E, Cotman SL (**2015**). Unbiased Cell-based Screening in a Neuronal Cell Model of Batten Disease Highlights an Interaction between Ca2+ Homeostasis, Autophagy, and CLN3 Protein Function. **J Biol Chem** 290(23): 14361-14380.

40. Toone WM, Kuge S, Samuels M, Morgan BA, Toda T, Jones N (**1998**). Regulation of the fission yeast transcription factor Pap1 by oxidative stress: requirement for the nuclear export factor Crm1 (Exportin) and the stress-activated MAP kinase Sty1/Spc1. **Genes & development** 12(10): 1453-1463.

41. Halova L, Du W, Kirkham S, Smith DL, Petersen J (**2013**). Phosphorylation of the TOR ATP binding domain by AGC kinase constitutes a novel mode of TOR inhibition. **J Cell Biol** 203(4): 595-604.

42. Schmidt A, Bickle M, Beck T, Hall MN (**1997**). The yeast phosphatidylinositol kinase homolog TOR2 activates RHO1 and RHO2 via the exchange factor ROM2. **Cell** 88(4): 531-542.

43. Madrid M, Jiménez R, Sánchez-Mir L, Soto T, Franco A, Vicente-Soler J, Gacto M, Pérez P, Cansado J (**2015**). Multiple layers of regulation influence cell integrity control by the PKC ortholog Pck2 in fission yeast. **J Cell Sci** 128(2): 266-80.

44. Sanchez-Mir L, Franco A, Martin-Garcia R, Madrid M, Vicente-Soler J, Soto T, Gacto M, Perez P, Cansado J (**2014**). Rho2 palmitoylation is required for plasma membrane localization and proper signaling to the fission yeast cell integrity mitogen- activated protein kinase pathway. **Mol Cell Biol** 34(14): 2745-2759.

45. Sanchez-Mir L, Soto T, Franco A, Madrid M, Viana RA, Vicente J, Gacto M, Perez P, Cansado J (**2014**). Rho1 GTPase and PKC ortholog Pck1 are upstream activators of the cell integrity MAPK pathway in fission yeast. **PLoS One** 9(1): e88020.

46. Nakano K, Arai R, Mabuchi I (**2005**). Small GTPase Rho5 is a functional homologue of Rho1, which controls cell shape and septation in fission yeast. **FEBS Lett** 579(23): 5181-5186.

47. Doi A, Kita A, Kanda Y, Uno T, Asami K, Satoh R, Nakano K, Sugiura R (**2015**). Geranylgeranyltransferase Cwg2-Rho4/Rho5 module is implicated in the Pmk1 MAP kinase-mediated cell wall integrity pathway in fission yeast. **Genes Cells** 20(4): 310-323.

48. Viana RA, Pinar M, Soto T, Coll PM, Cansado J, Perez P (**2013**). Negative functional interaction between cell integrity MAPK pathway and Rho1 GTPase in fission yeast. **Genetics** 195(2): 421-432.

49. Garcia P, Tajadura V, Sanchez Y (**2009**). The Rho1p exchange factor Rgf1p signals upstream from the Pmk1 mitogen-activated protein kinase pathway in fission yeast. **Mol Biol Cell** 20(2): 721-731.

50. Wang S, Xu B, Liou LC, Ren Q, Huang S, Luo Y, Zhang Z, Witt SN (**2012**). alpha-Synuclein disrupts stress signaling by inhibiting polo-like kinase Cdc5/Plk2. **Proc Natl Acad Sci USA** 109(40): 16119-16124.

51. Tuxworth RI, Vivancos V, O'Hare MB, Tear G (**2009**). Interactions between the juvenile Batten disease gene, CLN3, and the Notch and JNK signalling pathways. **Hum Mol Genet** 18(4): 667-678.

52. Chan D, Citro A, Cordy JM, Shen GC, Wolozin B (**2011**). Rac1 protein rescues neurite retraction caused by G2019S leucine-rich repeat kinase 2 (LRRK2). **J Biol Chem** 286(18): 16140-16149.

53. Namekata K, Harada C, Taya C, Guo X, Kimura H, Parada LF, Harada T (**2010**). Dock3 induces axonal outgrowth by stimulating membrane recruitment of the WAVE complex. **Proc Natl Acad Sci U S A** 107(16): 7586-7591.

54. Mocholi E, Ballester-Lurbe B, Arque G, Poch E, Peris B, Guerri C, Dierssen M, Guasch RM, Terrado J, Perez-Roger I (**2011**). RhoE deficiency produces postnatal lethality, profound motor deficits and neurodevelopmental delay in mice. **PLoS One** 6(4): e19236.

55. Bove J, Martinez-Vicente M, Vila M (**2011**). Fighting neurodegeneration with rapamycin: mechanistic insights. **Nat Rev Neurosci** 12(8): 437-452.

56. Sarkar S, Ravikumar B, Floto RA, Rubinsztein DC (**2009**). Rapamycin and mTOR-independent autophagy inducers ameliorate toxicity of polyglutamine-expanded huntingtin and related proteinopathies. **Cell death and differentiation** 16(1): 46-56.

57. Malagelada C, Jin ZH, Jackson-Lewis V, Przedborski S, Greene LA (**2010**). Rapamycin protects against neuron death in in vitro and in vivo models of Parkinson's disease. **J Neurosci** 30(3): 1166-1175.

58. Menzies FM, Huebener J, Renna M, Bonin M, Riess O, Rubinsztein DC (**2010**). Autophagy induction reduces mutant ataxin-3 levels and toxicity in a mouse model of spinocerebellar ataxia type 3. **Brain** 133(Pt 1): 93-104.

59. Zhao C, Yasumura D, Li X, Matthes M, Lloyd M, Nielsen G, Ahern K, Snyder M, Bok D, Dunaief JL, LaVail MM, Vollrath D (**2011**). mTOR-mediated dedifferentiation of the retinal pigment epithelium initiates photoreceptor degeneration in mice. **J Clin Invest** 121(1): 369-383.

60. Caccamo A, Maldonado MA, Majumder S, Medina DX, Holbein W, Magri A, Oddo S (**2011**). Naturally secreted amyloid-beta increases mammalian target of rapamycin (mTOR) activity via a PRAS40-mediated mechanism. **J Biol Chem** 286(11): 8924-8932.

61. Dehay B, Bove J, Rodriguez-Muela N, Perier C, Recasens A, Boya P, Vila M (**2010**). Pathogenic lysosomal depletion in Parkinson's disease. **J Neurosci** 30(37): 12535-12544.

62. Zhou J, Tan SH, Nicolas V, Bauvy C, Yang ND, Zhang J, Xue Y, Codogno P, Shen HM (**2013**). Activation of lysosomal function in the course of autophagy via mTORC1 suppression and autophagosome-lysosome fusion. **Cell Res** 23(4): 508-523.

63. Ries V, Henchcliffe C, Kareva T, Rzhetskaya M, Bland R, During MJ, Kholodilov N, Burke RE (**2006**). Oncoprotein Akt/PKB induces trophic effects in murine models of Parkinson's disease. **Proc Natl Acad Sci U S A** 103(49): 18757-18762.

64. Ling D, Song HJ, Garza D, Neufeld TP, Salvaterra PM (**2009**). Abeta42-induced neurodegeneration via an age-dependent autophagic-lysosomal injury in Drosophila. **PLoS One** 4(1): e4201.

65. Tuxworth RI, Chen H, Vivancos V, Carvajal N, Huang X, Tear G (**2011**). The Batten disease gene CLN3 is required for the response to oxidative stress. **Hum Mol Genet** 20(10): 2037-2047.

66. Bahler J, Wu JQ, Longtine MS, Shah NG, McKenzie A, 3rd, Steever

AB, Wach A, Philippsen P, Pringle JR (**1998**). Heterologous modules for efficient and versatile PCR-based gene targeting in *Schizosaccharomyces pombe*. **Yeast** 14(10): 943-951.

67. Sato M, Dhut S, Toda T (**2005**). New drug-resistant cassettes for gene disruption and epitope tagging in Schizosaccharomyces pombe. **Yeast** 22(7): 583-591.

68. Moreno S, Klar A, Nurse P (**1991**). Molecular genetic analysis of fission yeast *Schizosaccharomyces pombe*. **Methods Enzymol** 194:795-823.

69. Rorbach J, Richter R, Wessels HJ, Wydro M, Pekalski M, Farhoud M, Kuhl I, Gaisne M, Bonnefoy N, Smeitink JA, Lightowlers RN, Chrzanowska-Lightowlers ZM (**2008**). The human mitochondrial ribosome recycling factor is essential for cell viability. **Nucleic Acids Res** 36(18): 5787-5799.

70. Schonbrun M, Laor D, Lopez-Maury L, Bahler J, Kupiec M, Weisman R (**2009**). TOR complex 2 controls gene silencing, telomere length maintenance, and survival under DNA-damaging conditions. **Mol Cell Biol** 29(16): 4584-4594.

71. Tatebe H, Shiozaki K (**2003**). Identification of Cdc37 as a novel regulator of the stress-responsive mitogen-activated protein kinase. **Mol Cell Biol** 23(15): 5132-5142.

72. Rallis C, Lopez-Maury L, Georgescu T, Pancaldi V, Bahler J (**2014**). Systematic screen for mutants resistant to TORC1 inhibition in fission yeast reveals genes involved in cellular ageing and growth. **Biol Open** 3(2): 161-171.

73. Wagih O, Parts L (**2014**). gitter: a robust and accurate method for quantification of colony sizes from plate images. **G3 (Bethesda)** 4(3): 547-552.

74. Mi H, Muruganujan A, Thomas PD (**2013**). PANTHER in 2013: modeling the evolution of gene function, and other gene attributes, in the context of phylogenetic trees. **Nucleic Acids Res** 41(Database issue): D377-386.

75. Bosgraaf L, van Haastert PJ, Bretschneider T (**2009**). Analysis of cell movement by simultaneous quantification of local membrane displacement and fluorescent intensities using Quimp2. **Cell Motil Cytoskeleton** 66(3): 156-165.

76. Chen D, Toone WM, Mata J, Lyne R, Burns G, Kivinen K, Brazma A, Jones N, Bahler J (**2003**). Global transcriptional responses of fission yeast to environmental stress. **Mol Biol Cell** 14(1): 214-229.

The MAPKKKs Ste11 and Bck1 jointly transduce the high oxidative stress signal through the cell wall integrity MAP kinase pathway

Chunyan Jin[#], Stephen K. Kim, Stephen D. Willis and Katrina F. Cooper*

Department of Molecular Biology, Rowan University School of Osteopathic Medicine, Stratford, NJ, 08055 USA.

[#] Current address: Shaoxing Women & Children Hospital, 305 East Street, Shaoxing, Zhejiang, China, 312000.

* Corresponding Author: Katrina Cooper, E-mail: cooperka@rowan.edu

ABSTRACT Oxidative stress stimulates the Rho1 GTPase, which in turn induces the cell wall integrity (CWI) MAP kinase cascade. CWI activation promotes stress-responsive gene expression through activation of transcription factors (Rlm1, SBF) and nuclear release and subsequent destruction of the repressor cyclin C. This study reports that, in response to high hydrogen peroxide exposure, or in the presence of constitutively active Rho1, cyclin C still translocates to the cytoplasm and is degraded in cells lacking Bck1, the MAPKKK of the CWI pathway. However, in mutants defective for both Bck1 and Ste11, the MAPKKK from the high osmolarity, pseudohyphal and mating MAPK pathways, cyclin C nuclear to cytoplasmic relocalization and destruction is prevented. Further analysis revealed that cyclin C goes from a diffuse nuclear signal to a terminal nucleolar localization in this double mutant. Live cell imaging confirmed that cyclin C transiently passes through the nucleolus prior to cytoplasmic entry in wild-type cells. Taken together with previous studies, these results indicate that under low levels of oxidative stress, Bck1 activation is sufficient to induce cyclin C translocation and degradation. However, higher stress conditions also stimulate Ste11, which reinforces the stress signal to cyclin C and other transcription factors. This model would provide a mechanism by which different stress levels can be sensed and interpreted by the cell.

***Keywords**: cyclin C, oxidative stress, protein degradation, programmed cell death, MAPK signal transduction pathways, mitochondrial morphology.*

Abbreviations:
CWI - cell wall integrity,
IG - invasive growth,
HOG - high osmolarity glycerol,
MAPKs - mitogen activated kinase cascades.

INTRODUCTION

A conserved survival instinct in all living cells is the ability to adapt to changes in their extracellular environment. To achieve this, cells convey information from a sensory input via a signaling cascade to the genes in charge of orchestrating the cellular response. Among the signaling pathways, the mitogen activated kinase cascades (MAPKs) are very well conserved from yeast to mammals (reviewed in [1]). Each MAPK cascade consists of a canonical module of three protein kinases, a MAP kinase kinase kinase (MAPKKK/MEKK), a MAP kinase kinase (MAPKK/MEK) and a MAP kinase (MAPK). In *S. cerevisiae*, there are five known MAPK cascades (reviewed in [2, 3]) that control responses to different external signals stresses including nutrient starvation, high osmolarity, mating pheromone, and oxidative stress. Traditionally, the MAPK pathways have been depicted as insulated with each cascade being activated only in response to a particular extracellular signal. For example, the high osmolarity glycerol (HOG) pathway is predominantly (but not exclusively) activated in response to changes in osmolarity (reviewed in [4]), whereas the cell wall integrity (CWI) pathway is triggered by numerous stresses including cell wall deterioration, temperature shifts and oxidative stress (reviewed in [5, 6]). More recently, it has been suggested that in situations where signal strength exceeds pathway capacity, there is a crosstalk between the pathways [7, 8].

Ste11 is a shared MAPKKK that plays an important role in transducing several signals. Ste11 transduces signals important for mating, invasive growth (IG, also known as pseudohyphal pathway) and resistance to high osmolarity, even though each pathway responds to different external stimuli and produces different outputs (see Figure 1). Ste11 is activated by the yeast homolog of the p21-like kinase

(PAK) called Ste20 [9]. Dependent on the stimuli received, Ste11 phosphorylates a different MAPKK. How the cell determines which MAPKK to activate is only partially understood but thought to be dependent upon upstream components which route the signal to the appropriate MAPKK [3, 10]. Over the last two decades, a principle has emerged involving scaffold proteins, which physically assemble MAPK modules, in directing which MAPK module is activated following a particular stress (reviewed in [11]). However, the Ste5 scaffold mediating the mating pathway (see Figure 1) was shown to function as a conformational switch to gate the flow of information between the mating and IG pathways [12]. These results indicate that the role scaffolds play in sorting MAPK signaling may be more complicated.

In addition to signaling in the HOG, IG and mating pathways, Ste11 is also required for the expression of *FKS2* [13-15]. *FSK2* encodes one of two alternate stress-induced catalytic subunits of the beta-1,3 glucan synthase complex that is important for cell wall integrity under conditions of cell wall stress [16]. The role Ste11 plays in its activation is complex but it has been shown that its activation can be mediated by Ste11 cross talking with the mating, HOG, calcineurin and CWI pathways [13, 15]. Thus, although the molecular mechanism by which this cross-talk is achieved is not well defined, it seems clear that Ste11 has functional redundancy with other MAPKK's [14, 15] (see Figure 1).

The CWI pathway is normally triggered by factors that evoke cell wall damage including oxidative stress (reviewed in [6]). The cell wall sensors signal through Rho1 to protein kinase C [17, 18] that in turn stimulates the CWI MAPK module. In brief, the stress signal flows from Pkc1 to Bck1 (MEKK), the redundant MEKs Mkk1 or Mkk2 (written hereafter Mkk1/2) to finally the Slt2/Mpk1 MAP kinase (see [6] for review). One readout of the CWI pathway is the activation of two transcription factors, Rlm1 and SBF, which induce the expression of genes that remodel the cell wall [19-22]. Rlm1 is activated by Slt2 mediated phosphorylation [20] whereas Slt2, and its pseudokinase paralogue Kdx1, use a noncatalytic mechanism to activate SBF [23-25].

We have identified a third transcription factor called cyclin C that is regulated by the CWI pathway ([26-28] and reviewed in [29]). Unlike cyclins that regulate cell cycle progression, cyclin C and its cyclin dependent partner Cdk8, are conserved members of the Mediator complex associating with the RNA polymerase II holoenzyme (reviewed in [30]). Cyclin C-Cdk8 predominantly acts as a transcriptional

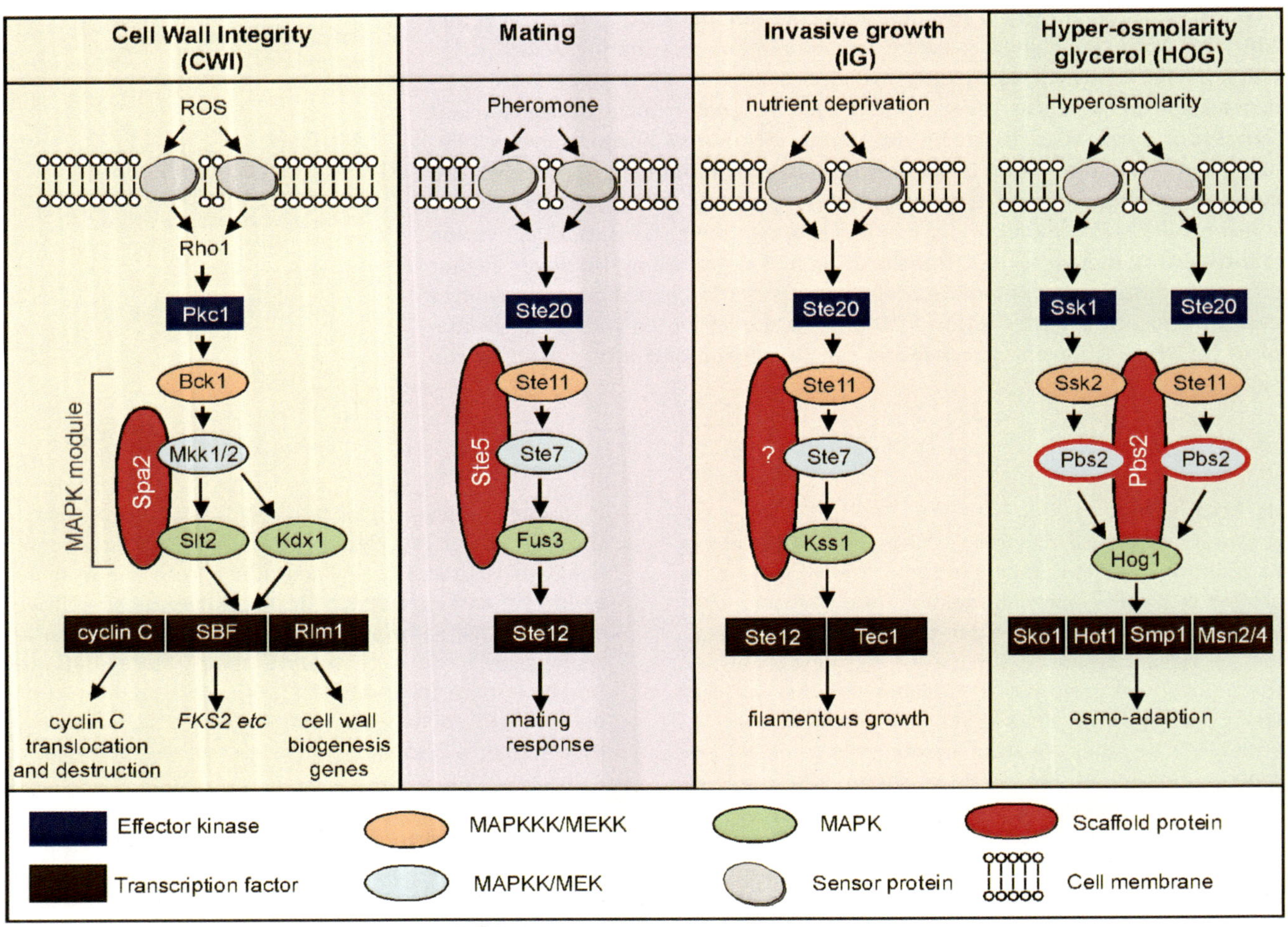

FIGURE 1: Schematic diagram of the MAPK modules of the cell wall integrity (CWI), mating, invasive growth (IG) and hyper-osmolarity glycerol (HOG) pathways in *S. cerevisiae*.

repressor, regulating over 100 loci including many stress response genes [31, 32]. To relieve this repression, cyclin C, but not Cdk8, is translocated to the cytoplasm where it is consequently degraded in an ubiquitin dependent manner [33, 34]. Importantly, before it is destroyed, cyclin C is required for stress-induced mitochondrial fission and programmed cell death [34, 35]. This cytoplasmic role of cyclin C in mediating mitochondrial fission and intrinsic apoptosis is conserved in mammals [36].

In this report, we show that under conditions of high oxidative stress, the capacity of the canonical CWI MAPK module is exceeded. In short, we provide evidence that Slt2 can be activated following high oxidative stress in the absence of Bck1. As a result, cyclin C is both translocated to the cytoplasm and degraded. This Bck1 independent translocation of cyclin C requires Ste11. Furthermore, we provide evidence that the normal stress-induced path for cyclin C translocation starts in the nucleus, then transits to the nucleolus and finally enters the cytoplasm. We also show that CWI MAPK activity is only required for nucleolus-cytoplasm translocation and not for the nuclear to nucleolus movement. These results suggest a mechanism by which high-stress conditions are recognized differently by the cell and transduced into changes in both gene expression and mitochondrial dynamics.

RESULTS

Bck1 is not required for MAPK mediated degradation of cyclin C under high stress

We have previously shown that the conserved CWI MAPK pathway transduces the oxidative stress signal from the cell wall sensors to induce cyclin C nuclear to cytoplasmic translocation and destruction [27]. In that report, a model was proposed in which the strength of the stress dictates which sensors are required. Here we examined whether the severity of the stress damage dictates the route the signal takes to trigger cyclin C relocalization.

To test this, we first examined whether the MEK kinase in the CWI pathway, Bck1, is required to transduce low and high oxidative stress signal to cyclin C. To address this question, we used functional cyclin C-myc and cyclin C-YFP reporters to examine cyclin C degradation kinetics and cytoplasmic localization, respectively. Consistent with our previously published work [26, 28, 34], Bck1 is required for cyclin C destruction and cytoplasmic relocalization of cyclin C following low oxidative stress (0.4 mM H_2O_2) exposure (Figure 2A, 2C and 2D). However, cyclin C destruction occurs with kinetics similar to wild type in a *bck1Δ* mutant exposed to 1.2 mM H_2O_2 (Figure 2B, quantitated in Figure 2C). As anticipated from these results, we find that cyclin C-YFP forms cytoplasmic foci in the *bck1Δ* mutant exposed to 1.2 mM H_2O_2 (Figure 2D, see Figure 2E for representative images). We also observed that cyclin C localization in 80% of *bck1Δ* cells is both just outside of the nucleus in a

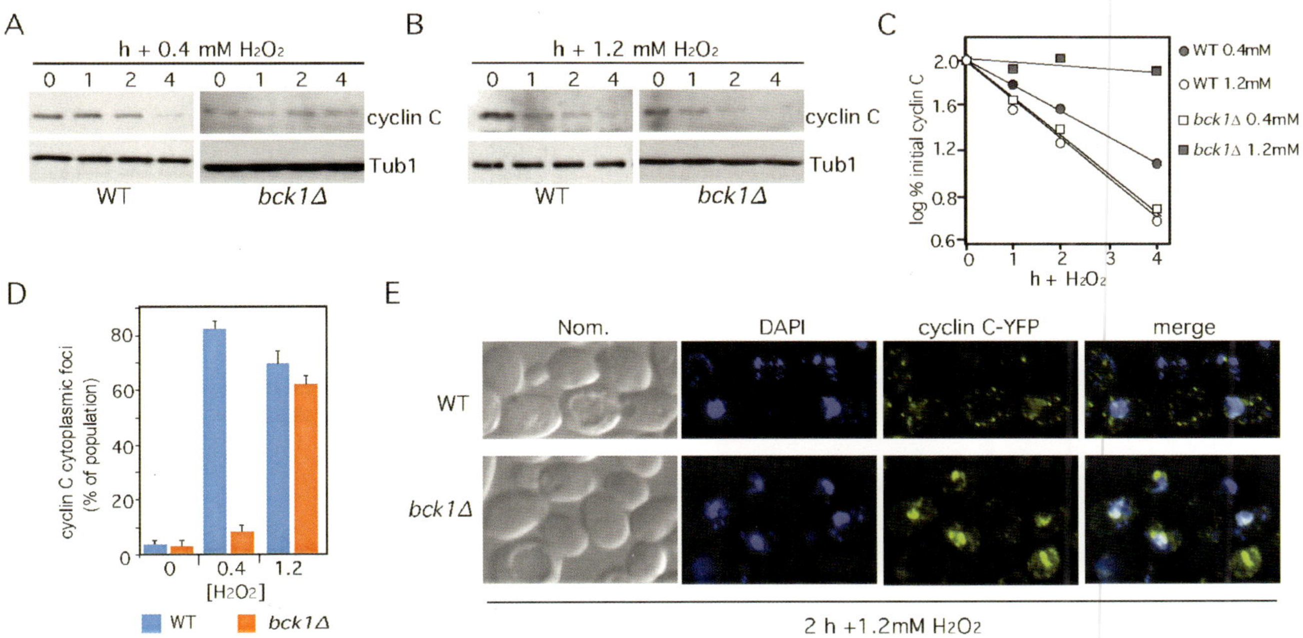

FIGURE 2: Bck1 is not required for cyclin C nuclear to cytoplasmic translocation and degradation following high (1.2 mM) H_2O_2 stress. Wild type (RSY10) and *bck1Δ* (RSY1050) cultures expressing myc-cyclin C (pLR337) were grown to mid-log phase (0 h) then treated with 0.4 mM **(A)** or 1.2 mM **(B)** H_2O_2 for the indicated times. Cyclin C levels were determined by Western blot analysis of immunoprecipitates. Tub1 levels were used as a loading control. **(C)** Quantification of the results obtained in (A) and (B). **(D)** The percent of cells (mean ± s.e.m.) within the population displaying at least 3 cytoplasmic cyclin C foci is given before and following H_2O_2 (0.4 and 1.2 mM) treatment for 2 h. At least 200 cells were counted per time point from 3 individual isolates. **(E)** Representative images (collapsed de-convolved 2 µm slices) of WT and *bck1Δ* cells harboring cyclin C-YFP and after 2 h 1.2 mM H_2O_2 stress.

perinuclear location as well as in the cytoplasm. This intermediate and cyclin C localization phenotype (see images in Figure S1A) is however also observed in 28% of wild type cells after 2 h stress and this number decreases to 5% after 3 h H_2O_2 stress. This suggests that the "intermediate phenotype" is a transient state in the kinetics of cyclin C export. Taken together these results indicate that an additional signaling pathway(s) is employed to transmit the oxidative stress signal when cells are exposed to elevated H_2O_2 concentrations.

Cyclin C destruction following high levels of oxidative stress requires both MAPKs of the CWI pathway

Both Slt2 and its pseudokinase Kdx1 mediate the export and degradation of cyclin C under conditions of low H_2O_2 stress [28]. The results described above indicate that a Bck1 independent pathway also transmits the high-stress signal to cyclin C. To determine whether downstream components of the CWI pathway are required for transducing the high-stress signal, cyclin C levels were monitored in a *slt2Δ kdx1Δ* strain. These results show that cyclin C is stable following 1.2 mM H_2O_2 exposure whereas it is degraded in the single mutants albeit with slower kinetics (Figure 3A, quantitated in Figure 3B). As cyclin C is normally destroyed following its translocation to the cytoplasm, we monitored cyclin C-YFP localization in the *slt2Δ kdx1Δ* strain subjected to 1.2 mM H_2O_2 treatment. Interestingly, fluorescence microscopy revealed that cyclin C-YFP did not form multiple cytoplasmic foci (Figure 3C and S1A). Rather, a single peri-nuclear focus was observed in most cells (quantitated in Figure 3D and see Figure S1B for field view). However, cyclin C was exported in *slt2Δ* cells after treatment with 1.2 mM H_2O_2 (Figure S1C and D). These results indicate that cyclin C is able to translocate from diffuse nuclear localization to a single peri-nuclear focus but was unable to enter the cytoplasm.

Cyclin C transits through nucleolus before it is exported into the cytoplasm

Our previous studies revealed that in oxidatively stressed *cdk8Δ* mutants, cyclin C-YFP formed a lone peri-nuclear focus that associated with the nucleolus [35]. In addition, Cdk8 also re-localized to the nucleolus in wild-type stressed cells but did not enter the cytoplasm [34]. These results led to the speculation that the normal stress-induced path for cyclin C starts in the nucleoplasm, then transits to the nucleolus and finally enters the cytoplasm. However, localization to the nucleolus was only observed in mutant cells raising the possibility that the nucleolar endpoint for cyclin C may not be the normal route in wild-

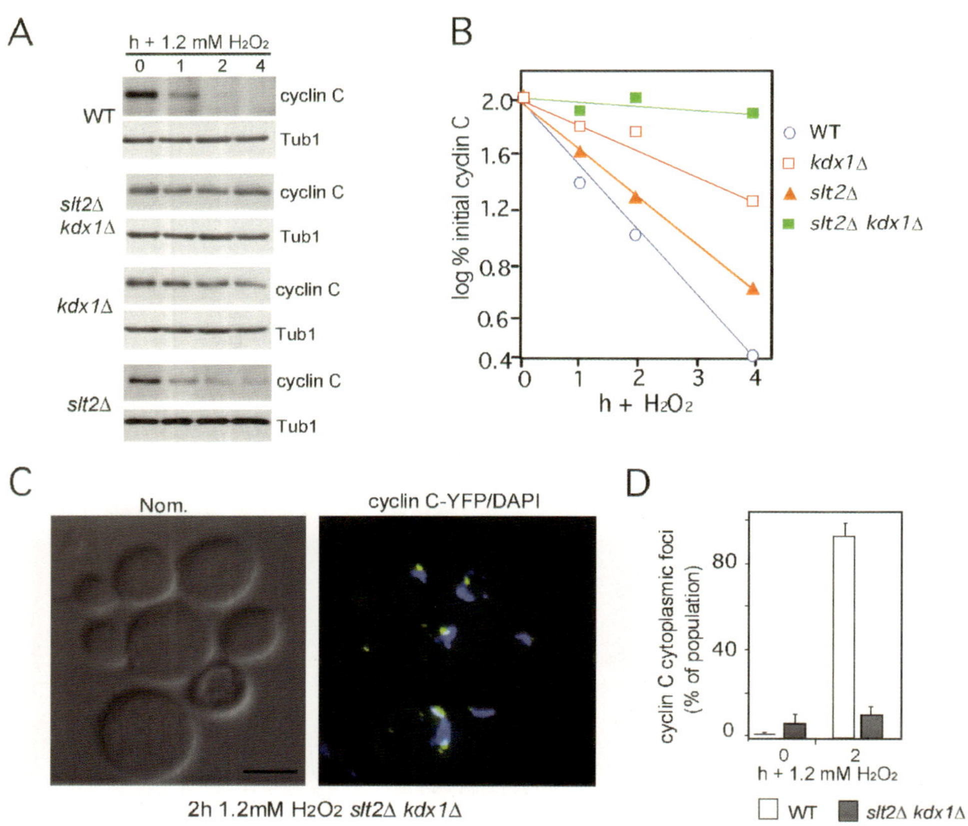

FIGURE 3: Kdx1 and Slt2 promote cyclin C relocalization and destruction in response to 1.2 mM H_2O_2 stress. (A) Cyclin C levels were monitored as described previously in wild type (RSY10), *slt2Δ kdx1Δ* (RSY1737), *slt2Δ* (RSY1006) and *kdx1Δ* (RSY1736) strains exposed to 1.2 mM H_2O_2 for the times indicated. Tub1 levels were used as a loading control. **(B)** Quantification of the results obtained in (A). **(C)** Cyclin C remains perinuclear in the *slt2Δ kdx1Δ* strain following 1.2 mM H_2O_2 stress. Fluorescence microscopy was conducted on mid-log phase on *slt2Δ kdx1Δ* cells expressing YFP-cyclin C (pBK37) following (2 h) 1.2 mM H_2O_2 treatment. The cells were fixed, stained with DAPI and then examined by fluorescence microscopy. Representative images (collapsed deconvolved 0.2 µm slices) of the results obtained are shown. **(D)** Quantification of the results in C. The percent of the population displaying at least 3 cyclin C-YFP foci in the cytoplasm is given (mean ± s.e.m.). At least 200 cells were counted per time point from 3 individual isolates.

type cells. To test this possibility, a wild-type culture expressing the nucleolar marker Nop1-CFP and cyclin C-YFP was grown to mid-log phase, harvested, then embedded in low temperature melting agarose containing 1.0 mM H_2O_2 (see Materials and methods for details). Images were collected immediately after the cells were immobilized in the agarose and every minute starting at 30 min for an additional 60 min. As expected in unstressed cells, cyclin C-YFP displayed diffuse nuclear staining that may partially overlap with the Nop1-CFP signal (0 min time point, Figure 4A). However, at minute 53, the cyclin C-YFP signal became more condensed and co-localized extensively with the nucleolar marker (bottom panels, Figure 4A). However, cyclin C-YFP co-localization with the nucleolus was transient, remaining nucleolar approximately seven minutes in this experiment before entry into the cytoplasm.

These results indicate that, in response to oxidative stress, cyclin C-Cdk8p normally transit to the nucleolus with only cyclin C continuing on to the cytoplasm. To further test this model, fixed wild-type cells harboring cyclin C-YFP and Nop1-RFP were examined 10, 20 and 60 minutes after treatment with 1.2 mM H_2O_2. After 10 minutes, a portion of cyclin C can be seen in the nucleolus (Figure 4B, see Figure S2 for complete image of cell). Cyclin C-YFP and Nop1-RFP co-localization becomes more pronounced with time and after 60 minutes of stress most of cyclin C has moved to the nucleolus. Taken together these results suggest a model in which cyclin C first transitions to the nucleolus before it is exported to the cytoplasm. However, the transient nature of cyclin C-nucleolar co-localization makes it difficult to capture this interaction for more detailed study.

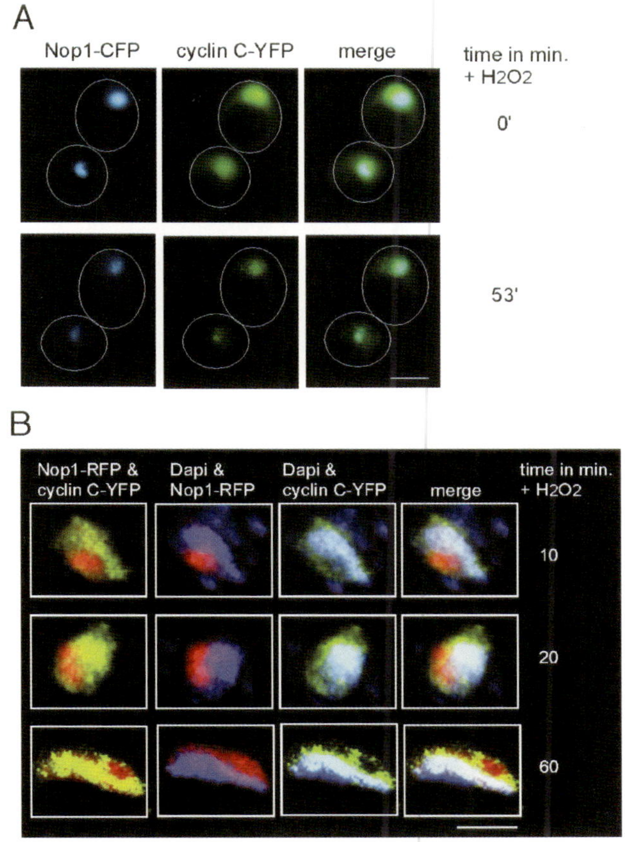

FIGURE 4: Cyclin C transits through the nucleolus in response to stress. (A) Live cell imaging of immobilized wild type cells (FLY1589) expressing cyclin C-YFP and Nop1-CFP before (0 min) and following (53 min) exposure to 1.0 mM H_2O_2. Arrows indicate restricted cyclin C-YFP signal coinciding with Nop1-CFP. **(B)** Enlarged fixed cell images of nucleus and surrounding area of wild type cells (RSY10) expressing cyclin C-YFP and Nop1-RFP. Cells were harvested and stained with DAPI following 1.2 mM H_2O_2 at the time points indicated.

Activated Rho1 signals cyclin C destruction through both CWI-dependent and independent pathways.

Our results indicate that exposure to high H_2O_2 concentrations leads to cyclin C destruction through both CWI-dependent and independent pathways. These results may indicate the existence of another pathway separate from the CWI system. Conversely, the CWI pathway itself may be modified in response to high stress conditions. To test

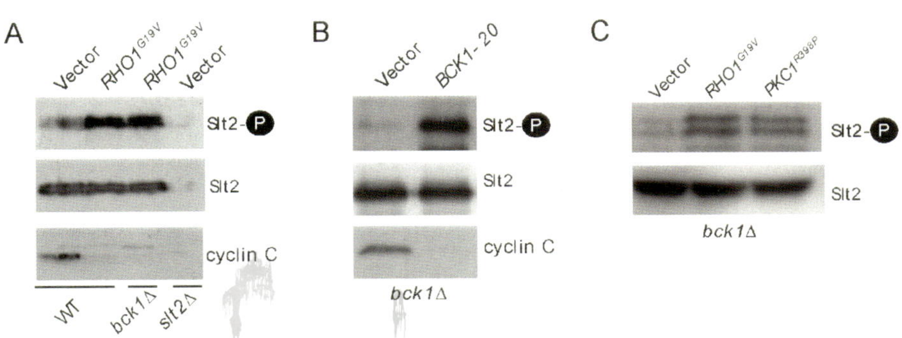

FIGURE 5: Bck1 is not required for Slt2 T-loop phosphorylation and cyclin C destruction following hyper-activation of the CWI pathway. (A) Slt2 T-loop phosphorylation (upper panel) was determined by Western blot analysis of protein extracts prepared from mid-log phase cultures (WT, RSY10, *bck1Δ*, RSY1050, and *slt2Δ*, RSY1057) expressing functional cyclin C-YFP (pBK37) and either $RHO1^{G19V}$ or a vector control. The blot was stripped and re-probed for Slt2 and cyclin C protein levels (middle and lower panels respectively). **(B)** As in (A) except that Slt2 and cyclin C were monitored in *bck1Δ* cells expressing either a vector or a plasmid harboring a constitutively active *BCK1* allele (*BCK1-20*). **(C)** As in (A) except that Slt2 phosphorylation (top panel) was also monitored in *bck1Δ* cells expressing a hyper-active allele of *PKC1* ($PKC1^{R398P}$).

these two models, we took advantage of previous studies demonstrating that cell wall stress induces sensor clustering [37] and enhanced Rho1 activation [38]. To genetically mimic this process, a constitutively activated *RHO1* allele, *RHO1^{G19V}* [39], was utilized. Protein extracts were prepared from log-phase cultures and T-loop phosphorylation was used to monitor Slt2 activation. As a downstream readout for Slt2 activation, cyclin C levels were also monitored. The results revealed that Bck1 is not required for Slt2 activation or cyclin C degradation in the presence of Rho1^{G19V} (Figure 5A). The *bck1Δ* mutant was acting as expected as introduction of a hyperactive *BCK1* allele (*BCK1-20* [40]) could efficiently induce Slt2 phosphorylation and trigger cyclin C degradation (Figure 5B). Taken together, these results indicate that hyperactivated Rho1, either by mutation, or by enhanced sensor clustering through elevated H_2O_2 exposure, can transmit the stress signal to cyclin C through Slt2 in a Bck1-independent manner.

GTP-bound Rho1 has six known effectors in yeast (see [41] for review) including Pkc1 (see Figure 1). Pkc1 also transmits the oxidative stress signal to Slt2 [42]. As *PKC1* is an essential gene [43], we monitored Slt2 T-loop phosphorylation in *bck1Δ* cells expressing a constitutively active form of Pkc1 (PKC1^{R398P} [43]). The results show that Slt2 is activated under these conditions (Figure 5C). In addition, these findings suggest that Pkc1 is the effector molecule that mediates cyclin C degradation under conditions of high stress.

Activated Rho1p induces mitochondrial fission and promotes PCD through both cyclin C-dependent and independent pathways.

A previous study found that introduction of a constitutively active Rho1, but not Bck1, could largely restore the normal H_2O_2-induced transcription program to cells deleted for the CWI pathway cell wall sensor Mtl1 [39, 44]. These results suggested to these authors that Mtl1 signals through Rho1 but not downstream components of the CWI pathway. We previously demonstrated that Mtl1 and the CWI signaling are required for normal cyclin C nuclear release and subsequent mitochondrial fragmentation [27, 28]. Therefore, we tested whether activated Rho1 alone was sufficient to regulate cyclin C translocation in the absence of additional stress signals. To address this question, we examined mitochondrial morphology by fluorescence microscopy in cells expressing either Rho1^{G19V} or a vector control. This experiment revealed that mitochondria were significantly more fragmented in cells expressing Rho1^{G19V} than the vector control (Figure 6A, quantitated in 6B). These results indicate that Rho1 signaling is sufficient to induce mitochondrial fragmentation. To determine if this phenotype is dependent on cyclin C, this experiment was repeated in a *cnc1Δ* strain. In the absence of cyclin C, mitochondrial fragmentation was significantly reduced in cells expressing Rho1^{G19V} compared to control. However, mitochondrial fragmentation was still observed above vector controls in the *cnc1Δ* strain indicating that Rho1 signaling is able to induce mitochondrial fragmentation in both cyclin C dependent and independent mechanisms.

We have previously demonstrated that activation of the CWI MAPK pathway using activated Rho1 is translated by the cells as a high stress situation [27, 28]. Therefore we addressed if the Rho1^{G19V} mediated mitochondria fragmentation seen is dependent upon Bck1. The results (Figure 6C and quantitated in 6D) show that this is indeed the case with over 60% of *bck1Δ* cells harboring Rho1^{G19V} showing cyclin C dependent mitochondrial fission compared to ~10% of the cells harboring a vector control. These results

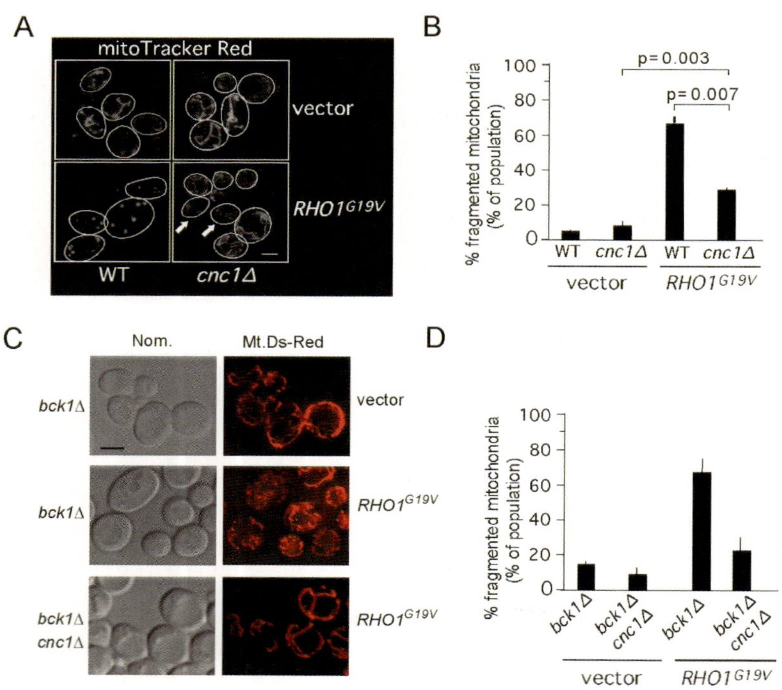

FIGURE 6: *RHO1^{G19V}* promotes cyclin C-dependent mitochondrial fission and programmed cell death in the absence of stress. (A) Mitochondrial morphology was followed by fluorescence microscopy in wild type (RSY10) and *cnc1Δ* (RSY391) cultures expressing mt-dsRed and either a vector control or *RHO1^{G19V}*. Representative images (0.2 µm slices) of each cell type are shown. The arrows indicate cells exhibiting fission in the *cnc1Δ* mutant. **(B)** Quantitation of three independent isolates exhibiting reticular or fragmented mitochondria were determined from ≥ 600 cells per culture (mean ± s.e.m.) p values for then indicated data sets are shown. **(C)** Mitochondrial morphology was followed by fluorescence microscopy in *bck1Δ* (RSY1050) and *bck1Δ cnc1Δ* (RSY1052) cultures expressing mt-dsRed and either a vector control or *RHO1^{G19V}* **(D)** Quantitation of three independent isolates exhibiting reticular or fragmented mitochondria were determined from ≥ 600 cells per culture.

support the model that, under conditions of high stress, the stress signal can reach cyclin C independent of Bck1.

Rho1-dependent cyclin C translocation induces oxidative stress hyper-sensitivity.

Extensive mitochondrial fission is an early cytological landmark in PCD initiation, although fragmentation itself does not insure cell death [45]. We have previously demonstrated that precocious release of cyclin C from the nucleus caused mitochondrial fission in the absence of stress and sensitized cells to H_2O_2 [46]. To determine whether Rho1-dependent cyclin C translocation also induced oxidative stress hyper-sensitivity, we monitored the viability of *bck1Δ* cells following treatment with 1.2 mM H_2O_2. As expected from our earlier results, loss of Bck1 activity did not alter the stress sensitivity compared to wild type (Figure 7A). This loss of viability was dependent on cyclin C as the *bck1Δ cnc1Δ* double mutant displayed resistance to H_2O_2. To determine whether the cell death observed on the plates were due to PCD, the strains just described were treated with H_2O_2 and PCD execution was measured by TUNEL assays. These experiments indicated that, similar to mitochondrial fission, activated Rho1 increased H_2O_2 sensitivity that was partially dependent on cyclin C (Figure 7B).

To understand the nature of this additional signal, the percentage of the population exhibiting high ROS concentrations was determined by DHE staining and fluorescence activated cell analysis. DHE oxidation measures internal ROS levels that can be elevated by damaged mitochondria and elevated DHE oxidation is casually related to the cell's likelihood to induce PCD. The results indicated that the presence of Rho1^{G19V} did not induce elevated ROS levels under normal growth conditions but the percent of the population of DHE positive cells increased following H_2O_2 treatment (Figure 7C). As observed with the TUNEL assays, much, but not all, of the increase in DHE positive cells is dependent on cyclin C. The cyclin C-independent ROS increase may be the result of continued mitochondrial fragmentation through another Rho1-dependent signaling pathway (see discussion).

The partial requirement of cyclin C for Rho1^{G19V}-dependent mitochondrial fission, H_2O_2-induced PCD and elevated internal ROS levels could be due to its transcriptional regulatory role, its mitochondrial function, or both. Given its reduced overall levels in cells expressing activated Rho1, we first asked whether cyclin C was still functioning as a transcription factor. Cyclin C represses the transcription of both stress response [33] and early meiotic genes (e.g., *SPO13*) [47] during mitotic cell division. Therefore, wild type and *cnc1Δ* strains were transformed with plasmids expressing either *RHO1^{G19V}* or a vector control and a *spo13*-lacZ reporter gene. Using a plate assay to monitor ß-galactosidase activity, this experiment revealed that cyclin C still repressed *spo13*-lacZ transcription even in the presence of Rho1^{G19V} (Figure 7D). These results indicate that although cyclin C levels are reduced, it must still pass through the nucleus with sufficient retention to repress

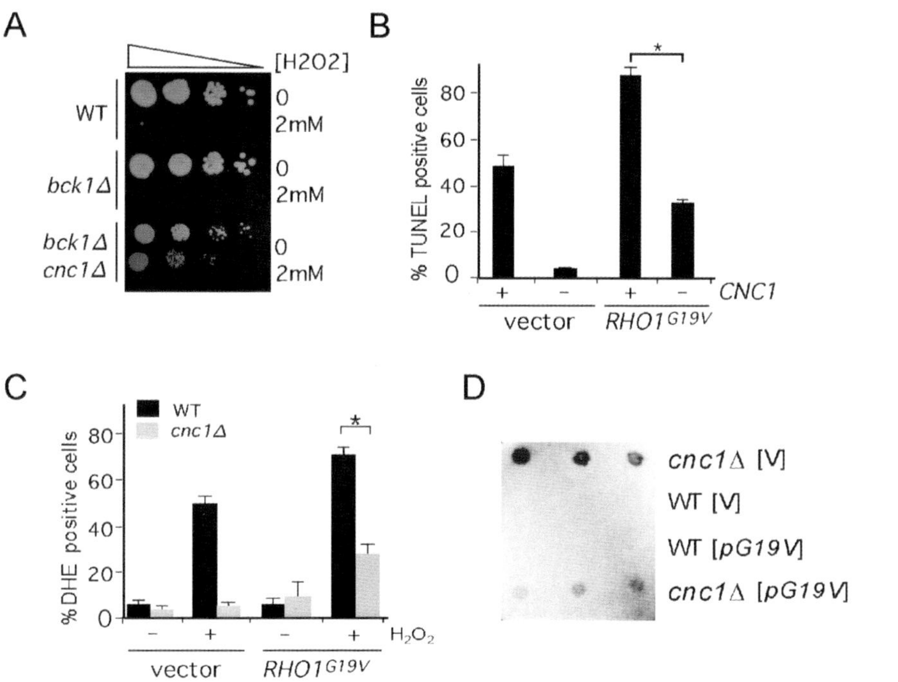

FIGURE 7: *RHO1^{G19V}* promotes cyclin C-dependent programmed cell death in the absence of stress. (A) Cell viability assays on wild type (RSY10), *bck1Δ* (RSY1050) and *bck1Δ cnc1Δ* (RSY1052) strains following treatment with 2 mM H_2O_2 for 2 h. Decreasing dilutions of the cells (represented by the arrow) were plated on YPDA media and the surviving colonies photographed after 2 days at 30°C. **(B)** Strains wild type (RSY10) and *cnc1Δ* (RSY391) cultures expressing mt-dsRed and either a vector control or *RHO1^{G19V}* were treated with 2 mM H_2O_2 for 20 h then assayed for double stranded breaks using TUNEL assays. The percent of TUNEL positive cells is given (mean ± s.e.m.). In all panels, the asterisk indicate $p < 0.05$ (Student's T test). **(C)** Strains described in (A) were treated with 2 mM H_2O_2 for 20 h and the percent of the population positive for DHE oxidation is shown (mean ± s.e.m.). **(D)** Cyclin C represses transcription in Rho1 hyper-activated strains. The aberrant vegetative expression of the meiosis-specific *spo13–lacZ* reporter gene (p42513Z) was analyzed in wild type (RSY10) and *cnc1Δ* (RSY391) strains harboring a vector control (V) or hyperactive *RHO1^{G19V}* allele. Three independent transformants were assayed on plates for ß-galactosidase expression by cleavage of the substrate 5-bromo-4-chloro-3-indolyl-ß-D-galactopyranoside (X-gal).

SPO13 transcription. In addition, these findings suggest that the role cyclin C plays in enhancing the activated Rho1 phenotypes can occur through its transcriptional and/or mitochondrial role (see Discussion).

Ste11 is required for cyclin C mitochondrial relocalization and destruction in response to high H_2O_2 induced stress.
Previous studies revealed a genetic interaction between the CWI pathway and Ste11 function. Specifically, these studies found that a *ste11Δ bck1Δ* double mutant displayed synthetic defects in cell wall integrity and that Slt2 is required for Ste11-dependent regulation of *FKS2* [14, 15]. To test whether Ste11 played a role in cyclin C destruction in response to high H_2O_2 treatment, a *ste11Δ* and a *bck1Δ ste11Δ* strain was constructed expressing myc-cyclin C. A time-course experiment revealed that cyclin C was protected from destruction in the double mutant whereas it was destroyed with similar kinetics to wild type cells in *ste11Δ* (Figure 8A, quantitated in 8B). Next, cyclin C localization was monitored under high H_2O_2 conditions. A *ste11Δ* single mutant or the *bck1Δ ste11Δ* double mutant was transformed with the cyclin C-YFP expression plasmid. In the absence of stress, cyclin C-YFP exhibited a diffuse nuclear signal in both strains similar to wild type (Figure 8C). In the *ste11Δ* mutant, exposure to high H_2O_2 concentrations resulted in cells displaying either cytoplasmic (light green arrows) or large peri-nuclear foci (red arrows). These results indicate that Ste11 is not essential for cyclin C relocalization but appears necessary for high efficiency translocation. In the double mutant, a higher percentage of the population exhibited a large focus associated with the nucleus and fewer cytoplasmic foci (Figure 8C, quantified in Figure 8D and see Figure S2 for field view of cells). These results indicate that Ste11 and Bck1 play partially redundant roles in directing cyclin C from the nucleolus to the cytoplasm in response to high H_2O_2 treatment.

DISCUSSION
Cyclin C and its kinase Cdk8 are components of the RNA polymerase II associated Mediator complex. This kinase functions primarily as a transcriptional repressor and targets a variety of genes most frequently those involved in the stress response. In response to stress, this repression is removed by nuclear to cytoplasmic relocalization of cyclin C. Previous studies found that the MAPKKK component of the CWI pathway (Bck1) is essential to trigger cyclin C relocalization in response to low-level H_2O_2 exposure. This report extends these observations by demonstrating that, in

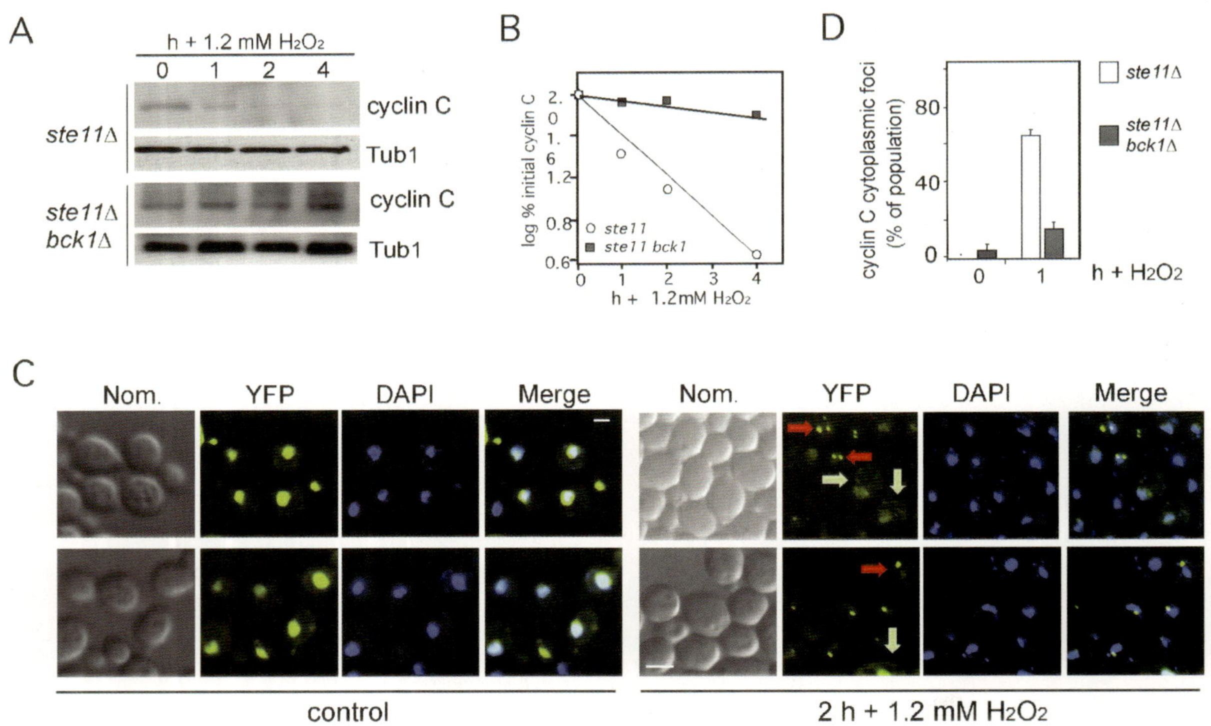

FIGURE 8: Ste11 is functionally redundant with Bck1 following 1.2 mM H_2O_2 stress. (A) Myc-cyclin C levels were monitored as described previously in *ste11Δ* (RSY1983) and *ste11Δ bck1Δ* strain (RSY1985) exposed to 1.2 mM H_2O_2 for the times indicated. Tub1 levels were used as a loading control. **(B)** Quantification of the results obtained in (A). **(C)** Ste11 is required for normal cyclin C nuclear release following 1.2 mM stress. Fluorescence microscopy was used to monitor cyclin C-YFP localization in *ste11Δ* (BLY21) and *bck1Δ ste11Δ* (BLY478) cells before (0 h) and following (2 h) 1.2 mM H_2O_2 treatment. Representative images (collapsed de-convolved 0.2 µm slices) of the results obtained are shown. Perinuclear and cytoplasmic localization of cyclin C-YFP are indicated by the red and yellow arrows, respectively. **(D)** Quantification of the results in (C). The percent of the population displaying at least 3 cyclin C-YFP foci in the cytoplasm is given (mean ± s.e.m.). At least 200 cells were counted per time point from 3 individual isolates.

response to either high H_2O_2 concentrations or in the presence of the constitutively active Rho1, Bck1 activity is combined with that of Ste11, the MAPKKK normally associated with mating type, pseudohyphal or high osmolarity environmental signaling. With respect to the cyclin C relocalization pathway, this study found that Ste11 and Bck1 are required for cyclin C translocation from the nucleolus to the cytoplasm. Taken together, these results revealed a mechanism by which the cell can distinguish low from high oxidative stress conditions by Ste11 activation. In addition, these results demonstrate a regulated step in the cyclin C relocalization pathway from the nucleolus to the cytoplasm.

MAP kinase signaling pathways are designed to recognize specific stimuli then transduce that signal to the nucleus to alter the transcription program. To achieve this goal, four general mechanisms (docking interactions, scaffold proteins, cross pathway inhibition and kinetic insulation) are in place to insulate the activation of one pathway from another (reviewed in [8]). However, recent studies have revealed an increasing important role for allowing MAP kinase pathway crosstalk. For example, in response to zymolyase cell wall stress, the sensors of the HOG pathway are required to upregulate genes controlled by the CWI MAPK module ([48, 49] and reviewed in [50]). Likewise, in strains deleted for either Pbs2 (MAPKK and scaffold) or Hog1 (see Figure 1), stimulation with high osmolarity stress results in crosstalk to the mating response [51]. These observations are consistent with the hypothesis of Murray and colleagues who posited that connectivity between different MAPK signaling pathways allows for multiple input sensing and decision making [52].

In this report, we identified a lateral signaling mechanism for the cell wall integrity (CWI) pathway that is activated in response to excessive oxidative stress. Specifically, Ste11 is required for normal degradation and cytoplasmic relocalization of cyclin C in response to high stress conditions (see Figure 9). The genetic, cellular and biochemical results presented here suggest that mechanistically this is achieved by Ste11 activating Mkk1/2. Mkk1/2 in turn activates Slt2 and Kdx1, an event that is required for cyclin C translocation [28]. Importantly, to date the redundant Mkks are the only reported MAPKKs that can transfer the activated phosphate group to Stl2 and Kdx1 (reviewed in [6]). Previous studies have also reported functional redundancy between Ste11 and Bck1 with regards to the activation of *FSK2* following stress [14]. Interestingly, Mkk1/2 is not the only kinase that Ste11 activates under high oxidative stress. Tyrosine phosphorylation of Hog1 has been observed following treatment of cells with 3 mM H_2O_2 [53]. Similarly, by using expression quantitative trait loci (eQTL) studies, Wang *et al.* proposed that Ste11 may also regulate the Pbs2-Hog1 signaling pathway under oxidative stress [54]. However, it is unlikely that H_2O_2 induced activation of Hog1 is required for cyclin C translocation to the cytoplasm as cyclin C is degraded in response to low H_2O_2 stress in *hog1Δ* cells [55].

One unanswered question that has arisen from this work is how the stress signal is transmitted to Ste11 following high H_2O_2 stress (blue arrows, Figure 9). Our previous

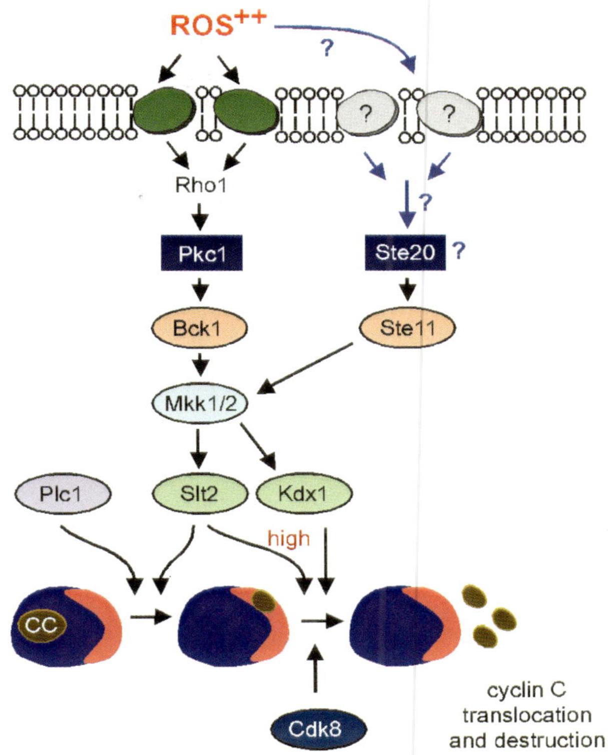

FIGURE 9: Proposed model for stress-activated signal transduction pathways mediating cyclin C nuclear export and destruction following high oxidative stress. Following high ROS stress cyclin C moves first to the nucleolus and then to the cytoplasm. Movement to the nucleolus requires Plc1 [55] whereas nucleolar export requires Slt2, Kdx1 activation and Cdk8 function [28, 34]. Under mild ROS stress the canonical CWI pathway is sufficient to transmit the stress signal to cyclin C leading to its nuclear export and destruction [28]. Under elevated ROS stress an additional MEKK (Ste11) is also required to transmit the stress signal via Mkk1/2 to cyclin C. Other sensors may also feed into the Ste11-Mkk1/2 pathway (see discussion for details).

work has suggested that additional cell wall sensors are activated in response to high stress [27]. One strong possibility is that similar to zymolyase stress, the sensors from the HOG pathway, Sln1 and/or Sho1, may transmit the oxidative stress signal to Ste11. Consistent with this, it has been reported that Ste11 can bind directly to the cytoplasmic tail of Sho1 [56] although it is not clear how this interaction directs kinase activation. Also, in support of this model is the observation that in pathogenic fungi *Aspergillus fumigatus* [57] and *Candida albicans* [58]), both Sho1 and Sln1 are cell wall sensors for environmental oxidative stress [59]. Consistent with the notion that both sensor proteins play a role, deletion of *SHO1* in the *bck1Δ* strain resulted in an "intermediate" phenotype with respect to cyclin C nuclear export and degradation (Figure S3). In other words, cyclin C was partially stabilized in the *bck1Δ sho1Δ* strain and was exported but not as robustly as in the *sho1Δ* mutant alone. However, the double mutant did not protect cells from dying following treatment with H_2O_2.

The nucleolus is the site of rRNA transcription and ribosome assembly. In response to many stressors including oxidative stress, rRNA transcription and ribosome assembly is halted. For example, the RNA polymerase I transcription factor Rrn3, relocalizes from the nucleolus to the cytoplasm to inhibit transcription following cellular damage [60]. In mammalian cells, the Rrn3 homolog TIF-1A is also inhibited in response to stress through phosphorylation by the stress activated Jnk MAPK [61]. This report demonstrated that cyclin C transits through the nucleolus in response to oxidative stress prior to its translocation to the cytoplasm. We have previously found that Cdk8 also exhibits a change from diffuse nuclear to the nucleolus following H_2O_2 treatment [34]. These results raise the possibility that cyclin C-Cdk8 performs a regulatory role in the stressed nucleolus. No target of cyclin C-Cdk8 in the RNA Pol I machinery has been identified but this kinase has been shown to modify components of the RNA Pol II Mediator complex, transcription factors and chromatin binding proteins [62]. Therefore, a role for cyclin C-Cdk8 in down regulating Pol I would not be surprising.

This study and previous reports [27, 28, 34, 35, 46] have begun to define the pathway that governs cyclin C translocation. Under normal conditions, cyclin C resides in the nucleoplasm as expected for a transcription factor (Figure 9). In response to stress, cyclin C relocalizes to the nucleolus through a process that requires Phospholipase C (Plc1) and a cis-acting element (A110) on the cyclin itself [34]. In addition, under low stress conditions, Slt2 is also required for this step and does so by directly phosphorylating cyclin C [28]. Increasing H_2O_2 concentrations or the presence of activated Rho1 relieves this requirement. Once cyclin C reaches the nucleolus, the combined activity of Cdk8 [35], Bck1 and Ste11 (this report) are necessary for efficient translocation to the cytoplasm where it interacts with the mitochondria. Taken together, these studies define a complicated, multi-step switch that controls subcellular cyclin C localization. However, it is important to note the consequences to the cell if cyclin C is mis-localized. For example, deleting *CNC1* results in no cytoplasmic cyclin C and significant protection from oxidative stress-induced PCD [26]. Conversely, loss of Med13, the nuclear anchor for cyclin C, allows continuous nuclear release of cyclin C resulting in constitutive mitochondrial fragmentation and hyper-sensitivity to oxidative stress [46]. Therefore, the cell must possess a responsive, interactive system to precisely control cyclin C subcellular localization.

MATERIALS AND METHODS

Yeast strains and plasmids

The haploid strains used in this study are listed in Table 1 and derived from RSY10, an *ade6* derivative of W303-1A [47]. BLY21 (*ste11Δ)* and BLY478 (*ste11Δ bck1Δ*) are isogenic W3031-A strains except that *ade3-1* is mutated instead of *ade6*. FLY1589 was a gift from F. Luca and is a diploid strain with integrated Nop1-CFP::*KANMX6*. *BCK1* was deleted from

TABLE 1. Yeast strains used in this study.

Strain	Genotype	Source
RSY10 [a]	*MAT**a** ade2 ade6 can1-100 his3-11,15 leu2-3,112 trp1-1 ura3-1*	[47]
RSY391 [a]	*cnc1Δ::LEU2*	[33]
RSY1006 [a]	*slt2Δ::his5+*	[26]
RSY1050 [a]	*bck1Δ::his5*$^+$	This study
RSY1052 [a]	*bck1Δ::his5*$^+$ *cnc1::LEU2*	This study
RSY1736 [a]	*kdx1Δ::KANMX6*	[28]
RSY1737 [a]	*kdx1Δ::KANMX6 slt2::his5+*	[28]
RSY1988 [a]	*sho1Δ::KANMX4*	This study
RSY1989 [b]	*sho1Δ::KANMX4 bck1:: his5*$^+$	This study
BLY21 [b]	*ste11Δ*	[14]
BLY478 [b]	*ste11Δ bck1::URA3*	[14]
FLY1589 [c]	*NOP1-CFP::KANMX6*	Frank Luca personal communication

[a]genotype: *MAT**a** ade2 ade6 can1-100 his3-11,15 leu2-3,112 trp1-1 ura3-1*

[b]genotype: *MATα ade2-1 ade3-1 can1-100 his3-11,15 leu2-3,112 trp1-1 ura3-1*

[c]genotype homozygous diploid *ura3-52 trp1Δ1 his3*

yeast strains RSY10 and RSY391 by using homologous recombination [63] to create RSY1050 (*bck1Δ*) and RSY1052 (*bck1Δ cnc1Δ*) respectively. Strains RSY1988 (*sho1Δ*) and RSY1989 (*bck1Δ sho1Δ*) were constructed by integrating the PCR amplified KANMX4 deleted *sho1* allele, obtained from the Research Genetics deletion strain collection, into RSY10 and RSY1050 respectively. In accordance with the Mediator nomenclature unification effort [64], we will use cyclin C (*SSN8/UME3/SRB11*) and Cdk8 (*SSN3/UME5/SRB10*) gene designations, respectively. Details about the plasmids used in this study can be found in Table 2. Plasmids pKC337 (*ADH1*-myc-cyclin C), and pBK37 (*ADH1*-cyclin C–YFP) are functional and have been previously described [27, 33, 34]. The initial *BCK1-20* construct (that was used to create pLR106) was a gift from D. Levin. The other constitutively activated alleles ($RHO1^{G19V}$ and $PKC1^{R398P}$) plasmids were gifts from Y. Ohya and M. Hall, respectively. The plasmid used to visualize mitochondria (mt-DsRed) was a gift from J. Shaw. The Nop1-RFP plasmid was a gift from M. Lisby. The *spo13*-lacZ fusion plasmid that exhibits meiosis-specific regulation similar to that of *SPO13* [65] was a gift from R. Strich.

Cell growth and culturing conditions

Cells were grown in either rich, non-selective medium (YPDA) or synthetic minimal medium (SC) to allow for plasmid selection as previously described [33]. The *spo13*-lacZ plate assays were conducted as previously described [47, 66]. In short, yeast cells were grown on filter paper (3MM, Midwest Scientific) placed on agar medium selective for the plasmids. Filters were then lifted and frozen in liquid nitrogen to rupture the cells then overlayed with agar containing 5-bromo-4-chloro-3-indolyl-β-D-galactopyranoside (X-Gal). After 2 to 36 h, color development was stopped by air drying.

Survival and stress assays

For all H_2O_2 stress assays, cells were grown to mid-log phase (6 x 10^6 cells/ml) in minimal media and then treated with H_2O_2 at the concentrations and times written. Cell survival, TUNEL and DHE assays were conducted exactly as previously described [35, 46]. For both DHE and TUNEL assays, positive cells were detected using flow cytometry analysis (30,000 cells counted per time point). All statistical analysis was performed using the Student's t test with $p < 0.05$ considered significant. All analyses were conducted with at least three independent cultures with 200 or more cells counted per time point.

Immunofluorescence microscopy and mitochondrial fission/fusion assays

YFP-cyclin C subcellular localization was monitored and scored as described previously [28]. For all experiments, the cells were grown to mid-log (5 x 10^6 cells/ml), treated with the concentration of H_2O_2 and for the time points indicated. The cells were harvested, fixed in 4% para-formaldehyde for 30 min, washed three times in water, stained with DAPI and then analyzed by fluorescence microscopy. Images were obtained using a Nikon microscope (model E800) with a 100 X objective (Plan Fluor Oil, NA 1.3) and a CCD camera (Hamamatsu model C4742). Data were collected using Autoquant® and processed using Image Pro software. All images of individual cells were optically sectioned (0.2 µm slices at 0.3 µm spacing) and deconvolved and the slices were collapsed to visualize the entire fluorescent signal within the cell. Cyclin C-YFP foci were scored as being cytoplasmic when 3 or more foci were observed outside of the nucleus. Mitochondrial fission assays were performed on live cells as described [35]. In brief, mitochondrial fission was scored positive if no reticular mitochondria were observed that transversed half the cell diameter. Fusion was scored when cells exhibited one or more reticular mitochondria the diameter of the cell. Fission and fusion was scored for 200 cells from three independent isolates. Statistical analysis was performed using the Student's T-test with $p < 0.05$ used to indicate significant differences. Live single cell imaging (Figure 4A) was accomplished by resuspending cells expressing cyclin C-YFP and Nop1-CFP in 1% low melting point agarose, 1 X complete minimal medium, 100 nM MitoTracker Red CMXRos (Molecular Probes) and 0.8 mM H_2O_2. Images were obtained immediately and every minute starting at 30 min. In all panels, the bar = 5 µm unless otherwise stated.

Western blot analysis

Extracts prepared for analyzing myc-cyclin C levels were prepared from mid-log cultures (6 x 10^6 cells/ml) as described

TABLE 2. Plasmids used in this study.

Plasmid Name	Gene	Epitope Tag	Marker	Promotor	2µ/CEN	Reference
pKC337	*CNC1*	1 myc	*TRP1*	*ADH1*	CEN	[33]
pBK37	*CNC1*	YFP	*TRP1*	*ADH1*	CEN	[27]
pLR106	*BCK1-20*	None	*HIS3*	*ADH1*	CEN	[26]
pYO964	$RHO1^{G19V}$	None	*URA3*	*RHO1*	CEN	[39]
PKC1*	$PKC1^{R398P}$	None	*URA3*	*PKC1*	2µ	[17]
Mt-DsRed	Mito-targeting	dsRed	*TRP1*	*ADH1*	2µ	[68]
Spo13-HIS3	*spo13*-LacZ	None	*HIS3*	own	CEN	[65]
Nop1-RFP	*Nop1*	RFP	*URA3*	own	2µ	[69]

previously [33] except that the lysis buffer used was 150 mM NaCl, 50 mM Tris-HCl pH 8.0, 1% NP-40, 0.15% deoxycholic acid sodium salt, 1 μg/ml pepstatin, 1 μg/ml leupeptin, 0.2% Protease Inhibitor cocktail (Sigma). In brief, 500 μg of soluble extract was immunoprecipitated using either α-myc or α-HA antibodies (Roche), collected on agarose A beads and analyzed by Western blot. Western blot signals were detected using goat α-mouse secondary antibodies conjugated to alkaline phosphatase (Sigma) and the CDP-Star chemiluminescence kit (Tropix). Signals were quantitated by phosphorimaging (Kodak Inc.). Half-life determinations were calculated by linear regression analysis with curves possessing r values > 0.9. Cyclin C levels were standardized to Tub1 levels before comparing to other values. For Slt2 phosphorylation assays, protein extracts were made using the LiOAc/NaOH extraction procedure exactly as described [67]. Slt2 phosphorylation was detected using α-phospho-p44/42 antibodies (Cell Signaling) as previously described [28]. Slt2 was detected using α-Mpk1 antibody (sc20168, Santa Cruz). Tub1 was visualized alpha-tubulin antibodies (12G10) were obtained from the Developmental Studies Hybridoma Bank (Univ. of Iowa).

ACKNOWLEDGMENTS

We thank E. Elion, M. Hall, D. Levin, M. Lisby, F. Luca, Y. Ohya, J. Shaw and R. Strich for strains and plasmids. We thank S. Beluch for help with western analysis of *sho1 bck1* strains. We also thank R. Strich for editorial help with this manuscript. This work was supported by the National Institutes of Health, USA (GM R15-113196).

CONFLICT OF INTEREST

The authors declare no conflict of interest.

REFERENCES

1. Widmann C, Gibson S, Jarpe MB, Johnson GL (**1999**). Mitogen-activated protein kinase: conservation of a three-kinase module from yeast to human. **Physiol Rev** 79(1): 143-180.

2. Chen RE, Thorner J (**2007**). Function and regulation in MAPK signaling pathways: lessons learned from the yeast Saccharomyces cerevisiae. **Biochim Biophys Acta** 1773(8): 1311-1340.

3. Gustin MC, Albertyn J, Alexander M, Davenport K (**1998**). MAP kinase pathways in the yeast Saccharomyces cerevisiae. **Microbiol Mol Biol Rev** 62(4): 1264-1300.

4. Brewster JL, Gustin MC (**2014**). Hog1: 20 years of discovery and impact. **Sci Signal** 7(343): re7.

5. Saito H, Posas F (**2012**). Response to hyperosmotic stress. **Genetics** 192(2): 289-318.

6. Levin DE (**2005**). Cell wall integrity signaling in Saccharomyces cerevisiae. **Microbiol Mol Biol Rev** 69(2): 262-291.

7. Junttila MR, Li SP, Westermarck J (**2008**). Phosphatase-mediated crosstalk between MAPK signaling pathways in the regulation of cell survival. **FASEB J** 22(4): 954-965.

8. Saito H (**2010**). Regulation of cross-talk in yeast MAPK signaling pathways. **Curr Opin Microbiol** 13(6): 677-683.

9. Wu C, Whiteway M, Thomas DY, Leberer E (**1995**). Molecular characterization of Ste20p, a potential mitogen-activated protein or extracellular signal-regulated kinase kinase (MEK) kinase kinase from Saccharomyces cerevisiae. **J Biol Chem** 270(27): 15984-15992.

10. Elion EA (**2001**). The Ste5p scaffold. **J Cell Sci** 114(Pt 22): 3967-3978.

11. Good MC, Zalatan JG, Lim WA (**2011**). Scaffold proteins: hubs for controlling the flow of cellular information. **Science** 332(6030): 680-686.

12. Zalatan JG, Coyle SM, Rajan S, Sidhu SS, Lim WA (**2012**). Conformational control of the Ste5 scaffold protein insulates against MAP kinase misactivation. **Science** 337(6099): 1218-1222.

13. Zhao C, Jung US, Garrett-Engele P, Roe T, Cyert MS, Levin DE (**1998**). Temperature-induced expression of yeast FKS2 is under the dual control of protein kinase C and calcineurin. **Mol Cell Biol** 18(1013-1022.

14. Lee BN, Elion EA (**1999**). The MAPKKK Ste11 regulates vegetative growth through a kinase cascade of shared signaling components. **Proc Natl Acad Sci U S A** 96(22): 12679-12684.

15. Wang X, Sheff MA, Simpson DM, Elion EA (**2011**). Ste11p MEKK signals through HOG, mating, calcineurin and PKC pathways to regulate the FKS2 gene. **BMC Mol Biol** 12:51.

16. Jung US, Levin DE (**1999**). Genome-wide analysis of gene expression regulated by the yeast cell wall integrity signalling pathway. **Mol Microbiol** 34(5): 1049-1057.

17. Nonaka H, Tanaka K, Hirano H, Fujiwara T, Kohno H, Umikawa M, Mino A, Takai Y (**1995**). A downstream target of RHO1 small GTP-binding protein is PKC1, a homolog of protein kinase C, which leads to activation of the MAP kinase cascade in Saccharomyces cerevisiae. **Embo J** 14(23): 5931-5938.

18. Kamada Y, Qadota H, Python CP, Anraku Y, Ohya Y, Levin DE (**1996**). Activation of yeast protein kinase C by Rho1 GTPase. **J Biol Chem** 271(16): 9193-9196.

19. Dodou E, Treisman R (**1997**). The Saccharomyces cerevisiae MADS-box transcription factor Rlm1 is a target for the Mpk1 mitogen-activated protein kinase pathway. **Mol Cell Biol** 17(4): 1848-1859.

20. Jung US, Sobering AK, Romeo MJ, Levin DE (**2002**). Regulation of the yeast Rlm1 transcription factor by the Mpk1 cell wall integrity MAP kinase. **Mol Microbiol** 46(3): 781-789.

21. Madden K, Sheu YJ, Baetz K, Andrews B, Snyder M (**1997**). SBF cell cycle regulator as a target of the yeast PKC-MAP kinase pathway. **Science** 275(5307): 1781-1784.

22. Baetz K, Moffat J, Haynes J, Chang M, Andrews B (**2001**). Transcriptional coregulation by the cell integrity mitogen-activated protein kinase Slt2 and the cell cycle regulator Swi4. **Mol Cell Biol** 21(19): 6515-6528.

23. Kim KY, Truman AW, Levin DE (**2008**). Yeast Mpk1 mitogen-activated protein kinase activates transcription through Swi4/Swi6 by a noncatalytic mechanism that requires upstream signal. **Mol Cell Biol** 28(8): 2579-2589.

24. Truman AW, Kim KY, Levin DE (**2009**). Mechanism of Mpk1 mitogen-activated protein kinase binding to the Swi4 transcription factor and its regulation by a novel caffeine-induced phosphorylation. **Mol Cell Biol** 29(24): 6449-6461.

25. Kim KY, Truman AW, Caesar S, Schlenstedt G, Levin DE (**2010**). Yeast Mpk1 cell wall integrity mitogen-activated protein kinase regulates nucleocytoplasmic shuttling of the Swi6 transcriptional regulator. **Mol Biol Cell** 21(9): 1609-1619.

26. Krasley E, Cooper KF, Mallory MJ, Dunbrack R, Strich R (**2006**). Regulation of the oxidative stress response through Slt2p-dependent destruction of cyclin C in Saccharomyces cerevisiae. **Genetics** 172(3): 1477-1486.

27. Jin C, Parshin AV, Daly I, Strich R, Cooper KF (**2013**). The cell wall sensors Mtl1, Wsc1, and Mid2 are required for stress-induced nuclear to cytoplasmic translocation of cyclin C and programmed cell death in yeast. **Oxid Med Cell Longev** 2013:320823.

28. Jin C, Strich R, Cooper KF (**2014**). Slt2p phosphorylation induces cyclin C nuclear-to-cytoplasmic translocation in response to oxidative stress. **Mol Biol Cell** 25(8): 1396-1407.

29. Strich R, Cooper K (**2014**). The dual role of cyclin C connects stress regulated gene expression to mitochondrial dynamics. **Microbial Cell** 1(10): 318-324.

30. Ansari SA, Morse RH (**2013**). Mechanisms of Mediator complex action in transcriptional activation. **Cell Mol Life Sci** 70(15): 2743-2756.

31. Holstege FC, Jennings EG, Wyrick JJ, Lee TI, Hengartner CJ, Green MR, Golub TR, Lander ES, Young RA (**1998**). Dissecting the regulatory circuitry of a eukaryotic genome. **Cell** 95(5): 717-728.

32. van de Peppel J, Kettelarij N, van Bakel H, Kockelkorn TT, van Leenen D, Holstege FC (**2005**). Mediator expression profiling epistasis reveals a signal transduction pathway with antagonistic submodules and highly specific downstream targets. **Mol Cell** 19(4): 511-522.

33. Cooper KF, Mallory MJ, Smith JB, Strich R (**1997**). Stress and developmental regulation of the yeast C-type cyclin Ume3p (Srb11p/Ssn8p). **EMBO J** 16(15): 4665-4675.

34. Cooper KF, Scarnati MS, Krasley E, Mallory MJ, Jin C, Law MJ, Strich R (**2012**). Oxidative-stress-induced nuclear to cytoplasmic relocalization is required for Not4-dependent cyclin C destruction. **J Cell Sci** 125(Pt 4): 1015-1026.

35. Cooper KF, Khakhina S, Kim SK, Strich R (**2014**). Stress-induced nuclear-to-cytoplasmic translocation of cyclin C promotes mitochondrial fission in yeast. **Dev Cell** 28(2): 161-173.

36. Wang K, Yan R, Cooper KF, Strich R (**2015**). Cyclin C mediates stress-induced mitochondrial fission and apoptosis. **Mol Biol Cell** 26(6): 1030-1043.

37. Heinisch JJ, Dupres V, Wilk S, Jendretzki A, Dufrene YF (**2010**). Single-molecule atomic force microscopy reveals clustering of the yeast plasma-membrane sensor Wsc1. **PLoS One** 5(6): e11104.

38. Philip B, Levin DE (**2001**). Wsc1 and Mid2 are cell surface sensors for cell wall integrity signaling that act through Rom2, a guanine nucleotide exchange factor for Rho1. **Mol Cell Biol** 21(1): 271-280.

39. Sekiya-Kawasaki M, Abe M, Saka A, Watanabe D, Kono K, Minemura-Asakawa M, Ishihara S, Watanabe T, Ohya Y (**2002**). Dissection of upstream regulatory components of the Rho1p effector, 1,3-beta-glucan synthase, in Saccharomyces cerevisiae. **Genetics** 162(2): 663-676.

40. Lee KS, Levin DE (**1992**). Dominant mutations in a gene encoding a putative protein kinase (BCK1) bypass the requirement for a Saccharomyces cerevisiae protein kinase C homolog. **Mol Cell Biol** 12(1): 172-182.

41. Levin DE (**2011**). Regulation of cell wall biogenesis in Saccharomyces cerevisiae: the cell wall integrity signaling pathway. **Genetics** 189(4): 1145-1175.

42. Vilella F, Herrero E, Torres J, de la Torre-Ruiz MA (**2005**). Pkc1 and the upstream elements of the cell integrity pathway in Saccharomyces cerevisiae, Rom2 and Mtl1, are required for cellular responses to oxidative stress. **J Biol Chem** 280(10): 9149-9159.

43. Levin DE, Fields FO, Kunisawa R, Bishop JM, Thorner J (**1990**). A candidate protein kinase C gene, *PKC1*, is required for the *S. cerevisiae* cell cycle. **Cell** 62:213-224.

44. Petkova MI, Pujol-Carrion N, Arroyo J, Garcia-Cantalejo J, Angeles de la Torre-Ruiz M (**2010**). Mtl1 is required to activate general stress response through Tor1 and Ras2 inhibition under conditions of glucose starvation and oxidative stress. **J Biol Chem** 285(25): 19521-19531.

45. Westermann B (**2010**). Mitochondrial fusion and fission in cell life and death. **Nat Rev Mol Cell Biol** 11(12): 872-884.

46. Khakhina S, Cooper KF, Strich R (**2014**). Med13p prevents mitochondrial fission and programmed cell death in yeast through nuclear retention of cyclin C. **Mol Biol Cell** 25(18): 2807-2816.

47. Strich R, Slater MR, Esposito RE (**1989**). Identification of negative regulatory genes that govern the expression of early meiotic genes in yeast. **Proc. Natl. Acad. Sci. USA** 86(10018-10022.

48. Bermejo C, Rodriguez E, Garcia R, Rodriguez-Pena JM, Rodriguez de la Concepcion ML, Rivas C, Arias P, Nombela C, Posas F, Arroyo J (**2008**). The Sequential Activation of the Yeast HOG and SLT2 Pathways Is Required for Cell Survival to Cell Wall Stress. **Mol Biol Cell** 19(3): 1113-1124.

49. Garcia R, Rodriguez-Pena JM, Bermejo C, Nombela C, Arroyo J (**2009**). The high osmotic response and cell wall integrity pathways cooperate to regulate transcriptional responses to zymolyase-induced cell wall stress in Saccharomyces cerevisiae. **J Biol Chem** 284(16): 10901-10911.

50. Rodriguez-Pena JM, Garcia R, Nombela C, Arroyo J (**2010**). The high-osmolarity glycerol (HOG) and cell wall integrity (CWI) signalling pathways interplay: a yeast dialogue between MAPK routes. **Yeast** 27(8): 495-502.

51. O'Rourke SM, Herskowitz I (**1998**). The Hog1 MAPK prevents cross talk between the HOG and pheromone response MAPK pathways in Saccharomyces cerevisiae. **Genes Dev** 12(18): 2874-2886.

52. Hartwell LH, Hopfield JJ, Leibler S, Murray AW (**1999**). From molecular to modular cell biology. **Nature** 402(6761 Suppl): C47-52.

53. Bilsland E, Molin C, Swaminathan S, Ramne A, Sunnerhagen P (**2004**). Rck1 and Rck2 MAPKAP kinases and the HOG pathway are required for oxidative stress resistance. **Mol Microbiol** 53(6): 1743-1756.

54. Wang L, Zheng W, Zhao H, Deng M (**2013**). Statistical analysis reveals co-expression patterns of many pairs of genes in yeast are jointly regulated by interacting loci. **PLoS Genet** 9(3): e1003414.

55. Cooper KF, Mallory MJ, Strich R (**1999**). Oxidative stress-induced destruction of the yeast C-type cyclin Ume3p requires phosphatidylinositol-specific phospholipase C and the 26S proteasome. **Mol Cell Biol** 19(5): 3338-3348.

56. Zarrinpar A, Bhattacharyya RP, Nittler MP, Lim WA (**2004**). Sho1 and Pbs2 act as coscaffolds linking components in the yeast high osmolarity MAP kinase pathway. **Mol Cell** 14(6): 825-832.

57. Ma Y, Qiao J, Liu W, Wan Z, Wang X, Calderone R, Li R (**2008**). The sho1 sensor regulates growth, morphology, and oxidant adaptation in Aspergillus fumigatus but is not essential for development of invasive pulmonary aspergillosis. **Infect Immun** 76(4): 1695-1701.

58. Roman E, Nombela C, Pla J (**2005**). The Sho1 adaptor protein links oxidative stress to morphogenesis and cell wall biosynthesis in the fungal pathogen Candida albicans. **Mol Cell Biol** 25(23): 10611-10627.

59. Brown AJ, Haynes K, Quinn J (**2009**). Nitrosative and oxidative stress responses in fungal pathogenicity. **Curr Opin Microbiol** 12(4): 384-391.

60. Lewinska A, Wnuk M, Grzelak A, Bartosz G (**2010**). Nucleolus as an oxidative stress sensor in the yeast Saccharomyces cerevisiae. **Redox Rep** 15(2): 87-96.

61. Mayer C, Bierhoff H, Grummt I (**2005**). The nucleolus as a stress sensor: JNK2 inactivates the transcription factor TIF-IA and down-regulates rRNA synthesis. **Genes Dev** 19(8): 933-941.

62. Conaway RC, Conaway JW (**2011**). Function and regulation of the Mediator complex. **Curr Opin Genet Dev** 21(2): 225-230.

63. Longtine MS, McKenzie A, 3rd, Demarini DJ, Shah NG, Wach A, Brachat A, Philippsen P, Pringle JR (**1998**). Additional modules for versatile and economical PCR-based gene deletion and modification in Saccharomyces cerevisiae. **Yeast** 14(10): 953-961.

64. Bourbon HM (**2008**). Comparative genomics supports a deep evolutionary origin for the large, four-module transcriptional mediator complex. **Nucleic Acids Res** 36(12): 3993-4008.

65. Buckingham LE, Wang H-T, Elder RT, McCarroll RM, Slater MR, Esposito RE (**1990**). Nucleotide sequence and promoter analysis of *SPO13*, a meiosis-specific gene of *Saccharomyces cerevisiae*. **Proc. Natl. Acad. Sci.** 87:9406-9410.

66. Washburn BK, Esposito RE (**2001**). Identification of the Sin3-binding site in Ume6 defines a two-step process for conversion of Ume6 from a transcriptional repressor to an activator in yeast. **Mol Cell Biol** 21(6): 2057-2069.

67. Zhang T, Lei J, Yang H, Xu K, Wang R, Zhang Z (**2011**). An improved method for whole protein extraction from yeast Saccharomyces cerevisiae. **Yeast** 28(11): 795-798.

68. Mozdy AD, McCaffery JM, Shaw JM (**2000**). Dnm1p GTPase-mediated mitochondrial fission is a multi-step process requiring the novel integral membrane component Fis1p. **J Cell Biol** 151(2): 367-380.

69. Torres-Rosell J, Sunjevaric I, De Piccoli G, Sacher M, Eckert-Boulet N, Reid R, Jentsch S, Rothstein R, Aragon L, Lisby M (**2007**). The Smc5-Smc6 complex and SUMO modification of Rad52 regulates recombinational repair at the ribosomal gene locus. **Nat Cell Biol** 9(8): 923-931.

7

In *Entamoeba histolytica*, a BspA family protein is required for chemotaxis toward tumour necrosis factor

Anne Silvestre[1, 2, 3, 4], Aurélie Plaze[1, 2], Patricia Berthon[3, 4], Roman Thibeaux[1, 2], Nancy Guillen[1, 2, *] and Elisabeth Labruyère[1, 2]

[1] Institut Pasteur, Unité Biologie Cellulaire du Parasitisme, F-75015 Paris, France.
[2] INSERM U786, F-75015 Paris, France.
[3] INRA, UMR1282 Infectiologie et Santé Publique, F-37380 Nouzilly, France.
[4] Université de Tours, UMR1282 Infectiologie et Santé Publique, F-37000 Tours, France.
* Corresponding Author: Nancy Guillén, Institut Pasteur, 28 rue du Dr Roux, F-75724 Paris cedex 15, France
E-mail: nguillen@pasteur.fr

ABSTRACT **Background:** ***Entamoeba histolytica*** **cell migration is essential for the development of human amoebiasis (an infectious disease characterized by tissue invasion and destruction). The tissue inflammation associated with tumour necrosis factor (TNF) secretion by host cells is a well-documented feature of amoebiasis. Tumour necrosis factor is a chemoattractant for *E. histolytica*, and the parasite may have a TNF receptor at its cell surface. Methods: confocal microscopy, RNA Sequencing, bioinformatics, RNA antisense techniques and histological analysis of human colon explants were used to characterize the interplay between TNF and *E. histolytica*. Results: an antibody against human TNF receptor 1 (TNFR1) stained the *E. histolytica* trophozoite surface and (on immunoblots) binds to a 150-kDa protein. Proteome screening with the TNFR1 sequence revealed a BspA family protein in *E. histolytica* that carries a TNFR signature domain and six leucine-rich repeats (named here as "cell surface protein", CSP, in view of its cellular location). Cell surface protein shares structural homologies with Toll-Like receptors, colocalizes with TNF and is internalized in TNF-containing vesicles. Reduction of cellular CSP levels abolished chemotaxis toward TNF and blocked parasite invasion of human colon. Conclusions: there is a clear link between TNF chemotaxis, CSP and pathogenesis.**

Keywords: *Entamoeba histolytica, chemotaxi, tumour necrosis factor, BspA protein.*

Abbreviations:
AS – antisense,
BspA – Bacteroides surface protein A,
CSP – cell surface protein,
LGT lateral gene transfer,
LRR – leucin rich repeats,
NGFR – nerve growth factor receptor,
TIR – toll/IL-1 receptor,
TLR – toll-like receptor,
TNF – tumour necrosis factor,
TNFR – TNF receptor,
WT – wild type.

INTRODUCTION

Entamoeba histolytica is the causative agent of human amoebiasis (a disease that targets intestinal and hepatic tissues). Parasite motility has an important role in invasive amoebiasis [1], since trophozoites cross the mucosal barrier and penetrate the intestinal epithelium. The motile trophozoites interact with extracellular matrix components and cells, leading to disruption of the intestinal architecture, cell death and acute inflammation [2].

During the host's amoeba-induced inflammatory response, human epithelial cells chemoattract neutrophils and macrophages to the site of invasion [3]. Trophozoite motility is related to chemotactic stimuli derived from components such as activated C5α in human serum, lysed red blood cells, bacteria, N-acetylneuraminic acid, fibronectin, and human tumour necrosis factor (TNF) [4, 5]. However, the molecular interplay between these compounds and *E. histolytica* has not been extensively characterized.

The TNF secreted by enterocytes and macrophages is a major component in the amplification of amoeba-related inflammation [6]. The levels of TNF secreted by the host is not sufficiently cleared through binding on trophozoites as this interaction does not influence the disease progress in experimental murine amoebiasis; only TNF depletion shows a reducing impact on the immunopathogenicity and the disease outcome [7]. We showed that human TNF is chemoattractant and chemokinetic for *E. histolytica* [4, 5, 8]. Chemotaxis of *E. histolytica* toward TNF is abrogated in presence of monoclonal anti-TNF antibody or with the soluble TNF receptor 1, demonstrating specificity of TNF chemotaxis [4]. Transmembrane tumour necrosis factor receptors (TNFRs) are characterized by extracellular cysteine-rich domains that are the hallmark of the TNFR superfamily [9]. To date, no members of the TNFR superfamily have been described in *E. histolytica*. Chemotaxis of *E. histolytica* [5, 8] results in motility toward the chemoattractant source [4, 10, 11]. The mechanism by which TNF induces *E. histolytica* chemotaxis is based on phosphatidylinositide-3-kinase signalling, reorgánization of the actin-rich cytoskeleton and galactose/N-acetyl-galactosamine lectin activity [10]. Our starting hypothesis was that TNF binds amoebic surface proteins capable of activating a signalling pathway involved in chemotaxis. Our goal was thus to determine whether *E. histolytica's* proteome contains TNFR homologues. In a BLAST search, we identified a protein presenting the TNFR signature and bearing six leucine-rich repeats (LRRs, structural motifs involved in protein interaction). Cell surface proteins with LRRs have already been identified as virulence factors in the anaerobic bacteria *Bacteroides forsythius* [12] and were named "*Bacteroides* surface protein A" (BspA). This type of protein has also been found in oral bacteria [13] and in anaerobic protozo ans as a predicted consequence of lateral gene transfer from bacteria into protozoan genomes [14]. It was shown that BspA-like proteins are involved in bacterial adherence, epithelial cell invasion [15, 16] and fibronectin and fibrinogen binding [12]. Direct evidence for an *in vivo* role of BspA in pathogenesis was provided by a study showing that BspA-defective bacteria were significantly less pathogenic than the wild type (WT) [17]. In *E. histolytica*, the BspA-like family is composed of 116 proteins [18], one of which has been found at the plasma membrane of trophozoites [19]. In the present work, we identified another member of the BspA-like family and named it "cell surface protein" (CSP). We found that CSP and TNF co-localized at the trophozoite surface; and are then internalized within vesicles. In order to gain insight into CSP's putative role in pathogenesis, we down-regulated expression of the *csp*-encoding gene by transfection of trophozoites with an antisense (AS) construct plasmid. Using biochemical and cellular analyses, we demonstrated that CSP has a key role in *E. histolytica's* ability to migrate toward a TNF gradient and to invade human colon explants. These findings suggest that CSP is significantly involved in TNF chemotaxis during the early stages of *E. histolytica* amoebiasis although the proof that CSP is a direct receptor for TNF is not established.

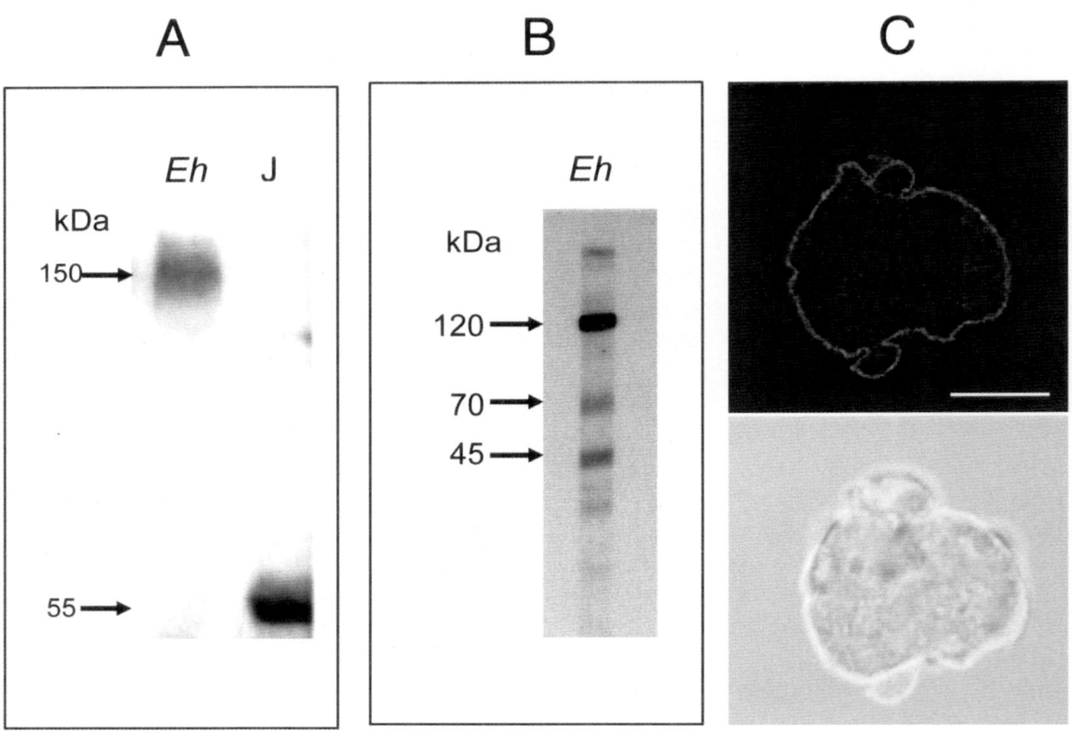

FIGURE 1: Screening of an *E. histolytica* protein extract with an antibody against human soluble hTNFR1. (A) Polyacrylamide gel electrophoresis under non-reducing conditions, and a Western blot analysis. Crude lysates of *E. histolytica* (Eh) and Jurkat (J) cells were loaded in the indicated lane (load: 20 µg of protein). Probing with the anti-hTNFR1 antibody revealed a single 150 kDa protein in *E. histolytica* and a single 55 kDa protein in Jurkat cells. **(B)** Polyacrylamide gel electrophoresis under reducing conditions, and a Western blot analysis. In a crude lysate from *E. histolytica*, three proteins (at approximately 120 kDa, 70 kDa and 45 kDa) were detected. **(C)** Entire trophozoites were fixed and stained with goat anti-hTNFR1 and anti-goat AlexaFluor488 in order to reveal the localization of amoebic proteins sharing an epitope with hTNFR1. Scale bar: 10 µm.

RESULTS

Identification of proteins with a TNFR signature

In order to determine whether proteins within the *E. histolytica* proteome shared common epitopes with human TNFR, a crude lysate from *E. histolytica* trophozoites was immunoblotted with an antibody against hTNFR1. A protein with a relative molecular mass of 150 kDa was detected by electrophoresis under non-reducing conditions (Figure 1A); whereas three proteins (120 kDa, 70 kDa and 45 kDa) were found upon treatment with reducing agents (Figure 1B). A crude lysate from Jurkat cells was used as a positive control; as expected, a signal at 55 kDa (corresponding to hTNFR1; accession number: AAA36754, [20]) was detected (Figure 1A). To determine the cellular localization of the homologous amoebic proteins, parasites in culture were fixed and stained with an anti-hTNFR1 antibody (Figure 1C). Labelling was observed at the surface of non-permeabilized trophozoites, suggesting that an *E. histolytica* surface protein shared a common epitope with the hTNFR1.

The amino acid sequence of hTNFR1 was used in a BLAST search of the amoebic proteome in the AmoebaDB. Of 8201 ORFs representing the amoebic proteome, a single sequence was identified (E-value: 3.5x10-7; entry EHI_016490; predicted molecular mass: 140 kDa) as comprising a TNFR domain (Figure 2A). EHI_016490 is annotated in AmoebaDB as a member of the LRR BspA-like family and we named it "cell surface protein" (CSP). The putative protein contains 1222 amino acids, shares 43% amino acid sequence identity with hTNFR1 (E-value=0.74) and does not contain a conventional signal peptide or a transmembrane domain. SignalP 4.1 server (used to predict the presence and location of signal peptide cleavage sites in amino acid sequences from Gram-positive prokaryotes, Gram-negative prokaryotes, and eukaryotes) was employed and conventional signal peptides were not found in CSP sequence, but we cannot rule out the fact that CSP amino acid sequence might contain a signal peptide different from those described for other organisms. Prosite predicted a cysteine-rich TNFR/nerve growth factor receptor (NGFR) domain between amino acids 957 to 998 (PS00652, TNFR_NGFR_1 family cysteine-rich region signature). Six LRR5 domains (Pfam13306, with E-values ranging from 4.03 x 10e-29 to 9.6 x 10e-17) were found. LRR5 domains account for 61% of the total CSP protein. Two of the domains show near-perfect identity with the consensus sequence CxxLxxLxLxxxL (at positions 707 and 1207), whereas four are degenerated sequences (at positions 162, 231, 509 and 1060). Three predicted prenylation sites (CaaX, where "a" is an aliphatic amino acid) were found (at positions 413, 909 and 1032, respectively). We used the Phyre2 server to predict alpha helix, beta strand and coiled domains with the CSP amino acid sequence and thus gain insight into the protein's secondary structure. The best score was obtained for the human extra cellular domain of toll-like receptor 3 (TLR3) [21] (Figure 2B), as characterized by a confidence score of 99.9% for 70% of CSP's total sequence and a secondary structure prediction score of 7.7 (with a peak at 9).

Further analysis on CSP structure was performed using I-TASSER [22] algorithms. The best structural prediction for CSP fit with the identified mouse TLR3 ectodomain (Accession number: 3ciy in Protein Database Bank) (supplemental Figure S1). The prediction emerge with a high confidence score (c-score = −1.5 from a theoretical value ranging from −5 to +2), indicating that more than 90 % of the predicted model was correct. A TM-score over 0.5 (theoretical value ranging from 0 to 1) supporting in addition the high quality of the structural alignment between CSP and known structures for TLR3 in the Protein Database Bank.

Because CSP showed high homology with TLRs, we investigated whether the entire BspA like family from *E. histolytica* (to which CSP belongs) has phylogenetic relatedness to TLR family of receptors. To identify the genes from the BspA family expressed in cultured *E. histolytica*, we took advantage of our precedent RNA seq approaches performed with *E. histolytica* virulent strain [23]. A total of 95 genes encoding BspA like proteins (over 100 formerly identified) were transcribed (Supplemental Table S1). To determine the genetic relatedness of CSP (EHI_016490) with its *Entamoeba* BspA like protein orthologues, a phylogenetic tree was build considering the 95 BspA predicted protein sequences from *E. histolytica* genome (supplemental Figure S2). The order of the sequences in the alignment reflects their relatedness with the most similar sequences typically being aligned next to each other. According to the topology of the evolutionary tree, EHI_016490 diverged early from the 94 other BspA like orthologues found in *E. histolytica*. Phylogeny of two other candidates (EHI_112030 and EHI_151330) was closely related to CSP (EHI_016490).

To gain insight into phylogenetic relatedness between CSP and mammalian TLR family of receptors, another phylogenetic tree was built with amino acid sequences from CSP (EHI_016490), its orthologues EHI_112030 and EHI_151330, human TLR (10 hTLR) and murine TLR (3 mTLR) (Figure 2C). The 16 sequences were dispatched in three distinct clusters, supported by high bootstrap values (72-94 %). The tree topology confirmed that CSP and its two closely related BspA like orthologues from *E. histolytica* (EHI_112030 and EHI_151330) were closely related to human TLR3.

Confronting EHI_112030 and EHI_151330 with PDB structural entries (using Phyre2 server) allowed proposing their secondary structure which showed similarities to these proposed for CSP: the top ten highest scoring alignments were with TLR domains. For the two-tested BspA like proteins, the best score was found with the extra cellular domain of human TLR3 (supplemental Figure S3). All the data from the bioinformatics analysis lead us to though that the TLR-like structure in CSP was mainly generated by LRR motifs, these represents 61 % of the total protein. As a consequence, we can expect that the distant prioritized BspA like proteins in the tree (also bearing LRR motifs) should have the same TLR 3D structure. Phyre 2 structural approaches performed on the two most distant proteins tell us that it was not the case: EHI_072070, EHI_152950 do not have the classical horseshoe-like structure of TLRs

A

```
   1 MKKLDSYSLL ICSKYFRYKS DFINVICVCK KFQETLEKFR YNPISISNLR LFPKIQTQCL YHKNEIRLPI ETYSFYYFLT
  81 YKEALNQMKN FNKCHQIVYT RSDREEFGID IPQNFAIKAL GDKCFESTPI QKIIIPNTIR KIGQEAFSQC TQLTQIQLPC
 161 TLKELPVCTF FNCIELEKIE IPSSVSIIDG ACFFCCSHLT EVKFPQNIVS IGYESFAFCA RLKEVVIQGT LYSLFNKSFF
 241 GCTALSSVHL PDTVKFISDS CFENCSSLOS INIPSTVVMI NOKVFKNCTS LKEIETPPSV DYIGERCFEN CYSLTRLKIS
 321 DTTVNISCNC FLNCTSLQTL EVPLKNNEYP FDVSYYDKQI LEKFGINCVH INFFSSGSVL TYNPLTHEPK IPDDALIIGK
 401 ECFKNIREIR SICIPTNIVI IDSNAFVGSF ITSIYIPTSV TYIISGAFSD CVRLKEIQLP SSISSIGCKL FMNCSALTSI
 481 TIPSTITSIN ASAFEFCINL STISLPPHLV KLKKNAFSGC VQLKEILLPS SLKRIEEKCF SDCHSLTFVS IPTTVTYIGK
 561 DICLNCRGLK NLIIPLEKDL SYKYKVSYQQ YQLFSSLNIH CTNIQFTDQD YLQRRNNNVD TIIPTDVDLH ISKLCFSKLV
 641 ENSFILPPNV ISLGKSCFQS SFNITSITLS TNITKIKSYA FNGCSSLKNL IIPSSVQYMG KYCFKNCDNL TSLSLPTNLL
 721 PYTSLVSYSE YLLLKRNNIE CLNIAQVNDD DIYDSKYLPS EIQTLNNTYF DFSSKELIVP SHITKIKVGV FCDCFQMSKI
 801 QIPSSVVSIK RNVFSNCPSL KSIELPPYLK KLSSSLFYYC ISLKSIEIPS KITKLSNNVF AECHSLSQIH FPNQLKRIKG
 881 CCFFNCKNLS SITIPSSVTK LGKRCFDFCL GLQKCKFEEP CQIKKIPENC FRMCDKLVSF NIPSSIEILD SSCFYKCFGL
 961 TSIHIPSNVK SIGQCCFKRC YFLKEVICDQ IQEIDKDCFS YCSRLESVIL PSSLKKIGQT AFSYCSALKE ICIPDSVEFI
1041 GGLCFSGCKQ LTRIALSSRL TSLSYDCFTN CHSLRSIIIN NTPISNYPFN VSLLQYIYFS KNKIPCYNIT LSQDEIYLLS
1121 TNIPHLVNCF NDNCFRNSVN LMNISIPSSV TSLGEYCFKN CINLTSITIP SSISSIPSHC FDSCYNLKSI ILPSTITSFG
1201 NHSFYGCSQL ESLNLIPKEC FE
```

B C

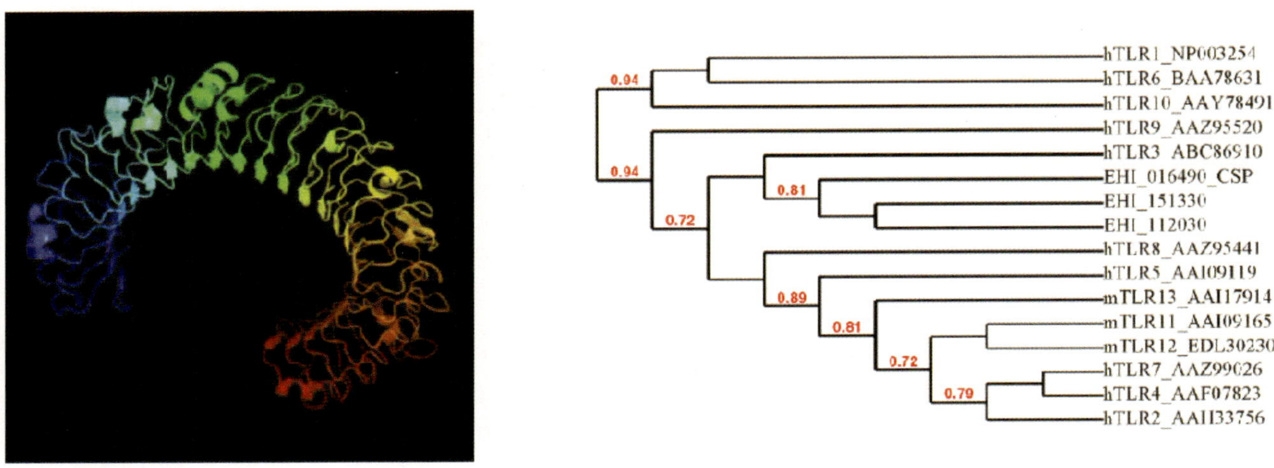

FIGURE 2: ***In silico*** **analysis of the CSP (EHI_016490) amino acid sequence. (A)** Protein domains found in the amino acid sequence of EHI_016490 after scanning with InterProScan and SMART online tools. The TNFR_NGFR_1 family's cysteine-rich region signature is indicated in bold. The leucine-rich repeats (xxxLxxLxx) are underlined with stars, the prenylation sites (CaaX, where each "a" is an aliphatic amino acid) are underlined and the two peptides used to produce polyclonal rabbit serum are bordered. **(B)** A three-dimensional structure model of EHI_016490, generated using Phyre2. The model is based on the c1ziwA template (the structure of the human toll-like receptor 3 extra-cellular domain). In all, 856 residues (70% of the full sequence) were modelled with a confidence score of 99.9% for the highest-scoring template. **(C)** Dendogram showing the relatedness between EHI_016490, EHI_112030, EHI_151330 from *Entamoeba histolytica*, human TLR (hTLR) and murine TLR (mTLR). The number at each branch point represents percentage bootstrap support, calculated from 1000 replicates. Bootstrap values below 0.7 are not represented.

(supplemental Figure S3). For these two last candidates, the best score (100 % alignment) fit with a leucine rich protein 2 from *Bacteroides ovatus*, with coverage of 82-86 %. Altogether, our bioinformatics analysis defined structural specificity of CSP and two orthologues among the large BspA like proteins family from *E. histolytica* and described homologies with TLRs from humans.

The amoebic CSP binds TNF and relocates to the uropod when a TNF gradient is present

In order to localize the CSP in the trophozoite, two potentially immunogenic peptides (Phe-92 to Glu-106 and Phe-606 to Thr-621) were synthesized and used to produce anti-CSP antibody. Probing *E. histolytica* crude lysates with the anti-CSP serum revealed three bands at 140, 70 and 45 kDa (supplemental Figure S4). CSP was predicted to be a roughly 140 kDa protein, which is the apparent molecular mass corresponding to the higher signal. The two smaller peptides recognized in reducing gel conditions might represent processed forms of CSP as we have already observed with the anti-sTNR human antibody (Figure 1B).

Anti-CSP and anti-TNFR antibodies were used to examine the proteins in trophozoites, using confocal microscopy. After staining trophozoites with both antibodies, the fluorescence signals were seen to co-localize in a continuous pattern on plasma membrane (Figure 3A). A control with pre-immune serum gave a low background signal and no signal at the trophozoite surface (data not shown). When

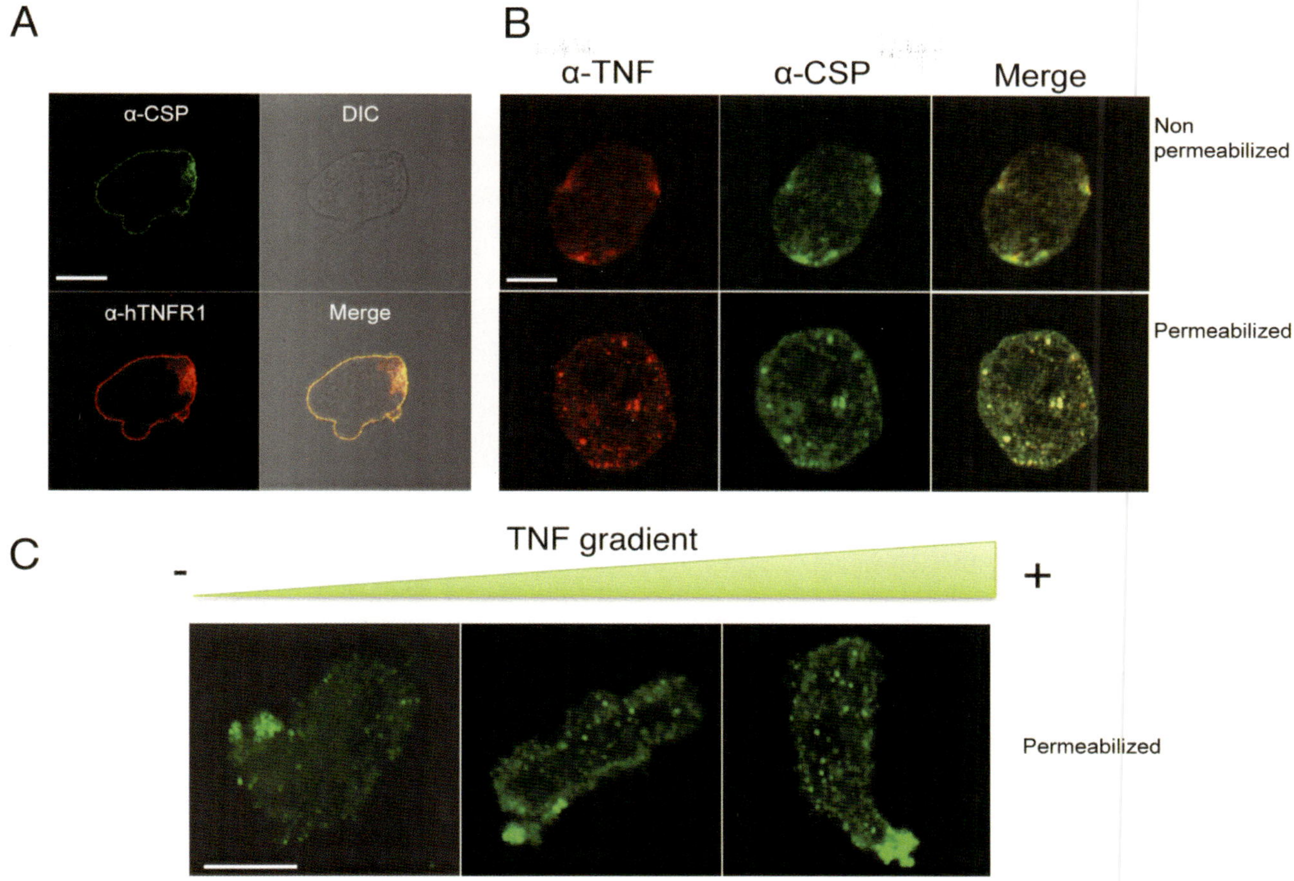

FIGURE 3: Cellular localization of CSP in *Entamoeba histolytica*. (A) Immunolocalization of CSP (green) and hTNFR1 (red) in fixed but non-permeabilized trophozoites revealed that both proteins are located at the plasma membrane. Scale bar: 10 µm. **(B)** Immunolocalization of CSP in *E. histolytica* incubated with a homogenous (non-gradient) concentration of TNF. Trophozoites were incubated with 70 nM TNF, fixed and permeabilized or not. CSP (green) and TNF (red) were co-localized at the trophozoite surface. CSP was detected as described for Fig. 3A. TNF was detected with 1E12 monoclonal antibody and mouse-Cy3 (dilution: 1:200). Scale bar: 10 µm. **(C)** Immunolocalization of CSP in *E. histolytica* incubated with a TNF gradient. Trophozoites exposed to a TNF gradient for 2 h were fixed and permeabilized. During trophozoite migration toward the TNF source, CSP (green) concentrated in the rear part of the cells (i.e. distal to the TNF source). The nucleus was stained with DAPI (blue). Scale bar: 10 µm.

trophozoites were incubated with recombinant human TNF and labelled with anti-CSP and anti-TNF antibodies, the trophozoite surface was stained by small dots of labelling, as a patched pattern. The fluorescence signals co-localize at the cell surface - showing that CSP and TNF bind to each other. When trophozoites were permeabilized, the fluorescence signals still co-localized as intra-cytoplasmic vesicules of various sizes (Figure 3B).

To investigate the potential influence of TNF on CSP localization during chemotaxis, trophozoites were incubated in a TNF gradient. CSP was detected in intracytoplasmic vesicles when the trophozoites were far from the TNF source but was concentrated in the uropod when the cells were close to the source (Figure 3C). These results indicate that CSP is usually located at the trophozoite surface but relocates to the uropod when trophozoites migrate up a TNF gradient. Several attempts were performed trying to express *csp* encoding gene in *Escherichia coli* to further analyze the physicochemical parameters of the possible interaction between TNF and CSP. Unfortunately, we did not succeed in cloning the entire gene into the bacterium, suggesting that this gene maybe toxic for *E. coli*.

Knocking down CSP with an AS RNA technique

To examine CSP's possible role in chemotaxis toward TNF, trophozoites were transfected with a plasmid carrying the *csp* gene inserted in an antisense orientation. The effect of CSP-AS RNA transcription on CSP levels was detected in immunoblot experiments (Figure 4); the level was 43% lower in CSP-AS trophozoites than in WT trophozoites (n=3; p=0.01). There was no significance between WT trophozoites and control GFP trophozoites in terms of protein levels of CSP.

The functional role of CSP in chemotaxis toward TNF

To determine whether CSP is involved in chemotaxis toward TNF, the displacement of trophozoites in which CSP expression had been modified was investigated *in vitro*. In

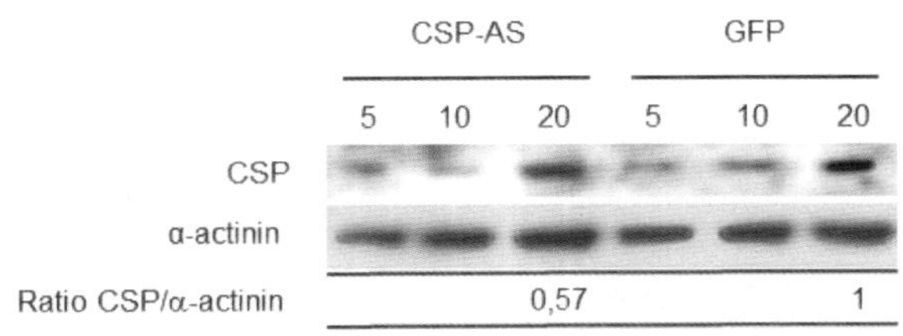

FIGURE 4: Quantification of CSP expression in modified *Entamoeba histolytica* trophozoites. Five, ten and twenty μg of protein from crude trophozoite extracts were loaded. CSP (140 kDa) and actinin (70 kDa) were revealed with specific antibodies. A decrease in CSP abundance was observed in trophozoites expressing the CSP-AS construct. Amounts of CSP and actinin in 20 μg of crude extract were quantified using AlphaEaseFC software. Levels of CSP were normalized against the amount of α-actinin. The relative abundance of CSP was 57% lower in CSP-AS trophozoites than in WT and GFP trophozoites (P = 0.04). Data are representative of three independent experiments.

the absence of chemoattractant, WT trophozoites were evenly distributed along the coverslip (p=0.531) (Figure 5). When TNF was injected into the agarose slice, trophozoites were significantly more abundant near the source of TNF when WT (p=0.042) and GFP (p=0.005) strains were tested. In contrast, similar numbers of CSP-AS trophozoites were located near to or far from the TNF source (p=0.249), indicating that most of these cells did not react to the presence of TNF. These results suggest that CSP is involved in TNF chemotaxis in trophozoites.

CSP is important for spreading within the human intestine

We used an *ex vivo* human model of intestinal amoebiasis to further investigate CSP's role during intestinal invasion by *E. histolytica*, [24]. This model has already enabled us to describe the early stages of pathogen invasion, including TNF secretion within the first 4 h of amoebic invasion [24].

Prior to infection, the viability of CSP-AS trophozoites in Krebs' buffer was determined in a trypan blue exclusion test. The mean ± SD percentages of dead cells were similar for GFP-expressing trophozoites and CSP-AS trophozoites (8.27 ± 0.56% and 8.40 ± 0.73%, respectively). Next, CSP-AS, WT and GFP trophozoites were incubated with human colon explants for 2, 4 and 7 h. Trophozoites were then detected on histological sections by immunostaining the Gal/GalNAc lectin (Figure 6). After 2 h of incubation, WT and GFP trophozoites adhered to the colonic epithelium. After 4 h, trophozoites detached the enterocytes and migrated into the lamina propria along the crypts. After 7 h of incubation, the tissue architecture was altered and trophozoites were found deep in the lamina propria, as observed in previous studies [24-27]. The destruction of the mucosa by WT and GFP trophozoites contrasted with the unaltered architecture of the control tissue after 7 h of culture.

The CSP-AS trophozoites differed from the WT in terms of intestinal invasion. Although no differences in tissue penetration were observed during the first 2 h trophozoites, the CSP-AS trophozoites' impairment in the invasive process appeared after 4 h; the colonic mucosa had not been invaded at this time point. Even after 7 h of incubation, CSP-AS trophozoites were never found into the lamina propria. Although some epithelial cells were detached, the overall tissue architecture was unaffected. The time point for the CSP-AS phenotype coincided with the TNF secretion previously observed after 4 h of explant incubation with *E. histolytica* (29). Taken as a whole, these data suggest that CSP has a role in amoebic chemotaxis toward TNF during the early steps of intestinal invasion.

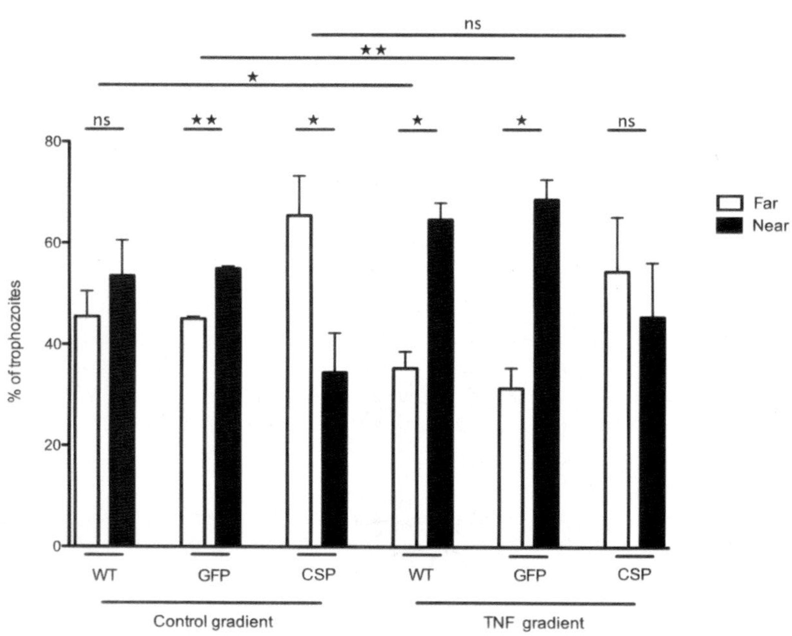

FIGURE 5: *In vitro* TNF chemotaxis is abolished in CSP-AS trophozoites. The displacement of the *E. histolytica* trophozoites was analyzed after 2 h of incubation in the presence of various compounds. Mean numbers of trophozoites in each group (either distal or proximal to the source) were compared in a Student's t-test. In the absence of chemoattractant (incomplete TY medium), WT trophozoites were distributed homogeneously along the coverslip. When the agarose slice was filled with TNF (50 nM), there were significant displacements of WT and GFP trophozoites up the gradient toward the TNF source. In contrast, the presence or absence of a TNF gradient did not significantly affect the distribution of CSP-AS trophozoites. Data are representative of three to four independent experiments. ** P < 0.01; * P < 0.05; ns, not significant.

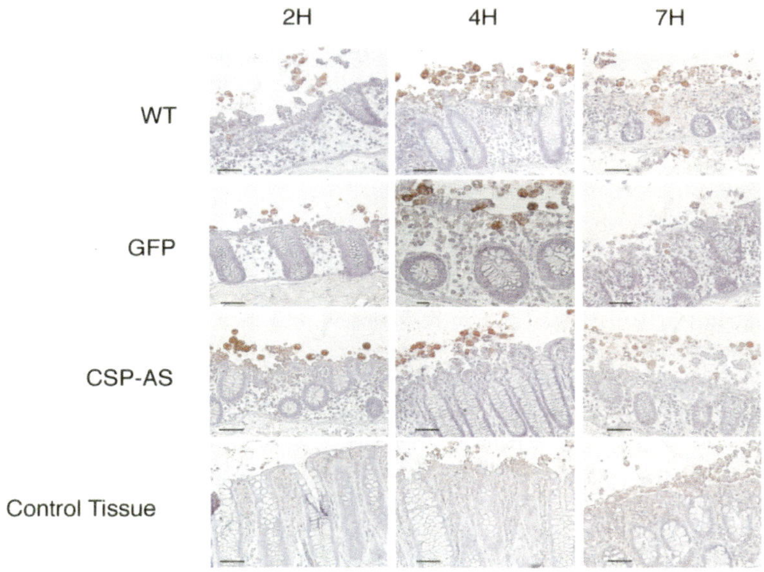

FIGURE 6: Invasion of *ex vivo* colonic explants was impaired for CSP-AS trophozoites. Histological examination of colonic tissue sections after 7 h of infection with *E. histolytica* WT, GFP, CSP-AS strains and in the absence of trophozoites (control). Tissue cross-sections were stained with hematoxylin/eosin reagent. Trophozoites were revealed with an antibody against Gal/GalNAc lectin and appear in red. WT and GFP trophozoites were able to invade the mucosa. In contrast, trophozoites expressing the CSP-AS construct were unable to invade the lamina propria and tended to remain at the surface of the mucosa. Images are representative of three individual experiments. Scale bar: 10 μm.

DISCUSSION

During human intestinal invasion, *E. histolytica* trophozoites degrade the mucus layer, adhere to epithelial cells, lyse cells and penetrate deep into the mucosa. Several parasitic proteins are known to be involved into these crucial events, i.e. the cysteine proteinase involved in mucus and extra cellular matrix degradation [28], the Gal/GalNAc lectin involved in cells and mucus adhesion and the amoebapore involved in cell death (for a review, see [29]). However, trophozoites with impaired Gal/GalNAc lectin signalling and impaired amoebapore A expression are still able to invade the human intestinal mucosa [24]. Trophozoites with impaired cysteine proteinase CP-A5 expression cross the mucus layer but are unable to penetrate deep into the mucosa, due to a lack of collagen remodelling via host matrix metalloproteinase activation [25]. None of these defective amoebic strains are impaired in cell adhesion or mucus adhesion [24]. In contrast, trophozoites with low levels of β-amylase are unable to deplete the mucus layer and thus unable to invade colonic explants [27]. In the present study, we characterized CSP (EHI_016490) and found that it is involved in tissue invasion by *E. histolytica* and chemotaxis toward TNF but not in mucus layer degradation.

Chemotaxis consists in the sensing of a chemoattractant molecule by cell surface proteins. The subsequent signal transduction leads to changes in the dynamics of the actin-rich cytoskeleton, which in turn induce trophozoite polarization [30]. These morphological changes result in directed cell movement.

The first step in TNF chemotaxis relies on TNF sensing by amoebic surface proteins. Our indirect immunofluorescence assay revealed that CSP was found associated to the trophozoite plasma membrane. However, the CSP sequence does not contain a conventional signal sequence and is not predicted to include a transmembrane domain, several other *E. histolytica* proteins (including the lysine- and glutamine-rich protein KERP1, peroxiredoxin and a BspA-like protein EHI_013940) lacking these features have been also found at the plasma membrane of trophozoites [19, 31], the mechanism of their traffic is not yet elucidated.

CSP might be bound to the *E. histolytica* membrane through a prenyl lipid anchor (as it has been suggested for other BspA-like proteins). Prenylation has not been described as a modification allowing protein anchoring to the external part of the plasma membrane. In contrast, the attachment of hydrophobic prenyl groups can anchor the proteins to intracellular membranes. Based on the power of genomics and computer predictions, our hypothesis is that following prenylation, CSP may bind ER-membranes and undergo CAAX box (Cys–aliphatic–aliphatic–Xaa) processing (CSP's sequence contains three prenylation sites, including two at the carboxyl terminus) with the cleavage of –AAX and carboxymethylation of the prenylated Cys residue. The eventual lipid modifications on CSP may favor the transport to the cell surface by vesicular trafficking, but this remains to be determined. Our hypothesis is supported by the detection of isoprenylation activity in *E. histolytica* [32, 33].

To determine the role of prenylation in anchoring CSP to the plasma membrane, it is possible to inhibit the enzymes necessaries for this phenomenon and that has been efficient in several protozoan parasites such as *Trypanosoma brucei*, *Leishmania mexicana*, *Plasmodium falciparum* [34-36]. Lastly, it is possible to inhibit vesicular trafficking with drugs such as Brefeldin A.

Although CSP seems to be devoid of transmembrane domain, CSP may acts as a receptor, in association with ancillary proteins (with transmembrane domains) which clustering with CSP would lead to TNF chemotaxis signaling. We showed that anti-CSP antibody stained the trophozoite surface, and TNF binding occurs upon addition into the extracellular milieu of this molecule, both data support that at least a CSP moiety is present at the trophozoite surface. Moreover, CSP shows a uniform distribution on the plasma membrane of the trophozoite, and in presence of TNF, CSP distribution becomes patched. This may correspond to clustering of CSP induced by TNF.

The second step in chemotaxis consists in the transduction of the surface signal toward the intracellular cytoplasm, leading to changes in the trophozoite's actin-rich cytoskeleton. Phosphatidylinositol 3,4,5-triphosphate and phosphatidylinositol 3-kinase (PI3K) have emerged as important components of the chemotactic signaling pathway in eukaryotic cells [37]. Previous work has demonstrated that *E. histolytica* chemotaxis toward TNF is abrogated after treatment with the PI3K inhibitor wortmannin [10]. Recently, it was shown that wortmannin directly affects amoebic motion and *in vivo* virulence by altering the structure of the cytoskeleton [38]. The effect of TNF signaling on amoebic gene expression has been analyzed in microarray experiments. Transcription of genes encoding proteins involved in cytoskeleton dynamics (microfilament dynamics, nucleation and bundling) was found to be modified [10, 39]. The third and last step in chemotaxis consists in the morphological changes that result in directional displacement of the cell.

CSP is a unique BspA-like amoebic protein that possesses a TNFR domain and a LRR motif (possibly involved in fibronectin/fibrinogen binding [12]) and is essential for invasion of the human colonic mucosa. The taxonomic distribution of genomes encoding BspA-like proteins indicates that most are bacterial lineages found in the gut of vertebrates, including humans [40]. Furthermore, BspA-like proteins have been found in three unicellular protozoans: *E. histolytica*, *E. dispar* [19] and *Trichomonas vaginalis* [40]. It was suggested that lateral gene transfer (LGT) from bacteria was responsible for BspA-like acquisition by the three protozoans. Indeed, LGT in *E. histolytica* genome may confer a direct, adaptive advantage on this species, since the functions encoded by the transferred gene should circumvent some constraints of the anaerobic, parasitic lifestyle. This hypothesis is supported by the fact that many amoebic genes encoding metabolic and fermentation enzymes (e.g. β-amylase) are derived from prokaryotes [41]. A growing body of evidence indicates that the donor organism may belong to the *Cytophaga-Flavobacterium-Bacteroides* group of the phylum *Bacteroidetes* [40, 42, 43]. The sequence of the CSP-encoding gene supports the hypothesis of CSP acquisition via LGT from a *Bacteroidetes*-group donor present in the commensal flora (microbiota) of the human gut.

We found that CSP had the same secondary structure as the ectodomain of the human toll-like receptor (TLR) 3. Toll-like receptors are sensors evolutionarily conserved between plants, fly and mammalians. TLRs recognize danger signals from diverse array of pathogens, which are called pathogen-associated molecular patterns (PAMPs), and also recognize molecules (including nucleic acids) released from intracellular stores of damaged or stressed cells, for review [44, 45]. TLR family has expanded greatly over the last few years, with 13 TLR now described, 10 of which are found in humans. All TLRs are composed of an ectodomain containing multiple LRRs directly involved in the recognition of PAMPs, a TM domain required for the subcellular localization of TLRs, and an intracellular domain with a conserved cytoplasmic signaling region called the Toll/IL-1 receptor (TIR), that triggers downstream complex signalling networks required to sustain appropriate immune response. In mammalian cells, TLRs recognize bacterial surface or cell wall components such as lipopolysaccharide, lipoteichoic acid and peptidoglycan [46]. The typical TLR (consisting of a LRR ectodomain and a cytoplasmic Toll/interleukin-1 receptor (TIR) domain) may result from the combination of a LRR-only protein and a TIR-only protein [47]. In this context, CSP may correspond to an ancestral (TIR-negative) form of the TLR present in higher eukaryotes.

In conclusion, we showed that CSP is located at trophozoite surface, the role of prenylation remains to be demonstrated. We hypothesize that it is associated to ancillary proteins with transmembrane domain to lead to TNF chemotaxis signaling. CSP-AS trophozoites may have low amoebic pathogenicity because they were impaired in invasion of the colonic mucosa. Further work is needed to characterize the interplay between CSP and the signaling pathway involved in TNF chemotaxis.

MATERIALS AND METHODS

Ethical aspects

Human intestinal tissues were processed according to the French Bioethics Act, following approval from two investigational review boards (Comité de Protection des Personnes Ile de France VII and Institut Pasteur Recherche Biomedicale: reference RBM/2009.50).

Amoebae strains and culture conditions

The HM1-IMSS pathogenic *Entamoeba histolytica* strain was cultured axenically in complete TYI-S-33 medium [48] at 37°C. Trophozoites were harvested during the exponential growth phase, collected by centrifugation at 1000 g for 10 min at 4°C, and then washed thoroughly with either complete TYI-S-33 or phosphate buffered saline (PBS) to remove all nutrients. Incubation with 75 nM human recombinant TNF (R&D Systems) was performed with 1.5 x 10^6 amoebae, in complete TYI-S-33 medium and at 37°C. The amoebae were then washed in PBS. Jurkat cells were cultured in RPMI (Gibco) at 37°C.

Plasmid construction and trophozoite transfection

An RNA AS technique was used to reduce CSP expression in a stable manner. The pEhNeo/CSP-AS plasmid was constructed by replacing the chloramphenicol acetyltransferase (CAT) coding sequence of pEhNEO/CAT [49] by the CSP AS (EHI_016490) sequence (generated by PCR using total RNA from trophozoites as a template). The primers were as follows: EhCSP1 5' GATCGGATCCTATTCGTTATTGATTTGTAGTAAATATTTTCGATA-TAAAAG 3' and EhCSP2 5' GATCGGTACCGATAACTAACTTTA-TATTTATATGACAAATCTTTTTCTAATG 3'. The PCR product was cloned, sequenced and subcloned into the KpnI and BamHI restriction sites of pEhNEO/CAT, flanked by untranslated 5' and 3' sequences of *E. histolytica* actin gene. The control transfection vector (pEhNEO/GFP) was constructed by cloning of GFP in the unique BamHI site of the pExEhNeo plasmid. GFP sequence was generated with primers: GFP1 5' GGATCCATGAGTAAAGGAGAAG 3' and GFP2 5'

TCATAGGATCCGGGTATCTTGA 3' [50]. After electroporation, transfected cells were isolated in the presence of G418 (30 µg/ml), as described previously [51].

Immunofluorescence staining of amoebae

The entire immunofluorescence procedure was carried out at 37°C, unless otherwise indicated. Amoebae were fixed in 3.7% paraformaldehyde for 30 min, permeabilized by adding 0.1% Triton X100 in PBS for 2 min at room temperature, and blocked in 1% bovine serum albumin in PBS for 30 min. Preparations were then incubated with the following antibodies: anti-CSP (I8, a custom rabbit polyclonal antibody raised against the peptides FNKCHOIVYTRSDREEE and FTDQDYLQRRNNNVDT (Eurogentec, Belgium)) and anti-TNF (1E12, anti-TIP domain, a kind gift from Dr Etienne Pays, Université Libre de Bruxelles, Belgium) [52]. Secondary antibodies coupled to AlexaFluor488 or AlexaFluor546 (Molecular Probes) were added (dilution 1:200) for 30 min at 37°C. The samples were mounted in Prolong medium containing DAPI (Invitrogen). Amoebae were examined by confocal microscopy (LSM510, Zeiss).

Western blot analysis

To evaluate the TORC1-dependent C-terminal phosphorylation Crude trophozoite extract (10-20 µg) was separated on 10-12% polyacrylamide gels, transferred onto nitrocellulose, and blocked with 5% dry milk in Tris-buffered saline/Tween 20. Proteins were probed with a goat anti-hTNFR1 (Sigma) (dilution 1:1000) and the anti-CSP (dilution 1:1000). Detection was performed with an anti-rabbit antibody (dilution 1:10000) (Jackson ImmunoResearch) and enhanced chemiluminescence reagents (Amersham Biosciences). Detected proteins were quantified using AlphaEaseFC software (Alpha Innotech).

***In silico* protein analysis**

The protein sequence of human TNFR1 (hTNFR1; GenBank reference AAA36754 [20]) was used to perform a BLASTp search of the *E. histolytica* AmoebaDB database (http://amoebadb.org/amoeba/). Signal peptide was predicted with SignalP algorithm [53], and transmembrane helix domains were predicted with the TMHMM-2.0 algorithm [54]. Post-translational modifications were identified with Prosite [55]. Conserved protein domains were identified using the PFAM [56], InterProScan [57] and SMART [58] online tools. Protein structure was predicted with the protein homology/analogy recognition engine (Phyre2) [59]. Phyre2 uses a library from the Structural Classification of Proteins (SCOP) Database, augmented with new depositions in the Protein Data Bank (PDB). Briefly, Psi-Blast was run to collect homologous sequences and built a statistical sequence profile. Three independent secondary structure prediction programs (PriPred, SSPro and JNet) were run, to generate a final consensus prediction. The confidence of the secondary structure prediction was determined by a prediction score (theoretical value ranging from 1 to 9). The highest the score, the highest the accurate secondary structure prediction. The top ten highest scoring alignments were used to construct full 3D models of the query. Phyre2 displayed the 3D models of the query, with the alignment coverage, the identity percentage and the confidence of the model. Phyre2 allows achieving high accuracy models at very low sequence identities (15-20 %).

The second freely available web server used for structure prediction was the iterative threading assembly refinement I-Tasser [22, 60]. I-TASSER fragments the query sequence into overlapping short stretches of amino acids. Candidate structures for those small fragments are generated as mentioned earlier, and assembled to construct a low-energy protein conformation.

A phylogenetic analysis was run on 95 Bsp A like proteins found in *E. histolytica* genome and expressed in trophozoites, to determine the genetic distance between EHI_016490 and its orthologs. The multiple alignment of 95 BspA like amino acids sequences was obtained by using ClustalX software [61]. The Neighbor-Joining algorithm was used to infer the topology based on multiple sequences alignment with identity percentage distance.

***In vitro* TNF chemotaxis**

To determine whether TNF chemotaxis was altered in *E. histolytica* trophozoites expressing the CSP-AS construct, we used the agarose chemotaxis-on-coverslip assay [4]. The coverslips were mounted onto glass slides and the cell distribution was examined under the microscope. Trophozoites were classified arbitrarily as groups that were near to or far from the source of test compound.

Preparation of human colon explants

Segments of human colon were obtained anonymously from three fully informed patients undergoing colon surgery. The conditions used for handling human colon explants have been published elsewhere [24]. Trophozoites ($8x10^5$) were added to the luminal face of the colon and incubated in Krebs' medium at 37°C for 2, 4 and 7 h. Amoebae-free segments served as controls for each time point.

Histological analysis

Tissue invasion by *E. histolytica* was monitored after incubation. Tissues were fixed in 4% paraformaldehyde in PBS for 48 h at 4°C and then embedded in paraffin. Three sections (thickness: 5 µm) were cut from paraffin blocks and stained with standard hematoxylin/eosin reagent. Trophozoites were stained with a 1:200 polyclonal antibody raised against the Gal/GalNAc lectin [62]. For each experiment, a representative histology image was taken.

Statistical analysis

Intergroup differences were evaluated in Student's unpaired t-test, using GraphPad software (http://www.graphpad.com). The threshold for statistical significance was set to P = 0.05.

ACKNOWLEDGMENTS

The authors greatly appreciate the support and enthusiasm of Dr Dominique Buzoni-Gatel as part of this collaboration between the INRA and the Institut Pasteur. We thank Guillaume Vogt for very helpful discussions in the early stages of this work, and Chung Chau-Hon for the RNASeq data. For all their support and advice, we thank Emmanuelle Perret and Pascal Roux (Plateforme Imagerie Dynamique, PFID, Imagopole IP), Laurence Fiette and Patrick Avé (Unité d'histologie humaine et modèles animaux IP), and Michèle Bernier and Marie Morcelet (Service d'Anatomopathologie, Hôpithal Foch, Suresnes, France). The authors gratefully acknowledge the Institut Pasteur's Pôle Intégré de Recherche Clinique for helping us to comply with guidelines on research on human tissues and with the French Bioethics Act. **Funding.** The work was funded by the National French Research Agency (ANR; grant MIE-08, Intestinalamibe), the French Parasitology consortium Labex ParaFrap (grant ANR-11-LABX0024), the Institut Pasteur (grant PTR 391) and the INRA Infectiologie et Santé Publique ISP Cluster (grant 32000266).

CONFLICT OF INTEREST

The authors do not have any commercial or other associations that might constitute a conflict of interest.

REFERENCES

1. Guillen N (**1996**). Role of signalling and cytoskeletal rearrangements in the pathogenesis of Entamoeba histolytica. **Trends Microbiol** 4(5):191-197.

2. Chelsea M and Petri WA (**2014**). Regulation of virulence of Entamoeba histolytica. **Annu Rev Microbiol** 68:493-520.

3. Stanley SL (**2001**). Protective immunity to amebiasis: New insights and new challenges. **J Infect Dis** 184(4):504-506.

4. Blazquez S, Zimmer C, Guigon G, Olivo-Marin JC, Guillen N and Labruyere E (**2006**). Human tumor necrosis factor is a chemoattractant for the parasite Entamoeba histolytica. **Infect Immun** 74(2):1407-1411.

5. Urban T, Jarstrand C and Austkettis A (**1983**). Migration of Entamoeba histolytica under agarose. **Am J Trop Med Hyg** 32(4):733-737.

6. Zhang Z, Mahajan S, Zhang XC and Stanley SL (**2003**). Tumor necrosis factor alpha is a key mediator of gut inflammation seen in amebic colitis in human intestine in the SCID mouse-human intestinal xenograft model of disease. **Infect Immun** 71(9):5355-5359.

7. Helk E, Bernin H, Ernst T, Ittrich H, Jacobs T, Heeren J, Tacke F, Tannich E and Lotter H (**2013**). TNF alpha-Mediated Liver Destruction by Kupffer Cells and Ly6C(hi) Monocytes during Entamoeba histolytica Infection. **Plos Pathogens** 9(1).

8. Leitch GJ, Dickey AD, Udezulu IA and Bailey GB (**1985**). Entamoeba histolytica trophozoites in the lumen and mucus blanket of rat colons studied in vivo. **Infect Immun** 47(1):68-73.

9. Hehlgans T and Pfeffer K (**2005**). The intriguing biology of the tumour necrosis factor/tumour necrosis factor receptor superfamily: players, rules and the games. **Immunology** 115(1):1-20.

10. Blazquez S, Guigon G, Weber C, Syan S, Sismeiro O, Coppee JY, Labruyere E and Guillen N (**2008**). Chemotaxis of Entamoeba histolytica towards the pro-inflammatory cytokine TNF is based on PI3K signalling, cytoskeleton reorganization and the Galactose/N-acetylgalactosamine lectin activity. **Cell Microbiol** 10(8):1676-1686.

11. Zaki M, Andrew N and Insall RH (**2006**). Entamoeba histolytica cell movement: A central role for self-generated chemokines and chemorepellents. **P Natl Acad Sci USA** 103(49):18751-18756.

12. Sharma A, Sojar HT, Glurich I, Honma K, Kuramitsu HK and Genco RJ (**1998**). Cloning, expression, and sequencing of a cell surface antigen containing a leucine-rich repeat motif from Bacteroides forsythus ATCC 43037. **Infect Immun** 66(12):5703-5710.

13. Sharma A (**2010**). Virulence mechanisms of Tannerella forsythia. **Periodontol** 54:106-116.

14. Hirt RP, Harriman N, Kajava AV and Embley TM (**2002**). A novel potential surface protein in Trichomonas vaginalis contains a leucine-rich repeat shared by micro-organisms from all three domains of life. **Mol Biochem Parasitol** 125(1-2):195-199.

15. Inagaki S, Onishi S, Kuramitsu HK and Sharma A (**2006**). Porphyromonas gingivalis vesicles enhance attachment, and the leucine-rich repeat BspA protein is required for invasion of epithelial cells by "Tannerella forsythia". **Infect Immun** 74(9):5023-5028.

16. Mengaud J, Lecuit M, Lebrun M, Nato F, Mazie JC and Cossart P (**1996**). Antibodies to the leucine-rich repeat region of internalin block entry of Listeria monocytogenes into cells expressing E-cadherin. **Infect Immun** 64(12):5430-5433.

17. Sharma A, Inagaki S, Honma K, Sfintescu C, Baker PJ and Evans RT (**2005**). Tannerella forsythia-induced alveolar bone loss in mice involves leucine-rich-repeat BspA protein. **J Dent Res** 84(5):462-467.

18. Lorenzi HA, Puiu D, Miller JR, Brinkac LM, Amedeo P, Hall N and Caler EV (**2010**). New Assembly, Reannotation and Analysis of the Entamoeba histolytica Genome Reveal New Genomic Features and Protein Content Information. **Plos Neglec Trop D** 4(6):e716.

19. Davis PH, Zhang Z, Chen MH, Zhang XC, Chakraborty S and Stanley SL (**2006**). Identification of a family of BspA like surface proteins of Entamoeba histolytica with novel leucine rich repeats. **Mol Biochem Parasitol** 145(1):111-116.

20. Himmler A, Maurerfogy I, Kronke M, Scheurich P, Pfizenmaier K, Lantz M, Olsson I, Hauptmann R, Stratowa C and Adolf GR (**1990**). Molecular cloning and expression of human and rat tumor necrosis factor receptor chain (P60) and its soluble derivative, tumor necrosis factor binding protein. **DNA Cell Biol** 9(10):705-715.

21. Jin MS and Lee JO (**2008**). Structures of TLR-ligand complexes. **Curr Opin Immunol** 20(4):414-419.

22. Roy A, Kucukural A and Zhang Y (**2010**). I-Tasser: a unified platform for automated protein structure and function prediction. **Nature Protoc** 5(4):725-738.

23. Hon CC, Weber C, Sismeiro O, Proux C, Koutero M, Deloger M, Das S, Agrahari M, Dillies MA, Jagla B, Coppee JY, Bhattacharya A and Guillen N (**2013**). Quantification of stochastic noise of splicing and polyadenylation in Entamoeba histolytica. **Nucleic Acids Res** 41(3):1936-1952.

24. Bansal D, Ave P, Kerneis S, Frileux P, Boche O, Baglin AC, Dubost G, Leguern A-S, Prevost M-C, Bracha R, Mirelman D, Guillen N and Labruyere E (**2009**). An ex-vivo Human Intestinal Model to Study Entamoeba histolytica Pathogenesis. **Plos Neglect Trop D** 3(11):e551.

25. Thibeaux R, Ave P, Bernier M, Morcelet M, Frileux P, Guillen N and Labruyere E (**2014**). The parasite Entamoeba histolytica exploits the activities of human matrix metalloproteinases to invade colonic tissue. **Nature Comm** 5:5142-5142.

26. Thibeaux R, Dufour A, Roux P, Bernier M, Baglin AC, Frileux P, Olivo-Marin JC, Guillen N and Labruyere E (**2012**). Newly visualized fibrillar collagen scaffolds dictate Entamoeba histolytica invasion route in the human colon. **Cell Microbiol** 14(5):609-621.

27. Thibeaux R, Weber C, Hon C-C, Dillies M-A, Ave P, Coppee J-Y, Labruyere E and Guillen N (**2013**). Identification of the Virulence Landscape Essential for Entamoeba histolytica Invasion of the Human Colon. **Plos Pathogens** 9(12):e1003824.

28. Lee YA, Saito-Nakano Y, Kim KA, Min A, Nozaki T and Shin MH (**2015**). Modulation of endogenous Cysteine Protease Inhibitor (ICP) 1 expression in Entamoeba histolytica affects amoebic adhesion to Extracellular Matrix proteins. **Exp Parasitol** 149:7-15.

29. Faust DM and Guillen N (**2012**). Virulence and virulence factors in Entamoeba histolytica, the agent of human amoebiasis. **Microbes Infect** 14(15):1428-1441.

30. Maugis B, Brugues J, Nassoy P, Guillen N, Sens P and Amblard F (**2010**). Dynamic instability of the intracellular pressure drives bleb-based motility. **J Cell Sci** 123(22):3884-3892.

31. Choi MH, Sajed D, Poole L, Hirata K, Herdman S, Torian BE and Reed SL (**2005**). An unusual surface peroxiredoxin protects invasive Entamoeba histolytica from oxidant attack. **Mol Biochem Parasitol** 143(1):80-89.

32. Maurer-Stroh S, Washietl S and Eisenhaber F (**2003**). Protein prenyltransferases. **Genome Biol** 4(4):2121-2129.

33. Shen PS, Sanford JC and Samuelson J (**1996**). Entamoeba histolytica: Isoprenylation of p21(ras) and p21(rap) in vitro. **Exp Parasitol** 82(1):65-68.

34. Yokoyama K, Trobridge P, Buckner FS, Scholten J, Stuart KD, Van Voorhis WC and Gelb MH (**1998**). The effects of protein farnesyltransferase inhibitors on trypanosomatids: inhibition of protein farnesylation and cell growth. **Mol Biochem Parasitol** 94(1):87-97.

35. Pendyala PR, Ayong L, Eatrides J, Schreiber M, Pham C, Chakrabarti R, Fidock DA, Allen CM and Chakrabarti D (**2008**). Characterization of a PRL protein tyrosine phosphatase from Plasmodium falciparum. **Mol Biochem Parasitol** 158(1):1-10.

36. Hancock JF, Paterson H and Marshall CJ (**1990**). A polybasic domain or palmitoylation is required in addition to the caax motif to localize p21ras to the plasma-membrane. **Cell** 63(1):133-139.

37. Devreotes P and Janetopoulos C (**2003**). Eukaryotic chemotaxis: Distinctions between directional sensing and polarization. **J Biol Chem** 278(23):20445-20448.

38. Lopez-Contreras L, Hernandez-Ramirez VI, Flores-Garcia Y, Chavez-Munguia B and Talamas-Rohana P (**2013**). Src and PI3 K inhibitors affect the virulence factors of Entamoeba histolytica. **Parasitology** 140(2):202-209.

39. de la Cruz OH, Muniz-Lino M, Guillen N, Weber C, Marchat LA, Lopez-Rosas I, Ruiz-Garcia E, Astudillo-de la Vega H, Fuentes-Mera L, Alvarez-Sanchez E, Mendoza-Hernandez G and Lopez-Camarillo C (**2014**). Proteomic profiling reveals that EhPC4 transcription factor induces cell migration through up-regulation of the 16-kDa actin-binding protein EhABP16 in Entamoeba histolytica. **J Proteomics** 111:46-58.

40. Noel CJ, Diaz N, Sicheritz-Ponten T, Safarikova L, Tachezy J, Tang P, Fiori PL and Hirt RP (**2010**). Trichomonas vaginalis vast BspA-like gene family: evidence for functional diversity from structural organisation and transcriptomics. **BMC Genomics** 11:1-26.

41. Samuelson J (**2002**). What Entamoeba histolytica and Giardia lamblia tell us about the evolution of eukaryotic diversity. **J Biosciences** 27(6):559-565.

42. Ali V and Nozaki T (**2006**). Biochemical and functional characterization of phosphoserine aminotransferase from Entamoeba histolytica, which possesses both phosphorylated and non-phosphorylated serine metabolic pathways. **Mol Biochem Parasitol** 145(1):71-83.

43. Horn M, Harzenetter MD, Linner T, Schmid EN, Muller KD, Michel R and Wagner M (**2001**). Members of the Cytophaga-Flavobacterium-Bacteroides phylum as intracellular bacteria of acanthamoebae: proposal of 'Candidatus Amoebophilus asiaticus'. **Environ Microbiol** 3(7):440-449.

44. Creagh EM and O'Neill LAJ (**2006**). TLRs, NLRs and RLRs: a trinity of pathogen sensors that co-operate in innate immunity. **Trends Immunol** 27(8):352-357.

45. Lee CC, Avalos AM and Ploegh HL (**2012**). Accessory molecules for Toll-like receptors and their function. **Nature Rev Immunol** 12(3):168-179.

46. Botos I, Liu L, Wang Y, Segal DM and Davies DR (**2009**). The Toll-like receptor 3:dsRNA signaling complex. **BBA-Gene Regul Mech** 1789(9-10):667-674.

47. Wu BJ, Huan TX, Gong J, Zhou P and Bai ZL (**2011**). Domain combination of the vertebrate-like TLR gene family: implications for their origin and evolution. **J Genet** 90(3):401-408.

48. Diamond LS, Harlow DR and Cunnick CC (**1978**). New medium for axenic cultivation of Entamoeba histolytica and other Entamoeba. **T Roy Soc Trop Med H** 72(4):431-432.

49. Hamann L, Nickel R and Tannich E (**1995**). Transfection and continuous expression of heterologous genes in the protozoan parasite Entamoeba histolytica. **P Natl Acad Sci USA** 92(19):8975-8979.

50. Jain R, Santi-Rocca J, Padhan N, Bhattacharya S, Guillen N and Bhattacharya A (**2008**). Calcium-binding protein 1 of Entamoeba histolytica transiently associates with phagocytic cups in a calcium-independent manner. **Cell Microbiol** 10(6):1373-1389.

51. Nickel R and Tannich E (**1994**). Transfection and transient expression of chloramphenicol acetyltransferase gene in the protozoan parasite Entamoeba histolytica. **P Natl Acad Sci USA** 91(15):7095-7098.

52. Magez S, Geuskens M, Beschin A, delFavero H, Verschueren H, Lucas R, Pays E and deBaetselier P (**1997**). Specific uptake of tumor necrosis factor-alpha is involved in growth control of Trypanosoma brucei. **J Cell Biol** 137(3):715-727.

53. Bendtsen JD, Nielsen H, von Heijne G and Brunak S (**2004**). Improved prediction of signal peptides: SignalP 3.0. **J Mol Biol** 340(4):783-795.

54. Krogh A, Larsson B, von Heijne G and Sonnhammer ELL (**2001**). Predicting transmembrane protein topology with a hidden Markov model: Application to complete genomes. **J Mol Biol** 305(3):567-580.

55. Sigrist CJA, Cerutti L, de Castro E, Langendijk-Genevaux PS, Bulliard V, Bairoch A and Hulo N (**2010**). PROSITE, a protein domain database for functional characterization and annotation. **Nucleic Acids Res** 38:D161-D166.

56. Finn RD, Tate J, Mistry J, Coggill PC, Sammut SJ, Hotz H-R, Ceric G, Forslund K, Eddy SR, Sonnhammer ELL and Bateman A (**2008**). The Pfam protein families database. **Nucleic Acids Res** 36:D281-D288.

57. Hunter S, Apweiler R, Attwood TK, Bairoch A, Bateman A, Binns D, Bork P, Das U, Daugherty L, Duquenne L, Finn RD, Gough J, Haft D, Hulo N, Kahn D, Kelly E, Laugraud A, Letunic I, Lonsdale D, Lopez R, Madera M, Maslen J, McAnulla C, McDowall J, Mistry J, Mitchell A, Mulder N, Natale D, Orengo C, Quinn AF, et al. (**2009**). InterPro: the integrative protein signature database. **Nucleic Acids Res** 37:D211-D215.

58. Letunic I, Doerks T and Bork P (**2009**). SMART 6: recent updates and new developments. **Nucleic Acids Res** 37:D229-D232.

59. Kelley LA and Sternberg MJE (**2009**). Protein structure prediction on the Web: a case study using the Phyre server. **Nat Protoc** 4(3):363-371.

60. Zhang Y (**2008**). I-Tasser server for protein 3D structure prediction. **Bmc Bioinformatics** 9.

61. Thompson JD, Gibson TJ, Plewniak F, Jeanmougin F and Higgins DG (**1997**). The CLUSTAL_X windows interface: flexible strategies for multiple sequence alignment aided by quality analysis tools. **Nucleic Acids Res** 25(24):4876-4882.

62. Tavares P, Rigothier MC, Khun H, Roux P, Huerre M and Guillen N (**2005**). Roles of cell adhesion and cytoskeleton activity in Entamoeba histolytica pathogenesis: a delicate balance. **Infect Immun** 73(3):1771-1778.

8

Two distinct and competitive pathways confer the cellcidal actions of artemisinins

Chen Sun[#], Jian Li[#], Yu Cao, Gongbo Long and Bing Zhou*
State Key Laboratory of Biomembrane and Membrane Biotechnology, School of Life Sciences, Tsinghua University, Beijing, China.
[#] These authors contribute equally to this work.
* Corresponding Author: Dr. Bing Zhou, School of Life Sciences, Tsinghua University; Beijing, 100084, China
E-mail: zhoubing@mail.tsinghua.edu.cn

ABSTRACT The biological actions of artemisinin (ART), an antimalarial drug derived from *Artemisia annua*, remain poorly understood and controversial. Besides potent antimalarial activity, some of artemisinin derivatives (together with artemisinin, hereafter referred to as ARTs), in particular dihydroartemisinin (DHA), are also associated with anticancer and other antiparasitic activities. In this study, we used baker's yeast *Saccharomyces cerevisiae* as cellular and genetic model to investigate the molecular and cellular properties of ARTs. Two clearly separable pathways exist. While all ARTs exhibit potent anti-mitochondrial actions as shown before, DHA exerts an additional strong heme-dependent, likely mitochondria-independent inhibitory action. More importantly, heme antagonizes the mitochondria-dependent cellcidal action. Indeed, when heme synthesis was inhibited, the mitochondria-dependent cellcidal action of ARTs could be dramatically strengthened, and significant yeast growth inhibition at as low as 100 nM ART, an increase of about 25 folds in sensitivity, was observed. We conclude that ARTs are endowed with two major and distinct types of properties: a potent and specific mitochondria-dependent reaction and a more general and less specific heme-mediated reaction. The competitive nature of these two actions could be explained by their shared source of the consumable ARTs, so that inhibition of the heme-mediated degradation pathway would enable more ARTs to be available for the mitochondrial action. These properties of ARTs can be used to interpret the divergent antimalarial and anticancer actions of ARTs.

***Keywords**: heme, mitochondria, artemisinin, yeast, cancer.*

Abbreviations:
ALA - δ-aminolevulinic acid,
ART - artemisinin,
DHA - dihydroartemisinin,
ETC - electron transport chain.

INTRODUCTION

Malaria is one of the most serious and prevalent infectious diseases worldwide. It remains a severe threat to human health; globally, an estimated 207 million cases of malaria and at least 627,000 deaths occurred in 2012 as reported by the World Health Organization (WHO) [1]. In the early 1970's, a large group of Chinese scientists, supported and organized by the Chinese government, extracted a highly effective and potent antimalarial drug, artemisinin (ART), from *Artemisia annua*, a traditional Chinese herb [2]. The next 40 years witnessed the synthesis of a series of artemisinin derivatives (ARTs), including dihydroartemisinin (DHA, Fig. 1A), artesunate, artemether and so on [3-6]. Due to the poor bioavailability of original ART, some artemisinin derivatives became more frequently used under clinical settings [7]. Because of emerging drug-resistances against other antimalarial drugs such as quinine, chloroquine and atovaquone [8-10], ARTs are currently becoming our last defensive line against this devastating disease. For this reason, as well as the high recrudescence rate in ARTs-treated patients, WHO recommended ART-Combination Therapy (ACT).

Chemically, ART is a sesquiterpene lactone containing an endoperoxide bond, which is essential for its antimalarial action. Deoxyartemisinin, which contains the same backbone structure as ART but missing the peroxide bond, is about three orders of magnitude less effective than ART [11]. Like H_2O_2, the peroxide bond in ART can be cleaved *in vitro* by ferrous iron, heme, or some other redox-active agents, similar to the Fenton reaction, to yield hydroxyl radicals. The majority of hypotheses concerning ART's mechanism were proposed based on this reaction. Specifically, three different theories exist explaining the mode of action of ARTs [12]: the protein target theory, heme theory and mitochondria theory. We previously established a yeast genetic model and found that ART interferes with mitochondrial function by depolarizing mitochondrial membrane potential [13]. This action was later confirmed with purified mitochondria from yeast and malaria parasites [11]. ARTs can quickly react with mitochondria of malaria parasites [11, 14]. In particular, ARTs directly induce ROS generation and membrane depolarization in purified yeast and malarial mitochondria, an effect that can be abrogated by ROS scavengers [11].

The specifics regarding anti-mitochondrial action of ARTs remain poorly understood. It was suggested that this anti-mitochondrial action may first involve ARTs activation by the electron transport chain (ETC), but direct evidence for this hypothesis is still lacking. We only know that ARTs' action on mitochondria is not through a direct inhibition of ETC because electron transport/respiration is not repressed [11]. Instead, the ETC may provide a reductive force for the activation of ARTs, which then interfere with mitochondrial functions [11, 13]. On the other hand, mammalian mitochondria do not activate ARTs, underlying the action specificity of these drugs. Therefore, although it is quite convincing that ARTs can be specifically activated by yeast and malarial but not mammalian mitochondria, leading to local damages and eventually cell death through resulting free radicals, the exact mechanism of mitochondrial activation of ARTs still remains uncertain and there is much that awaits to be explored.

In addition to the potent suppressing effect on malarial parasites, ARTs have also been reported to show inhibitory activities against other parasites [15, 16], viruses [17] and also cancer cells [18, 19]. The potential usage of ARTs in cancer therapies has recently drawn much attention. ARTs, especially dihydroartemisinin (DHA) and artesunate, which acts through DHA, were shown to inhibit cancer cell growth at low µM levels *in vitro*; *in vivo* experiments, using mouse cancer models, also produced some promising results [20]. Intensive efforts were devoted to exploring the mechanisms by which ARTs achieve the inhibitory effect against cancer cells. The cytotoxicity of ARTs against cancer cells was also demonstrated to be endoperoxide dependent [21]. Moreover, modulating intracellular iron concentrations altered the effect of ARTs [22]. The iron effect is consistent with the understanding that cancer cells contain more iron in order to maintain fast cell proliferation. Some recent studies further indicated that iron in the form of

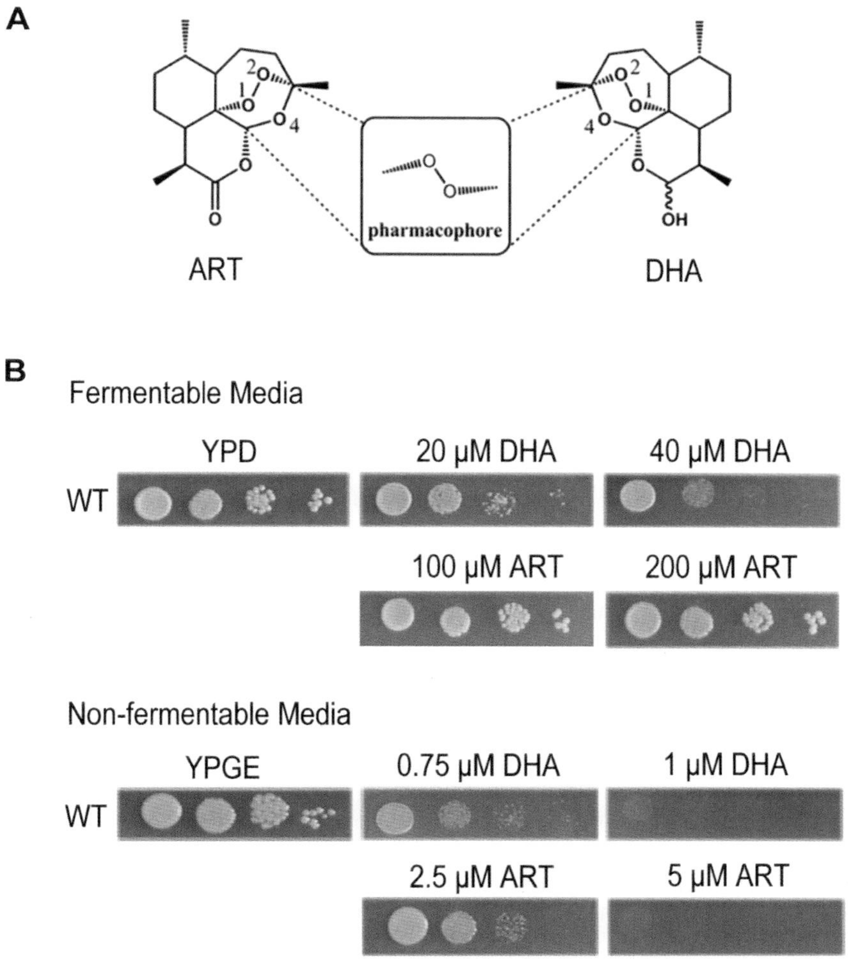

FIGURE 1: Dihydroartemisinin (DHA) inhibits yeast on both fermentable and nonfermentable media, albeit at much higher levels for the latter. Artemisinin (ART) has much weaker inhibition on the fermentable medium. **(A)** Molecular structures of ART and DHA. The endoperoxide bond in 1, 2, 4-trioxane is critical for the action of this class of drugs. **(B)** Inhibition of yeast growth on fermentable (YPD) and nonfermentable (YPGE) media with ART and DHA. DMSO was used when there was no drug added. ART: artemisinin; DHA: dihydroartemisinin.

heme may be the mediator of the ARTs' action [23, 24]. Regulating intracellular heme concentration of cancer cells altered the activity of ARTs correspondingly, while intracellular "free" iron did not directly participate in the action of ARTs but did increase heme synthesis [23].

Despite large efforts devoted to understanding how ARTs act biologically, it is still not known how exactly ARTs inhibit malarial parasites and cancer cells. In the past, medical chemistry studies helped reveal a large amount of *in vitro* properties of ARTs, but their biological properties in the context of a living cell were poorly understood. In order to decipher the biological actions of ARTs, we consider it essential to gain a comprehensive understanding of their molecular and cellular properties within a cell. In this report, we were able to reveal two modes of action of ARTs: the relatively nonspecific and general action as well as the mitochondrial action. We concluded that the heme-mediated nonspecific action and the specific mitochondrial pathway both exist *in vivo* and while both could be cell inhibitory, these two actions are also antagonistic likely due to their competition for the same source of the consumable ARTs.

RESULTS

High concentrations of ARTs, in particular DHA, can inhibit yeast growth even on fermentative media

Yeast is a facultative anaerobe and can grow on fermentable media (such as YPD medium wherein glucose is used as the carbon/energy source) and nonfermentable media (such as YPGE medium wherein ethanol and glycerol are used as the carbon/energy source) by either anaerobic or aerobic pathways. In the absence of mitochondrial respiration yeast cells adopt the anaerobic pathway and grow. This property makes it an ideal model organism to study the anti-mitochondrial properties of drugs. Our previous work demonstrated that yeast grown on nonfermentable media is a suitable model to study the mechanism of ARTs' actions on mitochondria, and showed that ART efficiently inhibits yeast growth on nonfermentable media, but not on fermentable YPD medium even at a 10-fold increased concentration [13]. At levels further up, inhibition by ARTs, in particular by DHA (Fig. 1A), on fermentable media becomes noticeable. As shown in Figure 1B, DHA can clearly inhibit yeast on YPD when about tenfold increase of drug concentration was applied. DHA is a more powerful antimalarial drug than ART, displaying about 5-fold higher antimalarial potency [11]. Consistently, it is also more powerful in inhibiting yeast growth on nonfermentable YPGE medium. In comparison to the strong suppression of 5 μM ART on normal yeast growth on YPGE plates, 1 μM DHA exhibited a similar level of inhibition. On fermentable YPD medium, growth inhibition was obviously observed when 20 μM DHA was used, but little growth inhibition was seen even when 200 μM of ART (more than ten times as used on nonfermentable media such as YPGE) was added to YPD plates (Fig. 1B). The much less potent and arguably insignificant inhibition by ART in fermentable media could be observed when growth in liquid media was monitored [13] (and also see below, Fig 2E). One thing worthy of mentioning is that although liquid inhibition is theoretically more quantitative, in reality we found using plate assays to examine the inhibition of ARTs on yeast growth not only more convenient, but also more reproducible.

In the yeast liquid growth assay for ARTs, the amount of cells inoculated, incubation duration and likely other factors dramatically affect the final result, possibly because ARTs are highly degradable, thus decreasing the actual drug concentration and significantly deviating it from the initial or designed one. When performed side by side and with well-controlled cell numbers and time, experimental errors are within a limited range. However, experiments performed at different times and under different conditions, can produce dramatically different absolute numbers (such as IC50), although with similar trends. In the plate assays, cell numbers inoculated are usually small and other likely contributing factors such as aeration might be less variable, making the assay more stable. We thus routinely used plate assays and resorted to well-controlled liquid assays, which are more quantitative but as explained above deliver relative and not absolute results, only when little difference could be noticed on the plate assays.

While DHA is somewhat more active than ART on nonfermentable media, it is much more effective than ART on fermentable media. This divergence of correlated activities for ART and DHA was reminiscent of findings found in cancer cell inhibition: while DHA can inhibit some cancer cells (IC50 normally in the range of a few to several dozens of μMs), ART is much less effective [25].

These results, in particular the inhibitory action of ARTs on yeast utilizing fermentable carbon sources, suggest to us the existence of a different and less potent type of inhibition besides the anti-mitochondrial action.

Heme mediates DHA's inhibitory action on fermentable media

Previous studies have shown that intracellular heme physiologically mediates the cytotoxicity of some ARTs in tumor cells [23, 24]. Direct interaction between heme and ARTs, as reported in numerous *in vitro* studies [26, 27], is well accepted in the medical chemistry field. To investigate the underlying mode of DHA's action on fermentable media, we tested whether DHA's inhibition on YPD medium can be affected by modulation of heme levels or not. Heme is indispensable for yeast growth on both fermentable and nonfermentable media. However, on fermentable media, heme requirement can be bypassed by the supplementation of Tween-80 and cholesterol to the medium, whose cellular synthesis presumably necessitates heme's presence [28]. Loss of function of *HEM1,* which catalyzes the first step of the heme synthesis pathway, drastically reduced heme synthesis (Fig. 2A, Table 1). In the conditional medium, which enables the growth of Δ*hem1,* the DHA inhibition on yeast was essentially completely eradicated (Fig. 2B). To be certain this was free of anti-mitochondrial action, we also examined whether petite strains, which lack respiration-functional mitochondria, behave similarly.

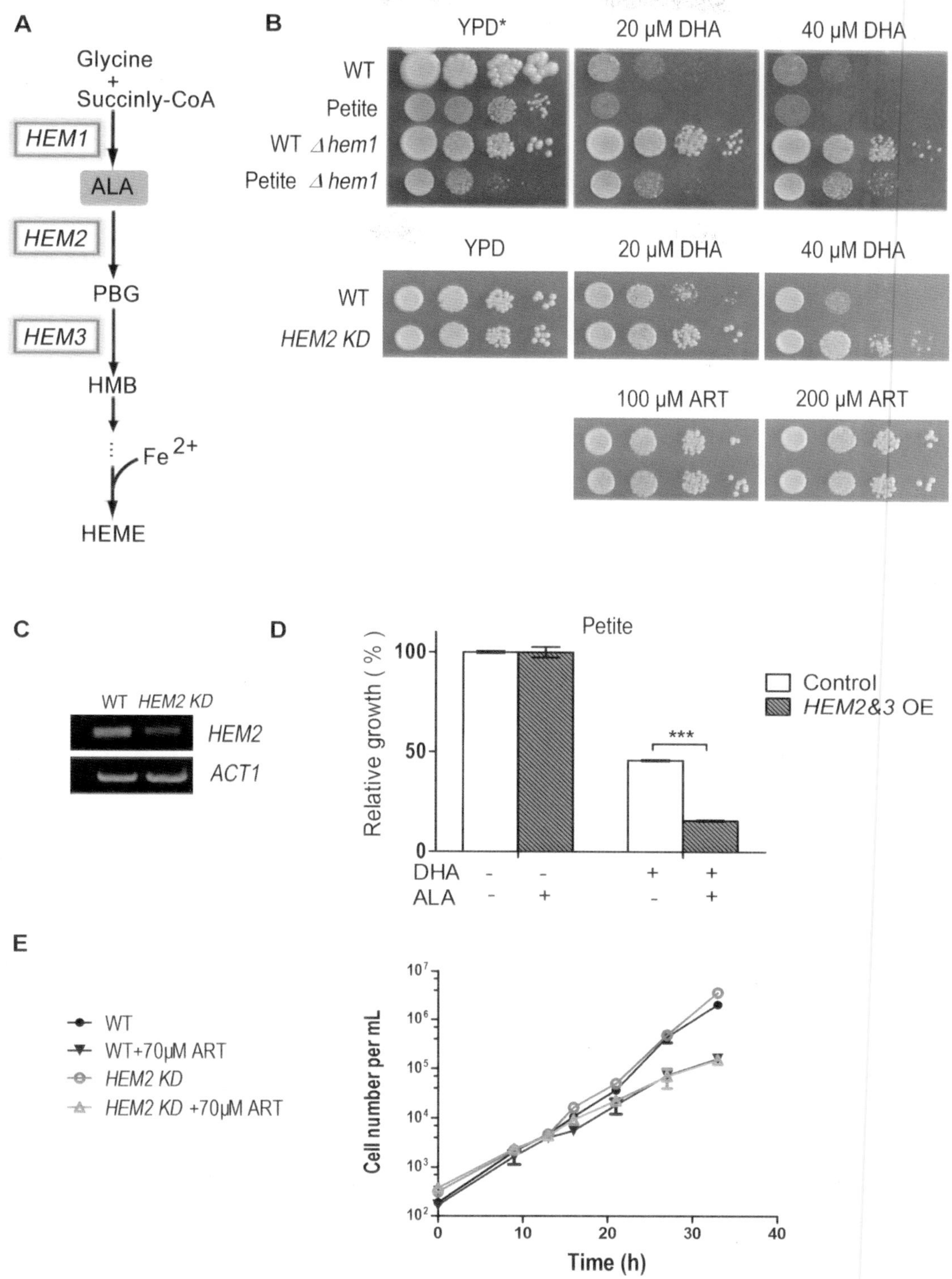

FIGURE 2: Heme mediates the inhibitory effect of DHA on fermentable media. (A) The heme biosynthesis pathway in *Saccharomyces cerevisiae*. **(B)** Deletion of *HEM1* essentially eradicates DHA inhibition of yeast on fermentable media. YPD medium supplemented with Tween-80 and cholesterol (YPD*) was used. *HEM2*-knockdown (*HEM2-KD*) strain is also resistant to DHA on fermentable media. **(C)** RT-PCR analysis of RNA extracted from wild type and *HEM2-KD* strain**.** Semi-quantitative RT-PCR reveals a down-regulation at the mRNA level of *HEM2* in the *HEM2-KD* strain. *ACT1* is the control. **(D)** Overexpression of *HEM2&3* (*HEM2&3* OE) sensitizes yeast to DHA in the presence of δ-aminolevulinic acid (ALA). The effect was not as dramatic and could not be detected with the plate assay and thus a liquid growth assay was adopted. A petite strain was used. The grey bar represents *HEM2&3* OE strain and white bar is the control strain with an empty vector. ***, $P < 0.001$. **(E)** The growth of the WT petite and *HEM2-KD* petite yeast in the presence and absence of ART in YPD liquid medium. *HEM2-KD* petite yeast is not more resistant than normal petite yeast to ART in fermentable media. Similar results were observed in 3 independent experiments.

TABLE 1. Heme content of yeast strains grown on different media.

Strains	YPD WT	YPD Petite	YPD WT Δ*hem1*	YPD Petite Δ*hem1*	YPGE WT	YPGE *HEM2*-KD	SD-Ura-Trp Petite	SD-Ura-Trp Petite *HEM2&3* OE	SGE-Ura-Trp WT	SGE-Ura-Trp WT *HEM2&3* OE
Heme content*	7.20 ±1.19	8.95 ±1.11	1.06 ±0.73	1.05 ±0.46	23.03 ±1.86	9.83 ±0.49	11.03 ±0.74	44.56 ±4.26	22.88 ±8.10	104.42 ±26.39

*ng/mg protein

Heme content was measured by a microplate reader (Fluoroskan Ascent) with excitation at 400 nm and emission at 660 nm. All of the results are expressed as the means ± S.D. calculated from four independent experiments.

DHA can similarly inhibit petite yeast in a heme-dependent manner (Fig. 2B). We then wondered whether up-regulating heme synthesis would show the opposite effect. With this consideration, we overexpressed the rate-limiting enzymes of heme synthesis [29] and supplemented this with the precursor/substrate δ-aminolevulinic acid (ALA). The effect due to heme increase was not as obvious as in the case of heme knockdown and was not noticeable on plate assays, so we had to use liquid growth to monitor the inhibition difference. In the liquid assay, overexpression of *HEM2&3* or supplementation of ALA into the medium alone did not obviously alter DHA sensitivity. However, when combined together, an apparent increase of DHA sensitivity was observed (Fig. 2D). Consistent with the improved inhibition efficacy, heme content was dramatically upregulated after the combined treatment (Table 1).

A *HEM2*-knockdown strain (*HEM2-KD*) was also utilized, wherein a KanMX4 coding sequence was inserted to the 3' end of *HEM2* gene [30]. This insertion leads to instability and degradation of *HEM2* mRNA, resulting in decreased heme synthesis (Table 1). RT-PCR indicated that the *HEM2* mRNA level was indeed down-regulated (Fig. 2C). Results showed that *HEM2-KD* yeast was significantly more resistant to DHA on fermentable media (Fig. 2B). We also tested whether heme is involved in ART's inhibition in fermentable media. As shown previously, ART has only weak inhibition of yeast in fermentable media. While DHA's non-mitochondrial effect is much affected by heme levels (Fig. 2B), *HEM2-KD* petite yeast is not more resistant than normal petite yeast to ART in fermentable media (Fig. 2E). This indicates that the much milder non-mitochondrial inhibition by ART, as observed in fermentable media, is not me-

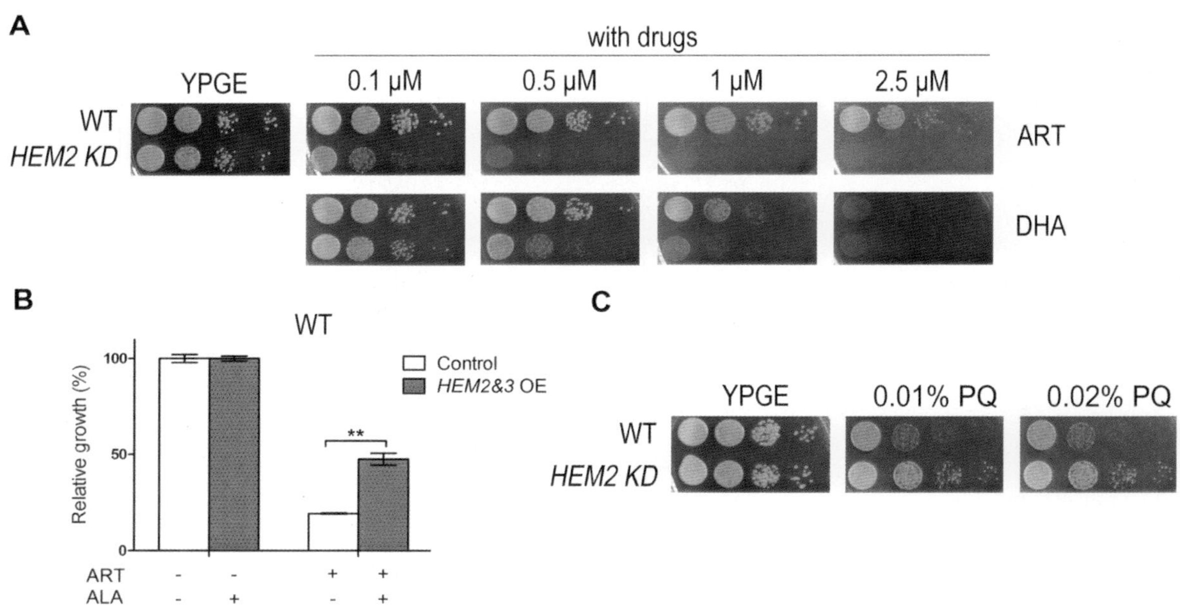

FIGURE 3: Modulation of heme synthesis dramatically alters the ART potency on nonfermentable media. (A) *HEM2-KD* strain is much more sensitive to both ART and DHA on YPGE medium. **(B)** ALA increases the resistance of *HEM2&3* overexpression (*HEM2&3* OE) strain to ART in nonfermentable media. Dropout medium SGE-Ura-Trp was used here. The grey bar represents *HEM2&3* OE strain and white bar is the control strain with an empty vector (**, P < 0.01). **(C)** While more sensitive to ARTs, *HEM2-KD* yeast is slightly more resistant to paraquat (PQ), a mitochondria ROS generator.

diated by heme. The underlying mechanism for the difference of inhibitory properties between DHA and ART remains to be explored, nevertheless it is beyond the scope of this investigation.

Heme antagonizes ARTs' action on nonfermentable media

Having shown that heme underlies DHA's action on fermentable medium, we next asked whether heme is in any way related to the mitochondrial mode of action of ARTs. ARTs are all potently cell-inhibitory in nonfermentable media [11] (Fig. 3A). Because heme is a component of ETC, complete removal of heme would cause respiration-defective phenotype. In fact, higher heme content was detected when yeast was grown on nonfermentable medium YPGE as compared with that on fermentable medium YPD (Table 1). To investigate how heme is related to the anti-mitochondrial action of ARTs, a *HEM2-KD* strain was utilized. Results of spotting assays are shown in Fig 3A. To our surprise, heme strongly antagonizes the mitochondrial mode of action. *HEM2-KD* yeast is much more sensitive to ARTs: whereas normal yeast is inhibited by 1 μM DHA on nonfermentable medium (YPGE), the growth of this heme synthesis reduction strain was greatly inhibited on YPGE with much lower DHA concentrations. Worth noting is that this potentiating effect is even more apparent for ART: on nonfermentable media (YPGE), the growth of this *HEM2-KD* strain was greatly inhibited with 0.1 μM ART as compared to about 5 μM ART for the normal isogenic control strain (Fig. 3A). In contrast, when heme levels were up-regulated by the overexpression of *HEM2&3* in addition to ALA, cell resistance to ART was increased on nonfermentable media (Fig. 3B). Again, this difference was less robust and we had to resort to liquid growth assays to detect the increase of drug resistance. Nevertheless, these results enabled us to conclude that heme plays a negative role in the mode of action of ARTs on nonfermentable media.

Heme reduction strengthens ARTs' anti-mitochondrial effects

The potentiating effect of heme down-regulation on ARTs' inhibitory activity in nonfermentable but not in fermentable media suggests that heme antagonizes ARTs' anti-mitochondrial action. However, because heme is also central to ETC activity in respiration, the increased sensitivity of *HEM2-KD* yeast to ARTs could alternatively be explained by the depleting effect of ARTs on heme in the ETC. In other words, it could be argued that when heme is down-regulated, heme in the ETC may be more vulnerable to the depleting effect of ARTs and therefore confers the increased sensitivity. If this holds true, we would expect that an addition of ARTs would reduce the ETC/respiration activity of yeast cells. We therefore tested the respiration ability of WT and *HEM2-KD* strains in nonfermentable medium YPGE with or without ART incubation. ART had little negative effect on O_2 consumption in both the WT and the heme-limited strain, indicating that the heme participating

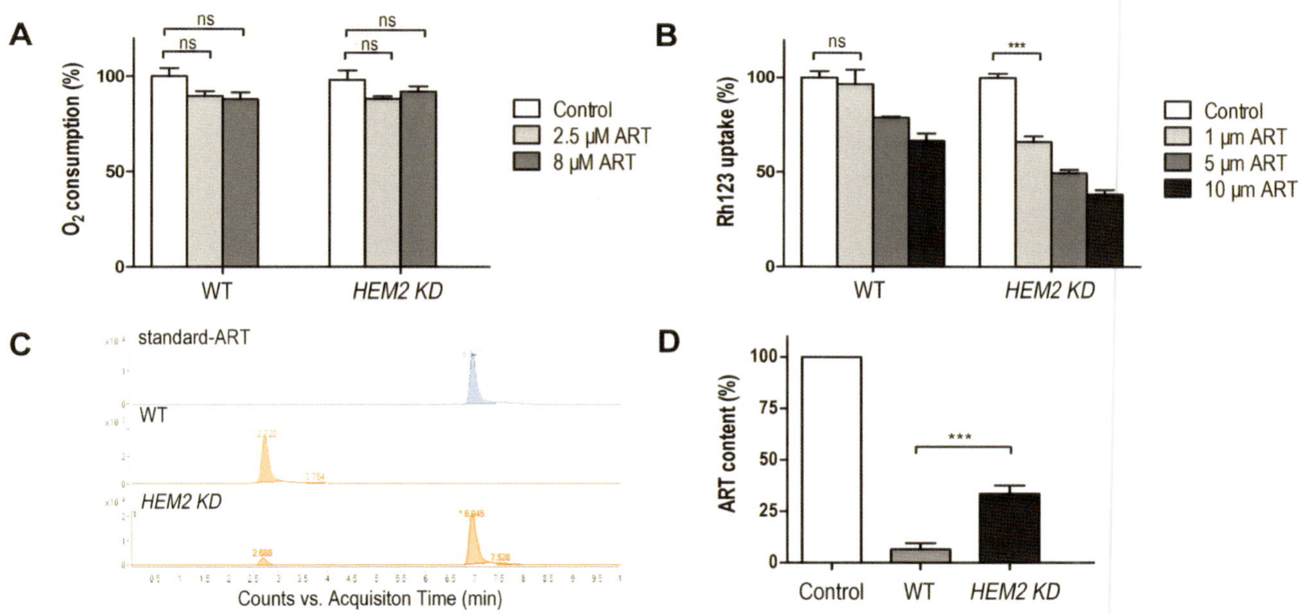

FIGURE 4: Heme reduction strengthens ARTs' anti-mitochondrial effect. (A) ART had little negative effect on O_2 consumption in both the WT and *HEM2-KD* strains (ns, no significance). **(B)** Similar levels of mitochondrial depolarization at much lower ART concentrations could be observed when heme is down-regulated. The membrane potential (Δψm) was assessed by measuring the Δψm-dependent uptake of Rh123. Lower uptake of the probe indicates lower Δψm. **(C, D)** *HEM2-KD* yeast has reduced ART's metabolism rate as compared to WT. **(C)** is the chromatogram of ART in LC-MS analysis. The peak of ART is at 6.94 min. **(D)** is quantitative measurement of results from three parallel experiments. ***, $p<0.001$

in respiration was not affected by ART (Fig 4A). Meanwhile, the mitochondrial membrane potential (monitored by rhodamine 123 uptake) of WT and *HEM2-KD* yeast strains were both greatly reduced upon ART treatment (Fig 4B). These results indicate that the potentiating effect of heme limitation on ARTs is not through affecting the heme in the ETC.

We observed similar levels of mitochondrial depolarization at much lower ART concentrations when heme was down-regulated (Fig. 4B). Therefore, heme limitation greatly strengthens ARTs' anti-mitochondrial effects.

ARTs' action is related to ROS generation. Is it possible that heme limitation increases ARTs' anti-mitochondrial potency because of a likely ROS scavenging activity of heme? To test whether heme could be a significant mitochondrial ROS scavenger *in vivo*, we tested whether heme down-regulation could increase the sensitivity to paraquat, a ROS generator in mitochondria [31]. *HEM2-KD* strain was found not more sensitive to paraquat, if anything, slightly more resistant to this agent (Fig. 3C) on nonfermentable medium YPGE, indicating that heme reduction possibly does not result in a significant loss of ROS scavenging activity.

Heme limitation reduces ARTs' intracellular consumption/metabolism rate

In vitro reactions of ARTs with non-heme or heme iron have been reported. For example, by using high-performance liquid chromatography (HPLC), heme-artemisinin adducts were isolated in a test tube reaction [32], and the physiological relevance of the interaction between heme and ART was further implied [33]. The capability of ART to inhibit the proteolytic activity of digestive vacuoles was observed in both *ex vivo* and *in vitro* experiments in which ART could potently inhibit heme polymerization [34]; ART was also demonstrated to be able to alkylate heme in infected mice [35]. We figured that the potentiating effect of heme limitation on ARTs' anti-mitochondrial action might be through an increase in the drugs' availability as a result of reduced consumption of ARTs. Indeed, heme levels play an important role in the metabolism of ARTs. After incubation with 10 μM ART in nonfermentable YPGE medium for 20 min, LC-MS analysis indicated that the ART in the medium was virtually all metabolized during wild type yeast growth, suggesting a rapid consumption. On the contrary, the remaining ART level in *HEM2-KD* yeast medium was much higher and readily detectable (Fig. 4C and 4D). We explain this by the fact that ART can quickly transverse cell membranes so that rapid

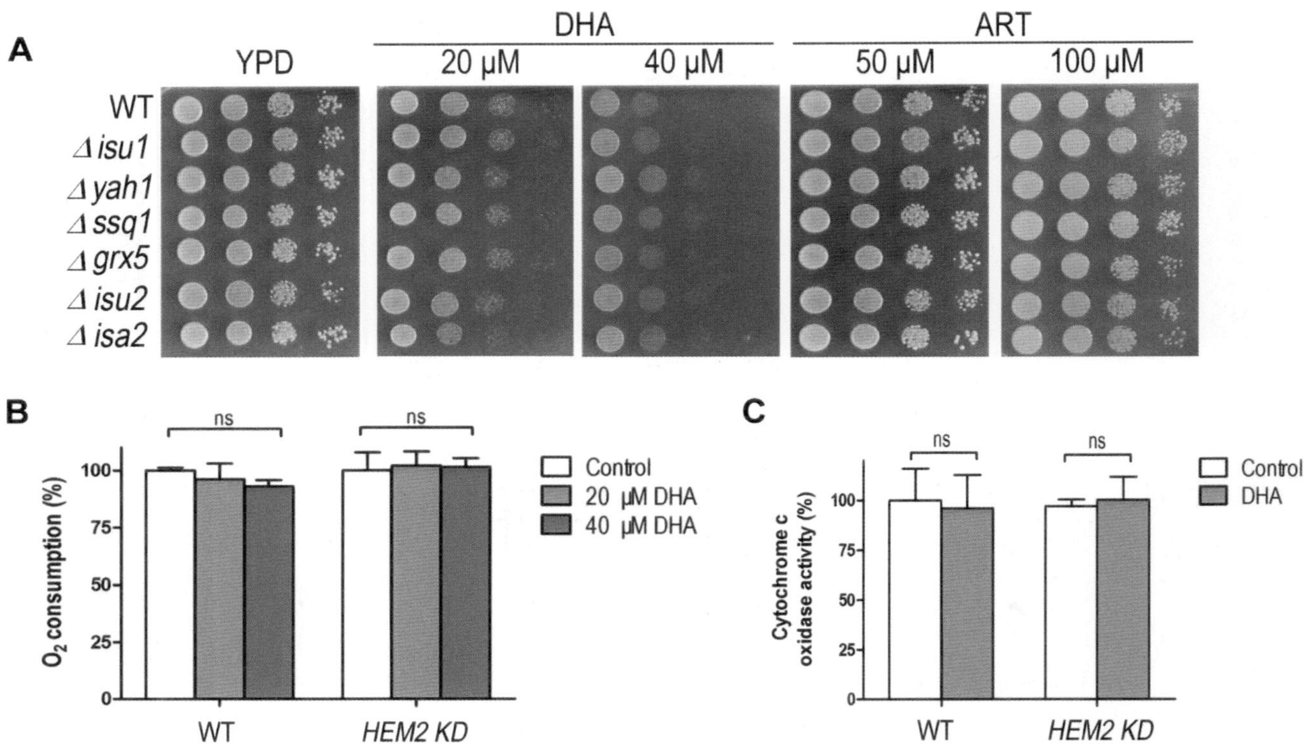

FIGURE 5: The heme-mediated inhibitory action of ARTs is not through interference of the iron-sulfur cluster biogenesis. (A) Compared with WT, the iron-sulfur cluster mutants do not exhibit extra sensitive phenotypes to both ART and DHA. All strains are in the BY4742 background. **(B)** DHA does not affect the respiration ability of either WT or *HEM2-KD* strains on fermentable media. **(C)** DHA does not affect the cytochrome *c* oxidase activity of either WT or *HEM2-KD* yeast strains. ns, no significance.

reduction of ART inside the cell can translate to rapid decrease of ART in the medium. *HEM2-KD* cells degrade ART more slowly so that depletion of ART in its growth medium takes longer. Indeed, it has been shown before that ART is enriched in most, if not all, cell membrane systems [36, 37], and we and others have shown that ART addition to intact cells or mitochondria can result observable effects within minutes [11, 14]. These all support that artemisinin can fast transverse membranes. Our results therefore indicated that ARTs are depleted more slowly when the heme level is lowered, making heme a major player in the ARTs metabolism, i.e. the process of ART reduction and subsequent steps that are not yet well characterized.

The heme-mediated inhibitory action of ARTs is not through interference of the iron-sulfur cluster biogenesis

To investigate whether the heme-mediated suppression effect could be related to iron-sulfur cluster biogenesis, we tested several mutants with impaired iron-sulfur (Fe-S) protein biogenesis for their DHA sensitivity. Isu1p and Isu2p act as scaffold proteins for the Fe-S cluster assembly; Isa2p is involved in the [4Fe-4S] transfer to mitochondrial target proteins; mitochondrial chaperone protein Ssq1p as well as a small glutaredoxin-like protein Grx5p, participate in the transfer of the different types of Fe-S cluster from Isu1p/Isu2p scaffold to apoproteins; Yah1p is a [2Fe-2S] mitochondrial ferredoxin involved in heme A biosynthesis and in Fe-S cluster assembly [38, 39]. As shown in Fig. 5A, these mutants do not exhibit extra sensitive phenotypes to DHA on fermentable medium compared to WT. To further examine whether DHA affect iron-sulfur protein functions, respiration ability and cytochrome *c* oxidase activity were assayed. Cytochrome *c* oxidase is an iron-sulfur protein and it is reported that the cytochrome *c* oxidase activity decreases upon the depletion of components of the iron-sulfur cluster assembly machinery [38]. However, no significant difference could be found in either the WT or heme knockdown strain with or without DHA incubation (Fig 5B, 5C), suggesting that the DHA toxicity in the fermentative growth condition is not due to the damage to iron-sulfur cluster biogenesis or inhibition of its functions.

DISCUSSION

The mode of action of ARTs has long been an intensely debated subject. In the present report, we used the yeast model to explore the biological properties of ARTs. We discovered the clear existence of two major types of pathways for ARTs' actions: the heme-mediated and non-heme-mediated cell inhibitory actions. On fermentable media, upregulating and downregulating intracellular heme content could respectively increase and decrease sensitivity of yeast to DHA. On the contrary, when ARTs' activity was assayed on nonfermentable media, which requires the mitochondrial respiratory function, modulating heme levels alters ART sensitivity in an opposite manner. Collectively, we propose a model of action for ARTs as illustrated in Fig 6. On one hand, ARTs can react with and be consumed by heme; on the other hand, ARTs can be activated by

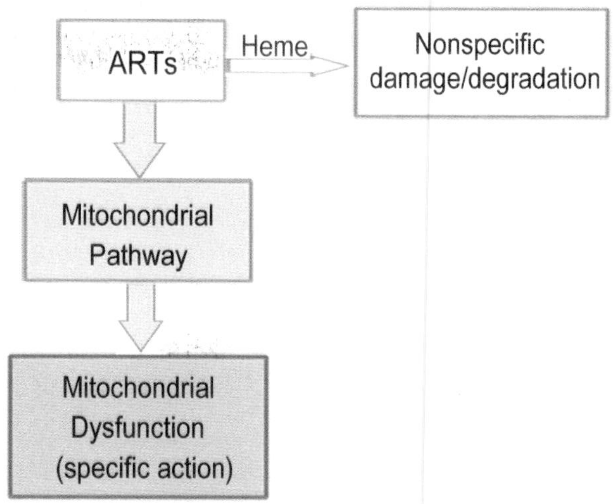

FIGURE 6: A model for the two distinct pathways of ARTs' actions. ARTs act in two distinct ways: the heme pathway and the mitochondrial pathway. ARTs can be consumed by intracellular heme, inducing general damages (in the case of DHA) or tolerated/wasted (ART). ARTs can also be activated by the mitochondrial pathway, causing mitochondrial dysfunction. The heme pathway is a much less efficient pathway for killing and explains the action in some tumor cells and probably other general cell types. The much more potent mitochondrial pathway is shared by ARTs, and can explain the action of ARTs on malaria parasites and yeast grown on nonfermentable media.

some unknown components (likely that of ETC) in mitochondria of certain organisms such as *S. cerevisiae* or malarial parasites. These pathways consume the same source of ARTs inside the cell. Between these two kinds of actions, the mitochondrial inhibition requires far less amounts of drugs and is greatly enhanced when the "wasting" pathway of heme is attenuated. Interaction of ARTs, in particular DHA, with heme likely generates nonspecifically ROS and some toxicity, underlying the mode of action of heme-dependent general cytotoxicity.

In our initial screen for ARTs that showed inhibitory activities against yeast cells on YPD medium, we tested several ARTs including ART, DHA, artelinic acid and artesunate, which are all potent antimalarial drugs. Except for DHA, which could suppress yeast growth at moderate concentrations, the other three drugs had significantly reduced effects. Interestingly, DHA also stands out among ARTs in inhibiting cancer cell growth [20, 40]. The divergence between antimalarial and anticancer potency and the correlated activity of ARTs on yeast in fermentable media and mammalian cell inhibition, from another perspective are consistent with the existence of two pathways of action for ARTs. Notably, yeast and cancer cells were reported to share many features, in particular, the "aerobic glycolysis", also known as the "Warburg effect". Likewise, comparative studies in previous and current works also found a high degree of correlation in cell inhibitory activities of these

ARTs between yeast grown on nonfermentable media and malarial parasites [11].

However, while both ART and DHA interact with heme, we are not sure why ART is far less toxic on fermentable medium YPD. One possibility might be that the resulted ART products are less cytotoxic. Another possibility is that DHA and ART have different chemical properties in terms of water solubility, affecting their intracellular distributions. Alternatively, their reaction profiles with heme could be different. The subtlety of biochemical/biophysical differences underlying their differing biological properties is certainly an issue worthy of future investigation.

Previous reports suggested that mitochondria were involved in the action of ARTs against cancer cells via apoptosis in a mitochondria-dependent manner [41, 42]. Recently, a detailed study further pointed out that active mitochondria are responsible for ARTs-induced cytotoxicity since HeLa rho^0 cells were more resistant to this class of drugs [24]. However, this role of mitochondria was not due to their reductive activation of ARTs to carbon-centered radicals, and indeed the same extent of activation took place in both the wild type and rho^0 cell lines. The mitochondrial involvement in cancer cell inhibition by ARTs is therefore likely an indirect one and secondary to its involvement in apoptosis induction.

The role of the hemoglobin degradation pathway for the action of ARTs in malarial parasites has been intensively studied [35, 43, 44]. Heme generated during hemoglobin catabolism is trapped in the vacuoles and has been proposed to be either the target [45] or the activating agent [46]. However, *Plasmodium falciparum* infecting alpha-thalassemic erythrocytes is resistant to ARTs *in vitro*, and heme and heme-containing proteins in erythrocyte are proposed to reduce ART's effectiveness [47, 48]. Furthermore, when malarial parasites are cultured under CO, which can block the interaction of ARTs with heme, inhibition by ARTs is more effective than when malarial parasites are cultured in the presence of O_2. Thus, the role of heme may be associated with degradation instead of activation of ARTs aiding in the killing of parasite [49]. Our results, showing that reduced intracellular heme content on nonfermentable media potentiated the effect of ART while increased heme levels attenuated the activity, provide supporting evidences that if there is any involvement of the hemoglobin catabolism pathway in the action of ARTs in malarial parasites, it likely acts more as a competing pathway for the consumption of ARTs. This is consistent with the malarial studies concluding that heme plays a deactivating role [47, 48].

Besides the heme molecules that are generated by the hemoglobin degradation and condensed/trapped in the vacuoles, malarial parasites also have a functional endogenous heme pathway, which might provide the synthesis of heme for the general need of the organisms. To date, the involvement of the endogenous heme pathway in the action of ARTs has not been clarified. In our preliminary experiments we found that neither inhibition of heme synthesis by succinylacetone, a *hem2* inhibitor, nor addition of ALA much altered the sensitivity of *Plasmodium falciparum* to ARTs, suggesting that the endogenous heme was not significantly associated with ARTs' activities. We suspect the activity of the endogenous heme pathway is rather weak so that not much heme is *de novo* synthesized.

Taken together, the yeast cell studies reveal to us two distinct types of biological actions for ARTs: both the mitochondrial and heme-dependent nonspecific pathways can mediate ARTs' damage. Emerging from our studies is this scenario: ARTs can react with a few components (largely heme or heme related) inside a cell, conferring a relatively general and nonspecific action in most, if not all, cells, and sometimes causing the general toxicity or side effects of the drugs; the mitochondrial interaction only occurs in those ARTs-sensitive cells. Our studies thus reconcile heme and mitochondria theories and further dissect their individual roles in the mechanism of action of ARTs. We suggest that the heme pathway could explain the antitumor properties of DHA, whereas the specific and potent mitochondrial pathway can explain the antimalarial action. The information provided by this study therefore not only helps us understand the biological properties of ARTs, but may also open a new avenue for anticancer ARTs designs. In addition, degradation by heme should also be a concern for new antimalarial endoperoxide drug development.

MATERIALS AND METHODS

Chemicals

ART and DHA were purchased from Chengdu Okay Medicine Co., Ltd (Chengdu, China). Hemin and ALA were purchased from Sigma. ARTs and hemin were prepared in DMSO, ALA was prepared in double distilled H_2O (ddH_2O). All the aliquots were used immediately or stored at -20°C for later use.

Yeast strains and growth

Standard yeast media and growth conditions were used. *Saccharomyces cerevisiae* BY4742 was used as the wild type control (BY4742: *MATα his3Δ1 leu2Δ0 lys2Δ0 ura3Δ0*). Petite strains in BY4742 background were made by ethidium bromide treatment. *Δhem1* and *HEM 2&3* overexpression strains for both wild type and petite strain, and the *HEM2-KD* strain were all kept in our lab and were based on the BY4742 background. Yeast was normally plated on YPD (2% glucose as carbon source) or YPGE (2% glycerol plus 2% ethanol as carbon source) agar plates supplemented with different concentrations of ART or DHA, and with 0.5% Tween-80 and 30 μg/ml cholesterol when necessary. For overexpression strains, uracil and tryptophan dropout synthetic medium was used.

Drug sensitivity/resistance assay

For growth testing on agar plates, yeast previously grown on YPD plates was 10-fold serial diluted with sterile ddH_2O, and then spotted on YPD or YPGE plates with or without drugs (Spotting assay). For tests in liquid media, yeast was cultured overnight in synthetic dropout media as follows: after being spun down and washed with ddH_2O, cells were resuspended with either SD-Ura-Trp or SGE-Ura-Trp medium to an initial A_{600} of 0.01; cells were then cultured with or without ALA (final concentration 250 μg/mL), DHA (0.5 μM) and ART (5 μM).

A_{600} was measured after 36 hours. Incubation was all performed at 30°C.

Heme content determination

Heme content was measured as previously described [50]. Briefly, 20 ml yeast cultures grown in different media (YPD, YPGE, SD-Ura-Trp or SGE-Ura-Trp) were harvested at the exponential phase, washed two times with ddH_2O and resuspended in 150 μl extraction buffer (PBS plus protease inhibitor cocktail). After the addition of isometric glass beads, the mixtures were vortexed for 15 minutes and centrifuged at 15,000 g for 30 min at 4°C. The supernatant was transferred to a new microcentrifuge tube, and the yeast's total protein concentration was measured by a BCA protein assay kit (Thermo Scientific). To prepare samples for a heme content assay, an aqueous oxalic acid solution (500 μl, 2 M) was added to each sample. The samples were mixed thoroughly before boiling for 30 min. Standard solutions of hemin (0.01 - 10 mM) were prepared in water/methanol (1:1, v/v) and boiled with oxalic acid as described above. These standard solutions and samples (200 μl) were then added into a 96-well plate and fluorescent intensities were measured by a microplate fluorometer (Fluoroskan Ascent) with the excitation wavelength setting at 400 nm and emission at 662 nm. Results were corrected by preparing samples without boiling to eliminate the influence of endogenous non-heme porphyrins.

Yeast mitochondria isolation

Mitochondria were purified from yeast cells by Zymolyase digestion and differential centrifugation. The preparation was performed as previously described with little modification [51]. Cells were collected and homogenized in a buffer containing 5.0 mM HEPES, 70 mM surcose, 0.22 M mannitol, proteinase inhibitor cocktail, 0.2% (wt/vol) BSA, pH 7.2 at 4°C. To remove nuclei and large membrane fragments, the homogenate was first centrifuged at 900 g for 10 min, and the resulting supernatant was subsequently centrifuged at 10000 g for 10 min to obtain crude mitochondrial fractions. The mitochondrial fractions were further purified by a sucrose gradient.

Membrane potential determination

The membrane potential of yeast cells was determined by measuring the fluorescence intensity of rhodamine 123 (Rh123). Yeast cells were first incubated in YPGE with the indicated concentration of ART for 30 min, then 2 μM Rh123 was used to incubate cells for 30 min at 30°C before washing and resuspension in PBS. The membrane potential of yeast cells was determined using a flow cytometer (FACScan, BD), with the excitation at 480 nm and emission at 530 nm. For each sample, 20,000 events were counted at the same flow cytometer setting.

Oxygen consumption

After grown to OD 1.00, yeast cells were treated with ART or DHA for 30 min, and then the cells were centrifuged, washed and resuspended by PBS. Oxygen consumption was measured using a Clark-type oxygen electrode in a total volume of 1 ml.

ART content analysis

Liquid chromatography–mass spectrometry (LC-MS) was applied to detect the ART in samples. HPLC was performed using an Agilent 1290 system (Palo Alto, CA, USA). The HPLC system was coupled to an Agilent 6460 triple-quadrupole mass spectrometer (Palo Alto, CA, USA) via a electrospray ionization (ESI) interface for mass analysis and detection. Chromatographic separation was achieved on an Acquity UPLC® BEH C18 column (2.1 × 50 mm, 1.7 μm; Waters Corp., Milford, USA). SecurityGuard C18 (5 μm) guard column (Waters Corp., Milford, USA). The chromatography was performed at room temperature. The mobile phase consisted of acetonitrile/ water containing 0.1% (v/v) formic acid (30:70, v/v), delivered at a flow rate of 0.2 mL/min. The mass spectrometer was operated in positive ion mode. The fragmentation transitions for the multiple reactions monitoring (MRM) were m/z 283.0 to 247.0 for ART.

For sample preparation, yeast cells were incubated with ART in YPGE medium for 20 min before centrifugation and the supernatant was collected and then extracted with 3 vol of dichloromethane for 3 times. The samples were air dried and resolved by 30% methanol before injection and chromatographic analysis.

Cytochrome *c* oxidase activity assay

Activities of cytochrome *c* oxidase were measured by the Cytochrome *c*-Oxidase Assay kit (# CYTOCOX1, Sigma, St. Louis, MO, USA) according to manufacturer's instructions.

Statistical analysis

Results are expressed as mean ± SD for the indicated number of experiments. Statistical analyses between parallel groups were calculated with Student's t-test. A value of P under 0.05 was considered to be statistically significant.

ACKNOWLEDGMENTS

This study was supported by National Science Foundation of China (31123004 and 30971568). We would like to thank Richard Jiang Yang (Dartmouth College, Hanover, US) for reading of the manuscript, Weihua Wang, Yu Tian, Yi Ding (Center of Biomedical Analysis, Tsinghua University) for technical assistance in ART content measurement by LC-MS and Ruifang Pang (Center of Biomedical Analysis, Tsinghua University) for technical assistance in heme content measurement.

CONFLICT OF INTEREST

The authors declare no conflict of interest.

REFERENCES

1. WHO (**2013**). World malaria report 2013. http://www.who.int/malaria/publications/world_malaria_report_2013/en/.

2. Qinghaosu Antimalaria Coordinating Research Group (**1979**). Antimalaria studies on Qinghaosu. **Chin Med J (Engl)** 92(12): 811-816..

3. O'Neill PM, Searle NL, Kan KW, Storr RC, Maggs JL, Ward SA, Raynes K, Park BK (**1999**). Novel, potent, semisynthetic antimalarial carba analogues of the first-generation 1,2,4-trioxane artemether. **J Med Chem** 42(26): 5487-5493.

4. Posner GH, Parker MH, Northrop J, Elias JS, Ploypradith P, Xie S, Shapiro TA (**1999**). Orally active, hydrolytically stable, semisynthetic, antimalarial trioxanes in the artemisinin family. **J Med Chem** 42(2): 300-304.

5. O'Neill PM, Rawe SL, Borstnik K, Miller A, Ward SA, Bray PG, Davies J, Oh CH, Posner GH (**2005**). Enantiomeric 1,2,4-trioxanes display equivalent in vitro antimalarial activity versus Plasmodium falciparum malaria parasites: implications for the molecular mechanism of action of the artemisinins. **Chembiochem** 6(11): 2048-2054.

6. Liu Y, Cui K, Lu W, Luo W, Wang J, Huang J, Guo C (**2011**). Synthesis and antimalarial activity of novel dihydro-artemisinin derivatives. **Molecules** 16(6): 4527-4538.

7. O'Neill PM, Posner GH (**2004**). A medicinal chemistry perspective on artemisinin and related endoperoxides. **J Med Chem** 47(12): 2945-2964.

8. Sidhu AB, Verdier-Pinard D, Fidock DA (**2002**). Chloroquine resistance in Plasmodium falciparum malaria parasites conferred by pfcrt mutations. **Science** 298(5591): 210-213.

9. Legrand E, Demar M, Volney B, Ekala MT, Quinternet M, Bouchier C, Fandeur T, Rogier C, Carme B, Puijalon OM, Esterre P (**2007**). First case of emergence of atovaquone resistance in Plasmodium falciparum during second-line atovaquone-proguanil treatment in South America. **Antimicrob Agents Chemother** 51(6): 2280-2281.

10. Foley M, Tilley L (**1997**). Quinoline antimalarials: mechanisms of action and resistance. **Int J Parasitol** 27(2): 231-240.

11. Wang J, Huang L, Li J, Fan Q, Long Y, Li Y, Zhou B (**2010**). Artemisinin directly targets malarial mitochondria through its specific mitochondrial activation. **PLoS One** 5(3): e9582.

12. Li J, Zhou B (**2010**). Biological actions of artemisinin: insights from medicinal chemistry studies. **Molecules** 15(3): 1378-1397.

13. Li W, Mo W, Shen D, Sun L, Wang J, Lu S, Gitschier JM, Zhou B (**2005**). Yeast model uncovers dual roles of mitochondria in action of artemisinin. **PLoS Genet** 1(3): e36.

14. Antoine T, Fisher N, Amewu R, O'Neill PM, Ward SA, Biagini GA (**2014**). Rapid kill of malaria parasites by artemisinin and semi-synthetic endoperoxides involves ROS-dependent depolarization of the membrane potential. **J Antimicrob Chemother** 69(4): 1005-1016.

15. Sen R, Bandyopadhyay S, Dutta A, Mandal G, Ganguly S, Saha P, Chatterjee M (**2007**). Artemisinin triggers induction of cell-cycle arrest and apoptosis in Leishmania donovani promastigotes. **J Med Microbiol** 56(Pt 9): 1213-1218.

16. Jones-Brando L, D'Angelo J, Posner GH, Yolken R (**2006**). In vitro inhibition of Toxoplasma gondii by four new derivatives of artemisinin. **Antimicrob Agents Chemother** 50(12): 4206-4208.

17. Romero MR, Serrano MA, Vallejo M, Efferth T, Alvarez M, Marin JJ (**2006**). Antiviral effect of artemisinin from Artemisia annua against a model member of the Flaviviridae family, the bovine viral diarrhoea virus (BVDV). **Planta Med** 72(13): 1169-1174.

18. Willoughby JA, Sr., Sundar SN, Cheung M, Tin AS, Modiano J, Firestone GL (**2009**). Artemisinin blocks prostate cancer growth and ce cycle progression by disrupting Sp1 interactions with the cyclin-dependent kinase-4 (CDK4) promoter and inhibiting CDK4 gene expression. **J Biol Chem** 284(4): 2203-2213.

19. Li PC, Lam E, Roos WP, Zdzienicka MZ, Kaina B, Efferth T (**2008**). Artesunate derived from traditional Chinese medicine induces DNA damage and repair. **Cancer Res** 68(11): 4347-4351.

20. Crespo-Ortiz MP, Wei MQ (**2012**). Antitumor activity of artemisinin and its derivatives: from a well-known antimalarial agent to a potential anticancer drug. **J Biomed Biotechnol** 2012:247597.

21. Beekman AC, Wierenga PK, Woerdenbag HJ, Van Uden W, Pras N, Konings AW, el-Feraly FS, Galal AM, Wikstrom HV (**1998**). Artemisinin-derived sesquiterpene lactones as potential antitumour compounds: cytotoxic action against bone marrow and tumour cells. **Planta Med** 64(7): 615-619.

22. Efferth T, Benakis A, Romero MR, Tomicic M, Rauh R, Steinbach D, Hafer R, Stamminger T, Oesch F, Kaina B, Marschall M (**2004**). Enhancement of cytotoxicity of artemisinins toward cancer cells by ferrous iron. **Free Radic Biol Med** 37(7): 998-1009.

23. Zhang S, Gerhard GS (**2009**). Heme mediates cytotoxicity from artemisinin and serves as a general anti-proliferation target. **PLoS One** 4(10): e7472.

24. Mercer AE, Copple IM, Maggs JL, O'Neill PM, Park BK (**2011**). The role of heme and the mitochondrion in the chemical and molecular mechanisms of mammalian cell death induced by the artemisinin antimalarials. **J Biol Chem** 286(2): 987-996.

25. Jiao Y, Ge CM, Meng QH, Cao JP, Tong J, Fan SJ (**2007**). Dihydroartemisinin is an inhibitor of ovarian cancer cell growth. **Acta Pharmacol Sin** 28(7): 1045-1056.

26. Zhang F, Gosser DK, Jr., Meshnick SR (**1992**). Hemin-catalyzed decomposition of artemisinin (qinghaosu). **Biochem Pharmacol** 43(8): 1805-1809.

27. Zhang S, Gerhard GS (**2008**). Heme activates artemisinin more efficiently than hemin, inorganic iron, or hemoglobin. **Bioorg Med Chem** 16(16): 7853-7861.

28. Gollub EG, Liu KP, Dayan J, Adlersberg M, Sprinson DB (**1977**). Yeast mutants deficient in heme biosynthesis and a heme mutant additionally blocked in cyclization of 2,3-oxidosqualene. **J Biol Chem** 252(9): 2846-2854.

29. Hoffman M, Gora M, Rytka J (**2003**). Identification of rate-limiting steps in yeast heme biosynthesis. **Biochem Biophys Res Commun** 310(4): 1247-1253.

30. Breslow DK, Cameron DM, Collins SR, Schuldiner M, Stewart-Ornstein J, Newman HW, Braun S, Madhani HD, Krogan NJ, Weissman JS (**2008**). A comprehensive strategy enabling high-resolution functional analysis of the yeast genome. **Nat Methods** 5(8): 711-718.

31. Cocheme HM, Murphy MP (**2008**). Complex I is the major site of mitochondrial superoxide production by paraquat. **J Biol Chem** 283(4): 1786-1798.

32. Robert A, Coppel Y, Meunier B (**2002**). Alkylation of heme by the antimalarial drug artemisinin. **Chem Commun (Camb)** (5): 414-415.

33. Hong YL, Yang YZ, Meshnick SR (**1994**). The interaction of artemisinin with malarial hemozoin. **Mol Biochem Parasitol** 63(1): 121-128.

34. Pandey AV, Tekwani BL, Singh RL, Chauhan VS (**1999**). Artemisinin, an endoperoxide antimalarial, disrupts the hemoglobin catabolism and heme detoxification systems in malarial parasite. **J Biol Chem** 274(27): 19383-19388.

35. Robert A, Benoit-Vical F, Claparols C, Meunier B (**2005**). The antimalarial drug artemisinin alkylates heme in infected mice. **Proc Natl Acad Sci U S A** 102(38): 13676-13680.

36. Li ZL (**1984**). Use of nuclear techniques in the study of the antimalarial action of Qinghaosu and dihydro-Qinghaosu. **Nucl Techniq** (2): 23-25.

37. Maeno Y, Toyoshima T, Fujioka H, Ito Y, Meshnick SR, Benakis A, Milhous WK, Aikawa M (**1993**). Morphologic effects of artemisinin in Plasmodium falciparum. **Am J Trop Med Hyg** 49(4): 485-491.

38. Lange H, Muhlenhoff U, Denzel M, Kispal G, Lill R (**2004**). The heme synthesis defect of mutants impaired in mitochondrial iron-sulfur protein biogenesis is caused by reversible inhibition of ferrochelatase. **J Biol Chem** 279(28): 29101-29108.

39. Alves R, Herrero E, Sorribas A (**2004**). Predictive reconstruction of the mitochondrial iron-sulfur cluster assembly metabolism: I. The role of the protein pair ferredoxin-ferredoxin reductase (Yah1-Arh1). **Proteins** 56(2): 354-366.

40. Woerdenbag HJ, Moskal TA, Pras N, Malingre TM, el-Feraly FS, Kampinga HH, Konings AW (**1993**). Cytotoxicity of artemisinin-related endoperoxides to Ehrlich ascites tumor cells. **J Nat Prod** 56(6): 849-856.

41. Mercer AE, Maggs JL, Sun XM, Cohen GM, Chadwick J, O'Neill PM, Park BK (**2007**). Evidence for the involvement of carbon-centered radicals in the induction of apoptotic cell death by artemisinin compounds. **J Biol Chem** 282(13): 9372-9382.

42. Mao H, Gu H, Qu X, Sun J, Song B, Gao W, Liu J, Shao Q (**2013**). Involvement of the mitochondrial pathway and Bim/Bcl-2 balance in dihydroartemisinin-induced apoptosis in human breast cancer in vitro. **Int J Mol Med** 31(1): 213-218.

43. Klonis N, Crespo-Ortiz MP, Bottova I, Abu-Bakar N, Kenny S, Rosenthal PJ, Tilley L (**2011**). Artemisinin activity against Plasmodium falciparum requires hemoglobin uptake and digestion. **Proc Natl Acad Sci U S A** 108(28): 11405-11410.

44. Kannan R, Kumar K, Sahal D, Kukreti S, Chauhan VS (**2005**). Reaction of artemisinin with haemoglobin: implications for antimalarial activity. **Biochem J** 385(Pt 2): 409-418.

45. Kannan R, Sahal D, Chauhan VS (**2002**). Heme-artemisinin adducts are crucial mediators of the ability of artemisinin to inhibit heme polymerization. **Chem Biol** 9(3): 321-332.

46. Meshnick SR (**2003**). Artemisinin and heme. **Antimicrob Agents Chemother** 47(8): 2712; author reply 2712-2713.

47. Vattanaviboon P, Siritanaratkul N, Ketpirune J, Wilairat P, Yuthavong Y (**2002**). Membrane heme as a host factor in reducing effectiveness of dihydroartemisinin. **Biochem Pharmacol** 64(1): 91-98.

48. Ponmee N, Chuchue T, Wilairat P, Yuthavong Y, Kamchonwongpaisan S (**2007**). Artemisinin effectiveness in erythrocytes is reduced by heme and heme-containing proteins. **Biochem Pharmacol** 74(1): 153-160.

49. Coghi P, Basilico N, Taramelli D, Chan WC, Haynes RK, Monti D (**2009**). Interaction of artemisinins with oxyhemoglobin Hb-FeII, Hb-FeII, carboxyHb-FeII, heme-FeII, and carboxyheme FeII: significance for mode of action and implications for therapy of cerebral malaria. **ChemMedChem** 4(12): 2045-2053.

50. Sassa S (**1976**). Sequential induction of heme pathway enzymes during erythroid differentiation of mouse Friend leukemia virus-infected cells. **J Exp Med** 143(2): 305-315.

51. Meisinger C, Pfanner N, Truscott KN (**2006**). Isolation of yeast mitochondria. **Methods Mol Biol** 313:33-39.

9

Improvement of biochemical methods of polyP quantification

Samuel Bru[1], Javier Jiménez[1], David Canadell[2,#], Joaquín Ariño[2,*], Josep Clotet[1,*]
[1] Department of Basic Sciences, Faculty of Medicine and Health Sciences. Universitat Internacional de Catalunya. Barcelona, Spain.
[2] Departament de Bioquímica i Biologia Molecular and Institut de Biotecnologia i Biomedicina, Universitat Autònoma de Barcelona. Cerdanyola del Vallès, Spain.
[#] Present Address: Cell Signaling Research Group, Departament de Ciències Experimentals i de la Salut, Universitat Pompeu Fabra, Barcelona, Spain.
* Corresponding Author:
J. Ariño, E-mail: joaquin.arino@uab.cat
J. Clotet, E-mail: jclotet@uic.cat

ABSTRACT Polyphosphate (polyP) is an abundant and physiologically important biomolecule for virtually any living cell. Therefore, determination of changes in cellular content of polyP is crucial for its functional characterization. Determination of cellular polyP has been performed by many different methods, and the lack of a standardized procedure is possibly responsible for the large dispersion of results found in the relevant literature. For a relatively simple organism, such as the yeast *Saccharomyces cerevisiae*, this variation can be up to 12-fold. polyP extraction and determination of free phosphate released by enzymatic degradation of the polymer is a method quite common and relatively straightforward for polyP determination. By using the yeast *S. cerevisiae* as model, we have experimentally evaluated the different steps in this procedure in order to identify critical issues that might explain the disparate reported results. As the main output of this evaluation we propose a straightforward and robust procedure that can be used as gold standard protocol for cellular polyP purification and determination from unicellular organisms, thus providing consistency to measurements and facilitating inter-laboratory comparisons and biological interpretation of the results.

***Keywords**: polyphosphate, yeast, Saccharomyces cerevisiae, neutral-phenol.*

Abbreviations:
DAPI - 4',6-diamidino-2-phenylindole, polyP - polyphosphate.

INTRODUCTION

Inorganic polyphosphate (polyP) is a linear polymer of orthophosphate with many biological functions in both prokaryotic and eukaryotic organisms. In microorganisms, polyP plays an important function as stores of phosphorus and energy, in cation homeostasis, and in adaptation to stress conditions [1]. Very recently, a role for polyP in post-translational modification of Lys residues in proteins has been reported [2]. In complex organisms, as humans, this polymer is involved in many diverse functions such as blood coagulation, bone formation, immune response, regulation of calcium level in mitochondria, or neuro-transmission (see [3] and references herein). In addition, polyP is used in diverse industrial processes, including water treatment, and as fertilizer in agriculture [4]. Therefore, research on polyP is of great relevance in a variety of fields, ranging from the environmental to the biological sciences.

In the past, quantification of intracellular polyP content has been carried out by means of a diversity of approaches and methodologies, often yielding results that are difficult to compare. One of the most commonly employed methods to quantify polyP has been ^{31}P NMR [5-7]. While this technique has produced a foundational knowledge in the field of polyP research, it suffers of significant drawbacks. For instance, ^{31}P NMR only detects phosphorus-containing molecules on the basis of bond class. Therefore, the method is not ideally suited to distinguish between polyP and molecules that also can contain phosphoanhydride bonds, such as nucleotides. Moreover, the abundant presence of P-ester compounds can obscure the comparatively smaller polyP signal in the majority of biological samples. In addition, usually NMR equipment is not well suited for time-resolved experiments in which environmental conditions must be quickly changed.

Staining with relatively specific dyes has been widely used for monitoring polyP accumulation both *in vivo* and *in vitro*. An example is the utilization of the metachromatic interaction of polyP with toluidine blue [8], although the most widely used method is based on the interaction between polyP and the fluorochrome 4',6-diamidino-2-phenylindole (DAPI). DAPI is usually used for DNA detection, because blue fluorescence is apparent when the stained cells are viewed under UV light. However, the binding of DAPI to polyP also shifts the peak of DAPI and the fluorescence intensity at this shifted wavelength is proportional to the concentration of polyP [9, 10]. Although previous studies concluded that the presence of DNA do not complicate the fluorimetric quantification of polyP with DAPI, it has recently been described that RNA, inositol phosphates, amorphous calcium phosphates and nucleotides can cause significant interference at the wavelengths used to measure polyP, concluding that the DAPI-polyP interaction is sensitive to sample composition [7, 11-14] and can lead to high variation between samples. Recently a new method for measuring polyP based on spectromicroscopy (Raman microscopy) has been developed [13, 15]. Raman microscopy is a fast evolving technology, but currently its sensitivity is relatively low and requires equipment not widely available.

Protocols that determine polyP concentration biochemically offer an appealing alternative to the methods discussed above, as exemplified by the early determination of polyP levels in different mammalian cells and tissues performed by the Kornberg's laboratory [16]. Such enzymatic methods are based on the use of polyP kinase, followed by ATP measurement [17, 18], or that of exopolyphosphatase followed by determination of free orthophosphate [19-21], and have been used in a variety of cells, from microorganisms to mammalian tissues. These methods are widely used because of their high sensitivity (particularly if radioactive phosphate is used), but they are not devoid of problems. For instance, they generally require of purification procedures, which can lead to erratic and/or low yields and, currently, no commercial sources exist for the required enzymes. In addition, even for enzyme-based protocols, the disparity of procedures used for purification and assays complicates inter-laboratory comparisons.

Possibly, the lack of a truly validated method to quantify cellular polyP content has been a drawback that has hampered the advance in the field. In fact, after evaluation of the relevant literature concerning the budding yeast *Saccharomyces cerevisiae*, we noticed a large disparity in the polyP content reported for this yeast that was difficult to explain on the basis of different genetic backgrounds or growth conditions. We considered that, at least in part, such inconsistency could derive from the use of different methods for polyP determination. In this work we analyze and compare possible alternatives for the main steps of exopolyphosphatase-based methods for polyP quantification, discuss the relative advantages, and propose a unified

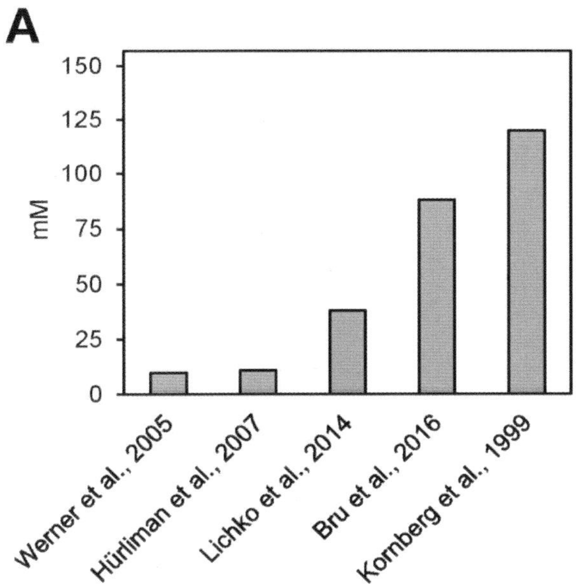

FIGURE 1: High variability on polyP determination in *S. cerevisiae*. (A) polyP content in mM concentration according to different authors. Values were converted to mM according to the following assumptions: i) The volume of the haploid cell is 42 fl, ii) One OD_{660} unit equals to 2.6 x 10^7 cells and represents 50 µg of RNA. The Mw for the [-PO_3^--] monomer is 79 g/mol. **(B)** Flow chart of this work, showing the different possibilities explored in each of the three parts needed for determining polyPs: extraction, purification and quantification.

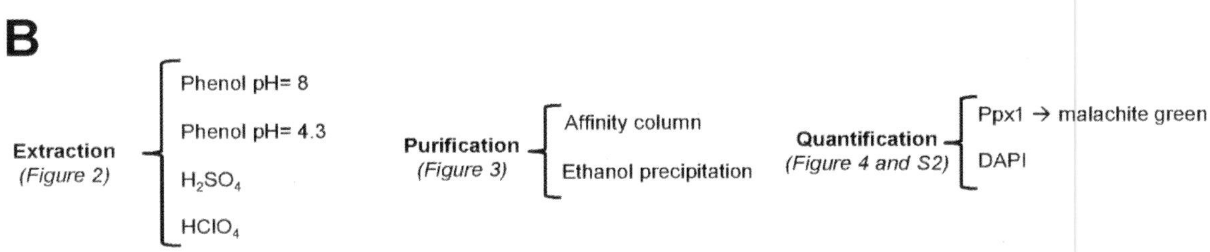

protocol for polyP determination that will help in the advance of the research of this polymer.

RESULTS AND DISCUSSION

Large discrepancies in determination of polyP cellular content claim for standardized determination methods

A major issue when any cellular metabolite from a given source is determined is to evaluate whether the values obtained are congruent with those previously reported under similar circumstances (growth conditions, genetic background, etc.) in the literature. In the case of polyP content in yeast cells, this is a difficult task because, in many cases, the units employed are different (i.e. μg/OD unit, μmol/mg RNA, nmol Pi/10^n cells, and so on). We have made an attempt to integrate data from the literature in a common frame of measurement units (mM Pi). The result of this transformation applied to five independent reports [22-26] is shown in Figure 1. As it can be seen, there are large differences (up to 12-fold) in the amount of cellular polyP determined. It might be argued that this disparity could result from the use of different genetic backgrounds in the experiments. However, in several of the examples included in Figure 1, the BY4741 strain, or very close derivatives, were employed. The use of different backgrounds, even if it can result in some fluctuations, can hardly explain the large variations observed (our own unpublished data, see also reference [25]). In addition, the Pi content in the media used and the growth phase of the cultures were fairly consistent among the different experiments. Therefore, it is reasonable to conclude that a substantial component of such variations must reside in the use of different protocols for polyP extraction, purification and quantification. In consequence, we considered necessary to carry out a comparative analysis of different options than can be proposed for each of these steps (Figure 1B).

Influence of the extraction method on polyP quantification

There is plenty of evidence that polyP polymers are labile to acidic conditions [27]. In fact, acid treatment at high temperature has been used for polyP quantification upon hydrolysis of the polymer [28]. However, some methods for polyP extraction from cells or tissues involve the use of strong acids (usually sulfuric or perchloric). Therefore, we considered the possibility that this step might influence in the recovery of polyP from its original source. To evaluate this, we incubated 4 μg of commercial polyP for different periods of time with the reagents more frequently used in the literature: a phenol-based extraction solution at pH 8.0 (Figure 2A), the same solution at pH 4.8 (Figure 2B), 1 M sulfuric acid (Figure 2C) or 1 M perchloric acid (Figure 2D). As it can be observed, treatment with strong acids results

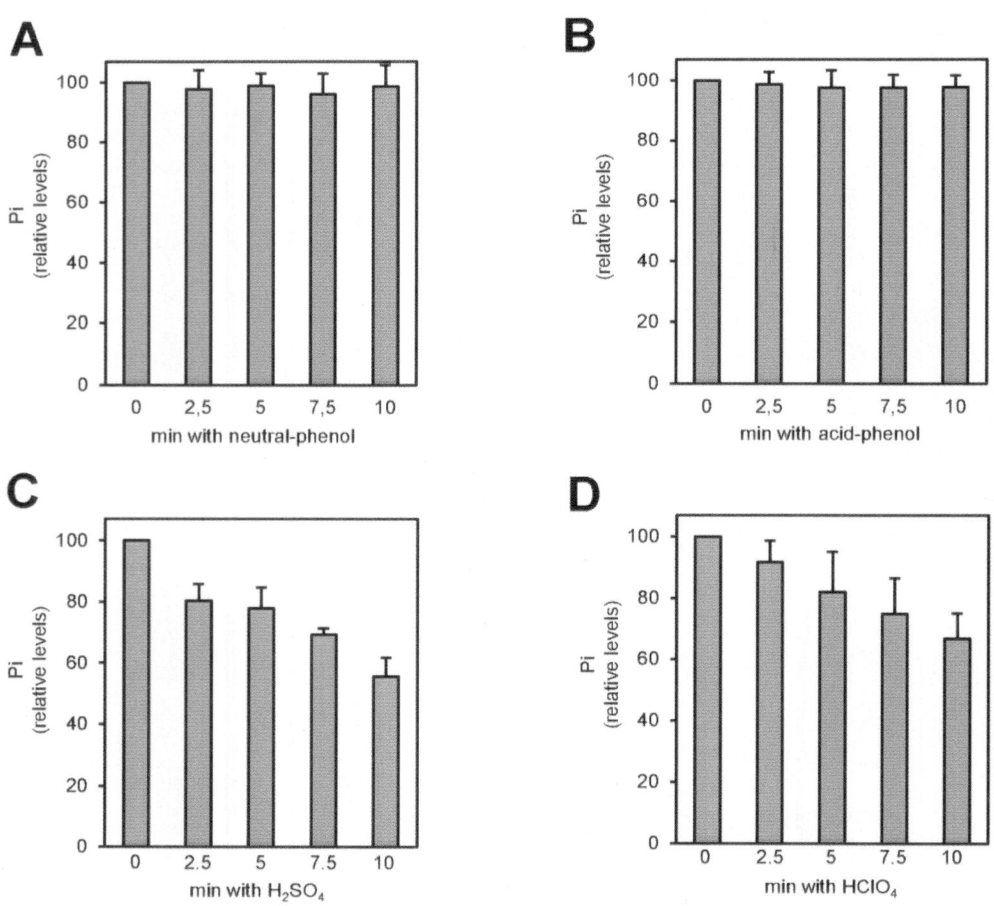

FIGURE 2: polyP is unstable when strong acids are employed during the extraction process. Relative amount of polyP after treatment with the different extraction solutions. Commercial polyP (4 μg) was incubated at different times with the different extraction solutions: **(A)** neutral-phenol at 4°C, **(B)** acid-phenol at 4°C, **(C)** 1 M H_2SO_4 at room temperature and **(D)** 1 M $HClO_4$ at 4°C. The mixes were neutralized, purified using affinity columns and the polyP eluted with MilliQ water. polyP amount was determined from the amount of Pi produced upon treatment with rPpx1. The graphs represent the percentage of polyP relative to time zero of each condition. Mean ± SEM from 3 independent experiments is shown.

in a time-dependent decrease in the amount of measured polyP, which was reduced up to one half of the untreated aliquot in as little as 10 min. This implies that some commonly used extraction methods would result in polyP degradation at this step, with the consequent impact in the final quantification.

The purification method strongly influences the size and amount of recovered polyP

Two of the most frequently used methods for polyP purification upon extraction are affinity chromatography (usually silica-based) and ethanol precipitation. Reported evidence indicated that different extraction methods may result in size fractionation of recovered polyP [29]. To test possible purification method-dependent differences in the length and recovery of polyP, we extracted polyP from two aliquots of yeast cultures by the phenol/chloroform (pH 8.0) method and samples were purified either by ethanol precipitation or by column chromatography (see Materials and Methods). Aliquots of the ethanol-precipitated material, as well as of the flow-through and the eluate of the chromatography were analyzed by polyacrylamide electrophoresis. As shown in Figure 3A, the chromatography step results in a dramatic size-dependent fractionation: polyP chains shorter than 60-80 residues were not retained by the column and appeared in the flow-through (which is usually discarded). This fractionation by size drastically decreased polyP recovery in the eluate which, upon quantification by degradation with rPpx1 and determination of the released phosphate, was estimated to be around 30-40% of the initial amount of polymer measured after the ethanol precipitation method (Figure 3B). Therefore, while extraction of polyP with the neutral phenol/chloroform method does not appear to introduce significant size selection bias (compare Figure 3A in this work with Figure 3 in reference [29]), subsequent column purification results in size fractionation leading to significant underestimation of the amount of cellular polyP.

Finally, we examined different factors that might affect the recovery in the process of precipitating polyP with ethanol. To this end, we added to the mixture various monovalent cations, magnesium or glycogen (as carrier). As deduced from Figure 3C, none of these components had significant effect on the recovery of polyP.

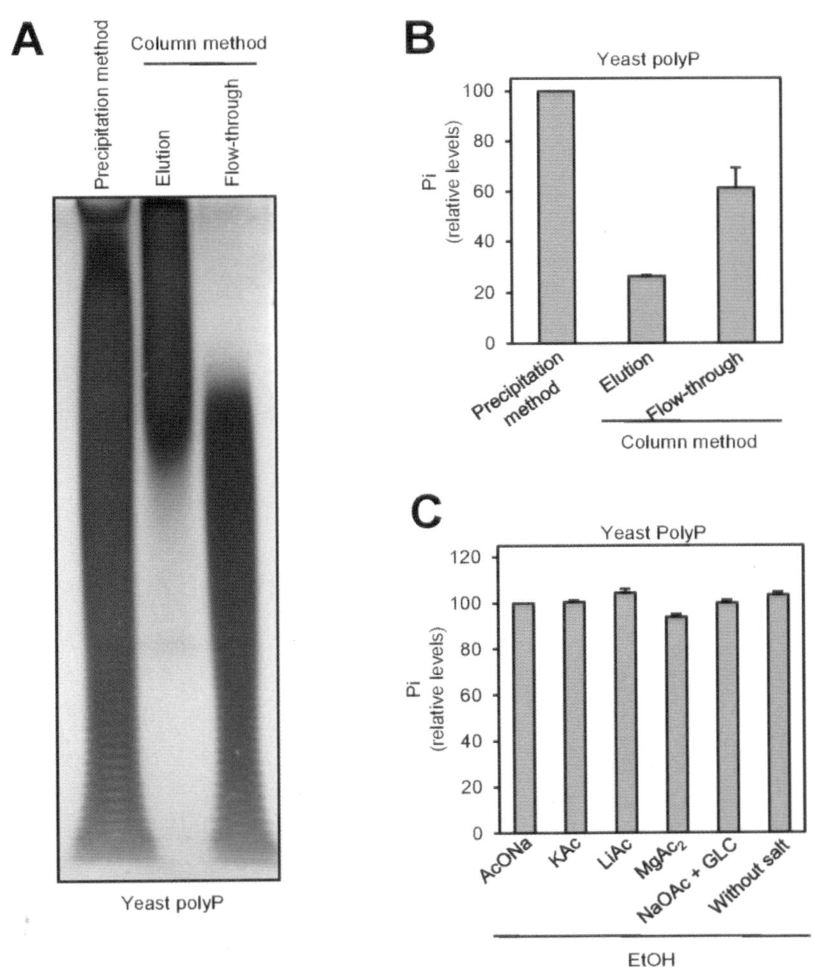

FIGURE 3: Ethanol precipitation method yields a broader spectrum of polyP sizes than affinity column purification. (A) PAGE and DAPI staining of polyP differently purified. polyP was extracted using the neutral phenol/chloroform procedure from a yeast pellet equivalent to 10^7 logarithmically growing yeast cells. The aqueous phase was treated with DNAse and RNAse solution, and purified by affinity columns or by ethanol precipitation. The resulting polyP fractions: precipitated (in the case of the ethanol) and eluted and flow-through (in the case of the affinity column) were analyzed by PAGE followed by DAPI staining. **(B)** Percentage of polyP relative to the purification method. polyP amount was determined using the same fractions obtained in panel A. The graph represents the Mean ± SEM from 3 independent experiments. **(C)** Relative amount of polyP obtained after the precipitation of polyP in the presence of different monovalent salts, divalent salts and a carrier. Aqueous phase from panel A was used. The graph represents the percentage of polyP relative to precipitation with NaOAc. Mean ± SEM from 3 independent experiments is shown. GLC, glycogen.

Characterizing the conditions for recombinant Ppx1-based polyP determinations

As described above, degradation of polyP by the exopolyphosphatase activity of yeast Ppx1 followed by measuring released phosphate is a common method for polyP determination. Therefore, it was considered of interest to test a number of practical aspects of this reaction suitable to be optimized. Ppx1 purified from yeast was reported to be fairly active up to 47°C [20], so we first examined the possibility to reduce the period of digestion with the enzyme by raising the incubation temperature. As shown in Figure S1A, the enzyme was highly active at 60°C but only for a relatively short period of time (20 min), followed by a sudden decline in activity, likely due to a thermal inactivation process (Figure S1B). In contrast, when incubated at 37°C, rPpx1 showed lesser, but constant activity during the experiment (60 min). As a result, incubation at 37°C resulted in a more exhaustive digestion of the polyP. However, since usually the amount of polyP to be measured is far smaller than the four µg used in this assay, it could be feasible to shorten the incubation time (and thus the duration of the assay) by raising the incubation temperature. We also tested the range of pH suitable for the use of the recombinant Ppx1 (Figure S1C). As observed, the enzyme was highly active in the pH range of 5.5 to 8.5, which fits well with the range reported for the native enzyme [20].

We observed that equivalents amounts of polyP either obtained by synthesis or purified from yeast (by neutral-phenol extraction followed by ethanol precipitation) displayed different degradation kinetics, being slower that of the polymer extracted from yeast (Figure 4A). Because certain charged molecules, such as heparin [30, 31] or

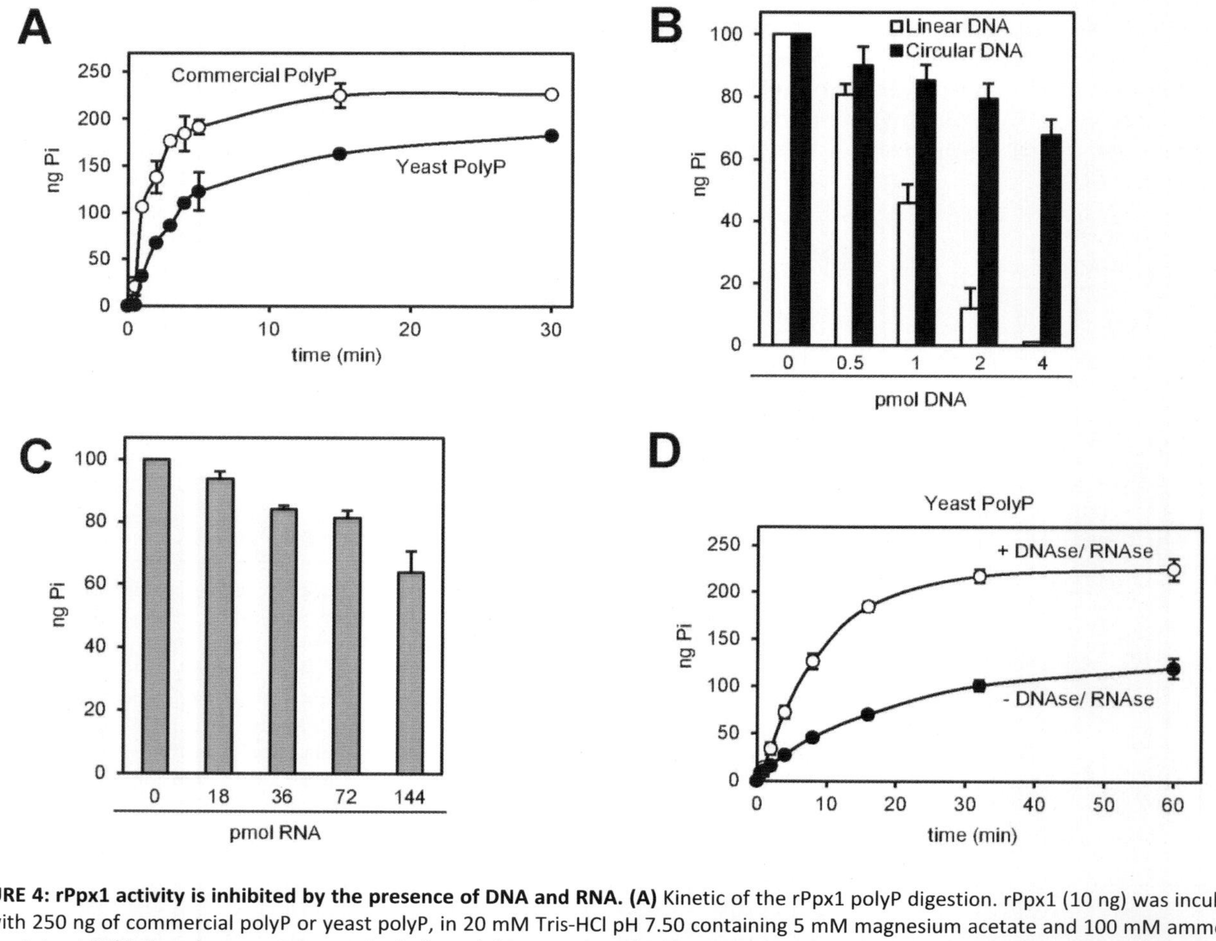

FIGURE 4: rPpx1 activity is inhibited by the presence of DNA and RNA. (A) Kinetic of the rPpx1 polyP digestion. rPpx1 (10 ng) was incubated with 250 ng of commercial polyP or yeast polyP, in 20 mM Tris-HCl pH 7.50 containing 5 mM magnesium acetate and 100 mM ammonium acetate at 37°C. Samples were taken at the indicated times to quantify the released Pi. Mean ± SEM from 3 independent experiments is shown. **(B)** rPpx1 activity on polyP in the presence of increasing amount of DNA. rPpx1 (10 ng) was incubated with 100 ng of commercial polyP in 20 mM Tris-HCl pH 7.50 containing 5 mM magnesium acetate and 100 mM ammonium acetate and at 37°C during 20 min with increasing concentrations of DNA (both circular and linear). The graph represents the released Pi in each condition. Mean ± SEM from 3 independent experiments is shown. **(C)** rPpx1 activity on polyP in the presence of increasing amount of RNA. Same experiment as in B, but with increasing concentrations of RNA. The graph represents the released Pi in each condition. Mean ± SEM from 3 independent experiments is shown. **(D)** Kinetics of rPpx1 polyP digestion in the presence of DNA and RNA. rPpx1 (1 ng) was incubated with 250 ng of yeast polyP previously treated or not with a DNAse/ RNAse solution in 20 mM Tris-HCl pH 7.50 containing 5 mM magnesium acetate and 100 mM ammonium acetate at 37°C. Samples were taken at the indicated times and the released Pi was quantified. Mean ± SEM from 3 independent experiments is shown.

spermidine [20], act as inhibitors of Ppx1 activity, we considered this behavior a symptom of the presence of inhibitory molecules. Ethanol precipitation is extensively used to concentrate nucleic acids, such as DNA and RNA, which are negatively charged. Given that the polyP samples used in this assay were not treated with RNAse or DNAse, and to test the possible interference of nucleic acids in the efficiency of polyP degradation, we incubated polyP with 10 ng (0.22 pmol) of rPpx1 in the presence of increasing amounts of linear or circular DNA. As showed in Figure 4B, linear DNA had a very strong inhibitory capacity, with total loss of enzymatic activity at an enzyme/DNA ratio of ~20. In contrast, circular DNA was less harmful to the process. Incubation with RNA resulted in some inhibition of the enzyme, although higher amounts of ribonucleic acid were needed. As documented in Figure 4D, treatment of the ethanol precipitated material extracted from yeast with a mixture of DNAse and RNAse not only accelerates the rate of the degradation reaction but also leads to a higher value for the amount of PolyP determined. These results would suggest a functional interaction between nucleic acids (mostly linear DNA) and Ppx1. Remarkably, poly(A) polymerase, an RNA-modifying enzyme, has been found to be inhibited by polyP [32]. Therefore, it must be concluded that accompanying nucleic acids can significantly interfere with the determination of polyP and that an excess of rPpx1 is recommend if it is likely the presence of nucleic acids in the assay.

rPpx1 degradation followed by free phosphate determination offers a better linear range than DAPI quantification

An alternative to the enzymatic degradation of polyP followed by quantification of the released phosphate is the electrophoretic resolution of polyP chains and subsequent staining of the polymers with DAPI. To compare the linearity of both methods we quantified a range of polyP from 0.5 to 8 μg by both enzymatic degradation and free Pi determination and DAPI staining of electrophoretically resolved polyP samples. As shown in Figure S2A the enzymatic method was linear within the entire tested range, whereas DAPI staining, due to saturation of the signal, was linear only for amounts of polyP up to 2 μg.

Experimental comparison of reported methods for polyP determination

Our precedent results indicate that the different steps in the determination of polyP from yeast cells can be a source of variation and could be at the origin of the large dispersions of values found in the literature (see Figure 1A). To directly test this possibility were carried out in our laboratory, for the same amount of cells, the determination of polyP using the procedures described by Neef and Kladde [33], Lichko and coworkers [23] and Hürliman *et al.* [25], in comparison with the one recently employed by our group [26]. All these methods have in common the enzymatic hydrolysis of polyP with exopolyphosphatase and chemical determination of the phosphate released, but differs in the upstream methodology.

As it can be seen in Figure 5A, the procedure described in Bru and coworkers, based in neutral phenol/chloroform extraction and ethanol precipitation after DNAse and RNAse treatment yielded an intracellular polyP concentration (expressed as Pi) of nearly 95 mM. This value is used here as reference. In contrast, the method described by Hürliman and coworkers gave the lower yield, likely because the extraction in this method is carried out by sulfuric acid treatment, which has a detrimental impact in the recovery of polyP (see Figure 2C), and because this step is followed by a Qiagen PCR column purification step, which is a second factor leading to polyP loss. Therefore, the accumulation of two detrimental steps in the entire process likely justifies the very low recovery observed for this method compared with the previous one (almost 20-fold lower). The method described by Neef and Kladde uses acid phenol and glass beads for extraction of polyP, which according to our results does not affect polyP recovery. However, these authors do not degrade nucleic acids with DNAses or RNAses prior enzymatic polyP degradation. Given the significant effect of these biomolecules on Ppx1 activity and polyP recovery (Figure 4), omission of this step could explain the somewhat lower recovery (about 30%) obtained with this procedure compared with that described in Bru *et al.* [26]. Finally, the method described by Lichko and coworkers is based in the extraction of polyP with perchloric acid, which also has a detrimental effect in polyP recovery (see Figure 2C). This might explain a polyP recovery of about one third compared with the reference value.

It is important to note that the procedure utilized for polyP determination can substantially affect the conclusions extracted from a given experiment. For instance, we have observed that the marked size-specificity of the chromatography column method used in the purification step can introduce a substantial bias when comparing different physiological conditions. As an example, we recently characterized the changes in cellular polyP in yeast cells exposed to sudden alkalization of the medium by using a column chromatography purification step [34]. Our results indicated a very rapid decrease, in a matter of few minutes, up to 15% of the initial polyP values. We were puzzled by such dramatic change, which was much faster than that obtained by depletion of Pi in the medium, and considered the possibility that it might be influenced by the determination procedure. Therefore, we repeated the experiment under identical conditions (using the neutral phenol extraction procedure) except that in one sample polyP was purified by affinity column and in the other the ethanol precipitation procedure was followed. As it can be observed in Figure S3, in untreated cells, utilization of the column step yields about 50% of the value obtained by the ethanol precipitation method. More importantly, according to the former method, alkalization of the medium leads to virtually complete depletion of polyP in about 2 min, whereas the later procedure yields a decrease of only 25% of the original value. Time-course electrophoretic analysis of the samples provides an explanation for such disparate results: high pH stress provokes rapid depletion of the largest polyP chains (the ones retained by the column). This result is not devoid of interest, as it might indicate that extracellular alkalization triggers the rapid activation of cellular endopolyphosphatases.

In conclusion, it is likely that most protocols described so far incur in one or more steps that negatively affect the yield of quantitated polyP and are at the basis of the dispersion of results obtained in many laboratories. Because there can be other (sometimes unavoidable) sources for variation in polyP content in yeast cells, such as the genetic background or the nature of the culture media, it is important to standardize as much as possible those factors that can be chosen at will. In this work we propose a protocol for total polyP extraction that integrates steps extracted from the literature with others shaped on the basis of the evidence presented here. This protocol is summarized in Figure 5B. Finally, our preliminary results (not shown) suggest that the protocol for polyP extraction and purification presented here can also be used for quantification of polyP from mammalian sources.

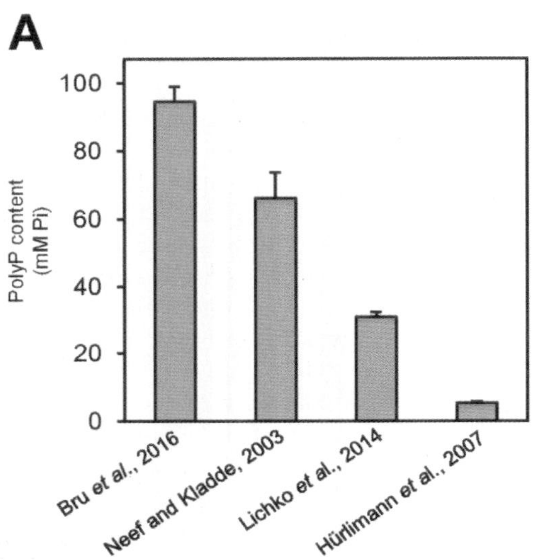

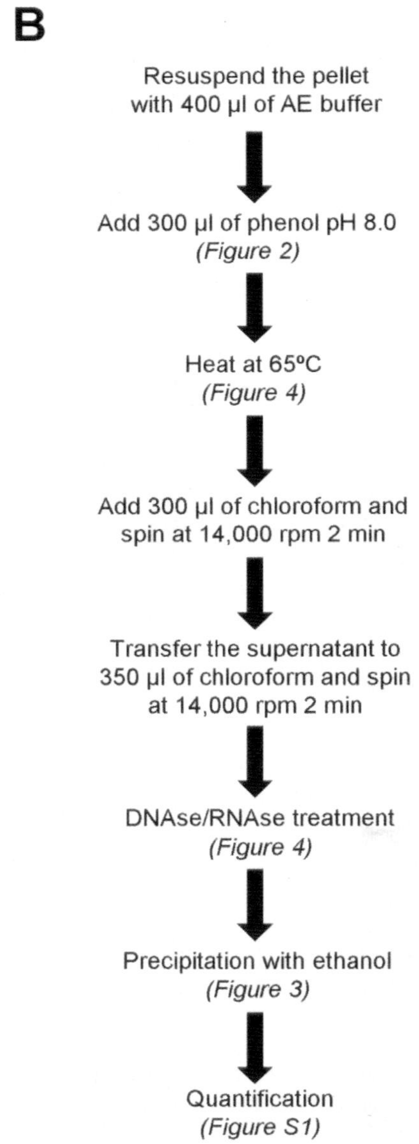

FIGURE 5: **Comparison of different polyP extraction and purification methods.** **(A)** Amount of polyP corresponding to 10^7 logarithmically growing yeast cells extracted and purified with the following methods: neutral-phenol/chloroform and ethanol precipitation [26], acid-phenol/chloroform and ethanol precipitation [33], perchloric acid [23], and sulfuric acid and affinity columns [25]. Mean ± SEM from 3 independent experiments is shown. **(B)** Scheme of the polyP extraction and purification protocol using neutral phenol and ethanol purification. In brackets appears the figure supporting this particular step. For details and tips see Materials and Methods.

MATERIALS AND METHODS

Yeast strains and growing conditions

Saccharomyces cerevisiae BY4741 yeast strain [35] was used in all the experiments. Yeast cells were grown in 5 ml of YPD medium (1% Yeast extract, 2% Peptone, 2% Dextrose) overnight at 30°C, diluted to OD (wavelength 660 nm) = 0.4 in 10 ml of fresh YPD and let them grow to OD = 1.5. The cells were harvested by centrifugation and instantly frozen by immersion in dry ice. In the alkaline stress experiments, cells were grown and harvested as in [36]. In all cases, pellets for polyP extraction were stored at -80°C.

PolyP extraction and purification methods

Method 1: Neutral phenol/chloroform and ethanol precipitation

1-2*10^7 exponentially growing cells were collected, the pellet was resuspended with 400 µl of AE buffer (50 mM sodium acetate (pH 5.3), 10 mM EDTA) at 4°C, transferred to a screw cap tube containing 300 µl phenol and 40 µl 10% SDS, mixed by inversion 4 times, vortexed 5 sec to homogenize, incubated at 65°C for 5 min and chilled for 1 min on ice. Three-hundred µl of chloroform were added, mixed by inversion 4 times, vortexed 5 sec to homogenize and centrifuged at room temperature for 2 min at 13,000 g. The top aqueous phase (around 450 µl) containing the polyP was transferred to a prepared 1.5 ml screw cap tube containing 350 µl chloroform (it is important not to carry over any phenol during pipetting by avoid touching the bottom phase or the white protein containing interphase), mixed by inversion 4 times, vortexed 5 sec to homogenize, centrifuged at room temperature for 2 min at 13,000 g and the aqueous phase was transferred to a new 1.5 ml microcentrifuge tube (it is important not to carry over any phenol). 2 µl of RNAse A 10 mg/ml (Sigma R6513) and 2 µl of DNAse I 10 mg/ml (Applichem, A3778.0100) were added, incubated 1 h at 37°C, transferred to a pre-cold at -20°C 1.5 ml microcentrifuge tube containing 1 ml of absolute ethanol and 40 µl of 3 M sodium acetate (pH 5.3), leaved 3 h at -20°C to precipitate polyP and centrifuged for 20 min at 13,000 g at 4°C. The supernatant was discarded by decantation, 500 µl of 70% ethanol were added, centrifuged for 5 min at 13,000 g at 4°C, the supernatant was discarded by decantation, centrifuged 1 min at 13,000 g and the last traces of ethanol were removed by pipetting. The tube was left open to dry the small translucent-white polyP pellet at room temperature for 5 min or until the pellet is completely dry. Finally, the polyP was resuspended in 50 µl of Milli-Q water. The polyP sample can be directly measured or stored at -20°C.

Method 2: acid phenol/chloroform and ethanol precipitation

PolyP extraction and purification was performed as described in [33]. Briefly, 10^7 logarithmically growing yeast cells were centrifuged at 9,000 g for 1 min. Pellets were resuspended in 350 µl of LETS buffer (0.1 M LiCl, 10 mM EDTA, 10 mM Tris pH 8.0, 0.5% SDS) and 350 µl of phenol pH 4.8, lysed with the addition of 500 µl glass beads and vortexed for 15 min at 4°C. After centrifugation of the mixture at 18,000 g for 15 min, the aqueous phase was transferred to a new Eppendorf tube and subjected to chloroform extraction as above. The supernatant was precipitated by adding 1 ml of absolute ethanol followed by overnight incubation at -20°C. The pellet containing the polyP was resuspended in 50 µl of 0.1% SDS, 1 mM EDTA, 10 mM, Tris-HCl, pH 7.4.

Method 3: sulfuric acid extraction and affinity columns purification

PolyP extraction and purification was performed as described in [22]. Briefly, 10^7 logarithmically growing yeast cells were centrifuged at 9,000 g for 1 min. The pellets were incubated with 50 µl of 1 M sulfuric acid for 5 min at room temperature. The suspension was neutralized by adding 50 µl of 2 M NaOH and cell debris removed by centrifugation. Finally, polyP was purified using Macherey-Nagel PCR affinity purification columns, and eluted in 50 µl of MilliQ water.

Method 4: perchloric acid extraction and purification

PolyP extraction and purification was performed as described in [23]. Briefly, 10^7 logarithmically growing yeast cells were centrifuged at 9,000 g for 1 min. The pellet was resuspended in 250 µl of 1 M perchloric acid, lysed by adding 500 µl of glass beads, vortex for 5 min at 4°C, and centrifuged at 18,000 g for 5 min at 4°C. The suspension was neutralized by adding 150 µl of KCE solution (1 M K_2CO_3, 5 mM EDTA), and cooled down in ice for 2 h. Cells debris was removed by centrifugation at 18,000 g. Finally, the supernatant was transferred to a new Eppendorf tube to be quantified.

polyP quantification

PolyP amount was determined as a measure of the inorganic phosphate produced by the complete digestion of the polyP by treatment with recombinant rPpx1 polyphosphatase protein [37]; briefly, *E. coli* BL21 cells transformed with pTrcPPX1 plasmid (kindly provided by A. Kornberg) containing yeast *PPX1* were grown over night at 37°C in 50 ml of LB (Luria Bertani medium), and the culture used as inoculum to a 500 ml culture in the same LB medium containing 0.5 mM IPTG as inducer. Growth was continued for 6 h at 25°C, *E. coli* cells were harvested by centrifugation, lysed and the recombinant proteins purified using Ni-nitrilotriacetic acid agarose (Qiagen, ID:30210). The purified polyP samples to be measured were diluted in 100 µl of a solution containing 20 mM Tris-HCl (pH 7.5), 100 mM NH_4 acetate, 5 mM Mg acetate, and 10 ng (measured by Bradford method) of rPpx1 for 1 h at 37°C. To quantify the released Pi, 86 µl of 28 mM ammonium heptamolybdate in 2.1 M sulfuric acid and 64 µl of 0.41 mM malachite green were added to the digested solution [38]. The OD_{595} was measured in a Synergy HT Elisa Reader and interpolation in a standard curve was used for obtaining absolute Pi amount values.

polyP detection by PAGE

Purified polyP was resolved electrophoretically using a 20% polyacrylamide gel (acrylamide 10:1 bisacrylamide) containing 7 M urea in TBE buffer pH 8.3, at 250 V/h for 5 h at 4°C. The dimensions of the gel were 200 mm height, 200 mm wide and 1.5 mm thick. The gel was stained by soaking it in the staining solution (25% methanol, 5% glycerol, 2 µg/ml DAPI, 50 mM Tris pH 10.5) for 30 min, and destained by soaking it in destaining solution (same as the staining solution but without DAPI) for 1 h. Finally, to visualize the polyP the gel was exposed to 254 nm UV light in Syngene G-BOX trans-illuminator.

Assessment of rPpx1 optimal activity

The optimal temperature, and pH value for rPpx1 activity was determined by incubating 1 ng of rPpx1 and 5 μg of commercial polyP (Shiba Regenetiss, Inc.) at several temperatures and pH values respectively in 20 mM Tris-HCl pH 7.50 (when temperature was varied) containing 5 mM magnesium acetate and 100 mM ammonium acetate. In the case of the pH optimal value assessment the incubation temperature was 37°C.

The influence of DNA or RNA presence in the activity of rPpx1 was assayed using 10 ng of rPpx1 and 100 ng of commercial polyP in the same buffer as above but adding increasing amounts of DNA (circular pUC19 vector or a lineal PCR fragment of 1.7 Kb) or RNA (yeast tRNA from Sigma).

rPpx1 activity is expressed as ng of Pi released/ min/ ng of enzyme.

ACKNOWLEDGEMENTS

We want to thanks A. Kornberg for the plasmid pTrcPPX1 and M. Pérez for invaluable technical assistance. This work has been funded by the Spanish Ministerio de Economía y Competitividad MINECO grants ref: BFU 2013-44189-P to JC ; BFU2011-30197-C3-01 and BFU2014-54591-C2-1-P to JA. SB was the recipient of a grant from the UIC.

CONFLICT OF INTEREST

The authors declare no conflict of interest.

REFERENCES

1. Rao NN, Gomez-Garcia MR, and Kornberg A (**2009**). Inorganic polyphosphate: essential for growth and survival. **Annu.Rev.Biochem.** 78: 605-647.

2. Azevedo C, Livermore T, and Saiardi A (**2015**). Protein polyphosphorylation of lysine residues by inorganic polyphosphate. **Mol.Cell.** 58(1): 71-82.

3. Angelova PR, Baev AY, Berezhnov AV, and Abramov AY (**2016**). Role of inorganic polyphosphate in mammalian cells: from signal transduction and mitochondrial metabolism to cell death. **Biochem.Soc.Trans.** 44(1): 40-45.

4. Kulakovskaya T, Vagabov V, and and Kulaev I (**2012**). Inorganic Polyphosphate in Industry, Agriculture and Medicine: Modern State and Outlook. **Process Biochemistry.** 47(1): 1-10.

5. Beauvoit B, Rigoulet M and Guerin P, and Canioni P (**1989**). Polyphosphates as a source of high energy phosphates in yeast mitochondria: a 31P NMR study. **FEBS Lett.** 252(1(2)): 17-21.

6. Castro CD, Meehan AJ, Koretsky AP, and Domach MM (**1995**). In situ 31P nuclear magnetic resonance for observation of polyphosphate and catabolite responses of chemostat-cultivated Saccharomyces cerevisiae after alkalinization. **Appl.Environ.Microbiol.** 61(12): 4448-4453.

7. Diaz J, Ingall E, Benitez-Nelson C, Paterson D, de Jonge MD, McNulty I, and Brandes JA (**2008**). Marine polyphosphate: a key player in geologic phosphorus sequestration. **Science.** 320(5876): 652-655.

8. Mullan A, Quinn JP, and McGrath JW (**2002**). A nonradioactive method for the assay of polyphosphate kinase activity and its application in the study of polyphosphate metabolism in Burkholderia cepacia. **Anal.Biochem.** 308(2): 294-299.

9. Tijssen JP, Beekes HW, and Van Steveninck J (**1982**). Localization of polyphosphates in Saccharomyces fragilis, as revealed by 4',6-diamidino-2-phenylindole fluorescence. **Biochim.Biophys.Acta.** 721(4): 394-398.

10. Aschar-Sobbi R, Abramov AY, Diao C, Kargacin ME, Kargacin GJ, French RJ and Pavlov E (**2008**). High sensitivity, quantitative measurements of polyphosphate using a new DAPI-based approach. **J.Fluoresc.** 18(5): 859-866.

11. Martin P, and Van Mooy BA (**2013**). Fluorometric quantification of polyphosphate in environmental plankton samples: extraction protocols, matrix effects, and nucleic acid interference. **Appl.Environ.Microbiol.** 79(1): 273-281.

12. Omelon S, Georgiou J, and Habraken W (**2016**). A cautionary (spectral) tail: red-shifted fluorescence by DAPI-DAPI interactions. **Biochem.Soc.Trans.** 44(1): 46-49.

13. Kolozsvari B, Firth S, and Saiardi A (**2015**). Raman spectroscopy detection of phytic acid in plant seeds reveals the absence of inorganic polyphosphate. **Mol.Plant.** 8(5): 826-828.

14. Kolozsvari B, Parisi F, and Saiardi A (**2014**). Inositol phosphates induce DAPI fluorescence shift. **Biochem.J.** 460(3): 377-385.

15. Omelon S, Georgiou J, Variola F, and Dean MN (**2014**). Colocation and role of polyphosphates and alkaline phosphatase in apatite biomineralization of elasmobranch tesserae. **Acta Biomater.** 10(9): 3899-3910.

16. Kumble KD, and Kornberg A (**1995**). Inorganic polyphosphate in mammalian cells and tissues. **J.Biol.Chem.** 270(11): 5818-5822.

17. Ault-Riche D, Fraley CD, Tzeng CM, and Kornberg A (**1998**). Novel assay reveals multiple pathways regulating stress-induced accumulations of inorganic polyphosphate in Escherichia coli. **J.Bacteriol.** 180(7): 1841-1847.

18. Motomura K, Hirota R, Ohnaka N, Okada M, Ikeda T, Morohoshi T, Ohtake H, and Kuroda A (2011). Overproduction of YjbB reduces the level of polyphosphate in Escherichia coli: a hypothetical role of YjbB in phosphate export and polyphosphate accumulation. **FEMS Microbiol.Lett.** 320(1): 25-32.

19. Ruiz FA, Rodrigues CO, and Docampo R (**2001**). Rapid changes in polyphosphate content within acidocalcisomes in response to cell growth, differentiation, and environmental stress in Trypanosoma cruzi. **J.Biol.Chem.** 276(28): 26114-26121.

20. Wurst H, and Kornberg A (**1994**). A soluble exopolyphosphatase of Saccharomyces cerevisiae. Purification and characterization. **J.Biol.Chem.** 269(15): 10996-11001.

21. Seidlmayer LK, Gomez-Garcia MR, Blatter LA, Pavlov E, and Dedkova EN (**2012**). Inorganic polyphosphate is a potent activator of the mitochondrial permeability transition pore in cardiac myocytes. **J.Gen.Physiol.** 139(5): 321-331.

22. Werner TP, Amrhein N, and Freimoser FM (**2005**). Novel method for the quantification of inorganic polyphosphate (iPoP) in Saccharomyces cerevisiae shows dependence of iPoP content on the growth phase. **Arch.Microbiol.** 184(2): 129-136.

23. Lichko LP, Eldarov MA, Dumina MV, and Kulakovskaya TV (**2014**). PPX1 gene overexpression has no influence on polyphosphates in Saccharomyces cerevisiae. **Biochemistry (Mosc).** 79(11): 1211-1215.

24. Kornberg A (**1999**). Inorganic polyphosphate: a molecule of many functions. **Prog.Mol.Subcell.Biol.** 23: 1-18.

25. Hurlimann HC, Stadler-Waibel M, Werner TP, and Freimoser FM (**2007**). Pho91 Is a vacuolar phosphate transporter that regulates phosphate and polyphosphate metabolism in Saccharomyces cerevisiae. **Mol.Biol.Cell.** 18(11): 4438-4445.

26. Bru S, Martinez JM, Hernandez-Ortega S, Quandt E, Torres-Torronteras J, Marti RR, Canadell D, Arino J, Sharma S, Jimenez J, and Clotet J (**2016**). Polyphosphate is involved in cell cycle progression and genomic stability in Saccharomyces cerevisiae. **Mol. Microbiol.** 101(3): 367-80.

27. Harold FM (**1966**). Inorganic polyphosphates in biology: structure, metabolism, and function. **Bacteriol.Rev.** 30(4): 772-794.

28. Lichko L, Kulakovskaya T, Pestov N, and Kulaev I (**2006**). Inorganic polyphosphates and exopolyphosphatases in cell compartments of the yeast Saccharomyces cerevisiae under inactivation of PPX1 and PPN1 genes. **Biosci.Rep.** 26(1): 45-54.

29. Vagabov VM, Trilisenko LV, Kulakovskaya TV, and Kulaev IS (**2008**). Effect of a carbon source on polyphosphate accumulation in Saccharomyces cerevisiae. **FEMS Yeast Res.** 8(6): 877-882.

30. Kulaev IS, Andreeva NA, Lichko LP, and Kulakovskaya TV (**1997**). Comparison of exopolyphosphatases of different yeast cell compartments. **Microbiol.Res.** 152(3): 221-226.

31. Lichko LP, Kulakovskaya TV, and Kulaev IS (**2003**). Nuclear exopolyphosphatase of Saccharomyces cerevisiae is not encoded by the PPX1 gene encoding the major yeast exopolyphosphatase. **FEMS Yeast Res.** 3(1): 113-117.

32. Holbein S, Freimoser FM, Werner TP, Wengi A, and Dichtl B (**2008**). Cordycepin-hypersensitive growth links elevated polyphosphate levels to inhibition of poly(A) polymerase in Saccharomyces cerevisiae. **Nucleic Acids Res.** 36(2): 353-363.

33. Neef DW, and Kladde MP (**2003**). Polyphosphate loss promotes SNF/SWI- and Gcn5-dependent mitotic induction of PHO5. **Mol.Cell.Biol.** 23(11): 3788-3797.

34. Serra-Cardona A, Canadell D, and Ariño J (**2015**). Coordinate responses to alkaline pH stress in budding yeast. **Microbial Cell.** 2(6): 182-196.

35. Brachmann CB, Davies A, Cost GJ, Caputo E, Li J, Hieter P, and Boeke JD (**1998**). Designer deletion strains derived from Saccharomyces cerevisiae S288C: a useful set of strains and plasmids for PCR-mediated gene disruption and other applications. **Yeast.** 14(2): 115-132.

36. Canadell D, Gonzalez A, Casado C, and Arino J (**2015**). Functional interactions between potassium and phosphate homeostasis in Saccharomyces cerevisiae. **Mol.Microbiol.** 95(3): 555-572.

37. Wurst H, Shiba T, and Kornberg A (**1995**). The gene for a major exopolyphosphatase of Saccharomyces cerevisiae. **J.Bacteriol.** 177(4): 898-906.

38. Cogan EB, Birrell GB, and Griffith OH (**1999**). A robotics-based automated assay for inorganic and organic phosphates. **Anal.Biochem.** 271(1): 29-35.

10

Integrative modules for efficient genome engineering in yeast

Triana Amen[1,*] and Daniel Kaganovich[1,*]

[1] Department of Cell and Developmental Biology, Hebrew University of Jerusalem, Givat Ram, Jerusalem 91904, Israel.

* Corresponding Authors:

Daniel Kaganovich, E-mail: dan@cc.huji.ac.il;

Triana Amen, E-mail: amentri@gmail.com

ABSTRACT We present a set of vectors containing integrative modules for efficient genome integration into the commonly used selection marker loci of the yeast *Saccharomyces cerevisiae*. A fragment for genome integration is generated via PCR with a unique set of short primers and integrated into *HIS3*, *URA3*, *ADE2*, and *TRP1* loci. The desired level of expression can be achieved by using constitutive (*TEF1p*, *GPD1p*), inducible (*CUP1p*, *GAL1/10p*), and daughter-specific (*DSE4p*) promoters available in the modules. The reduced size of the integrative module compared to conventional integrative plasmids allows efficient integration of multiple fragments. We demonstrate the efficiency of this tool by simultaneously tagging markers of the nucleus, vacuole, actin, and peroxisomes with genomically integrated fluorophores. Improved integration of our new pDK plasmid series allows stable introduction of several genes and can be used for multi-color imaging. New bidirectional promoters (*TEF1p-GPD1p*, *TEF1p-CUP1p*, and *TEF1p-DSE4p*) allow tractable metabolic engineering.

***Keywords**: vector, bidirectional promoter, integrative plasmid, genetic integration, yeast, Saccharomyces cerevisiae.*

Abbreviations:

MCS – multiple cloning site;

VHL - von Hippel Landau tumor suppressor.

INTRODUCTION

Saccharomyces cerevisiae is an indispensable tool for high throughput studies of biological processes. The ever-increasing diversity of tools for yeast genome manipulation allows systematic examination of biological pathways in a highly physiological, cost-effective, and tractable manner [1-7]. However, attempts to introduce multiple genes into chromosomal loci often encounter complications due to decreased efficiency and cloning limitations, such as overlaps in restriction site usage amongst multiple inserts [7, 8]. In order to overcome the difficulties of multiple gene integrations we sought to expand the variety of yeast integration cassettes with a specially designed pDK vector set.

Several methods enable rapid gene introduction. The most common ones are ectopic plasmid expression and chromosomal integration. Ectopic expression via multicopy or centromeric plasmids is often faster and easier than integration, but poses problems due to highly inhomogeneous expression [9, 10]. Hence, multi-color imaging and metabolic engineering in yeast often require stable integration of genes. Genome integration via homologous recombination has advantages over ectopic expression: e. g. stable strains, controlled copy number, and uniform expression [9].

Several integration strategies are commonly employed. (1) Yeast integrative plasmids are introduced into the genome in linearized form [1]. The linearized fragment contains the desired insert and selection marker, but also substantial superfluous genetic material including the bacterial selection marker and replication origins [1]. These extraneous sequences may reduce integration efficiency [8]. Additionally, the customized insert must not contain the restriction site used for plasmid linearization, adding further limitations. (2) Another frequently used strategy is PCR amplification using extended primers with homology regions allowing integration into any locus. Since the primers provide relatively short homology regions, the integration efficiency is comparatively low with this approach as well [8]. Insertion into loci rather than selective markers does not pose a particular advantage since the selective marker locus still has to be present in the integrated fragment. (3) Recent advances in CRISPR/Cas9 genome engineering allow highly efficient introduction of multiple fragments with short homology regions and often without the need for a selective marker, however the strain must

TABLE 1. pDK plasmid series with integrative modules.

Module	Locus			
	HIS3	*URA3*	*TRP1*	*ADE2*
TEF1p-MCS-*TEF1t*	pDK-HT	pDK-UT	pDK-TT	pDK-AT
CUP1p-MCS-*TEF1t*	pDK-HC	pDK-UC	pDK-TC	pDK-AC
GAL1p-MCS(1)-*CYC1t* *GAL10p*-MCS(2)-*ADHt*	pDK-HGG	pDK-UG	pDK-TG	pDK-AG
TEF1p-MCS(1)-*CYC1t* *GPD1p*-MCS(2)-*ADHt*	pDK-HTG	pDK-UTG	pDK-TTG	pDK-ATG
TEF1p -MCS(1)-*CYC1t* *CUP1p*-MCS(2)-*ADHt*	pDK-HTC	pDK-UTC	pDK-TTC	pDK-ATC
TEF1p -MCS(1)-*CYC1t* *DSE4p*-MCS(2)-*ADHt*	pDK-HTD	pDK-UTD	pDK-TTD	pDK-ATD

also carry Cas9 and gRNA expressing cassettes [4, 11, 12]. (4) I-SceI-assisted integration has a significantly increased efficiency, but this approach also requires the additional expression of the meganuclease I-SceI locus [13]. (5) Plasmids carrying integrative cassettes excised by restriction with extended homology regions flanking the insert significantly improve integration efficiency [6, 9, 14]. We were inspired by this last strategy and decided to create a set of yeast integration cassettes that contain extended homology regions corresponding to a common selective marker locus, which eliminates the need to have a marker locus inside the cassette. An advantage of not having a selective marker locus inside the cassette reduces its size and potentially increases integration efficiency [8]. To reduce multiple integration time and to maximize available markers we constructed novel bidirectional promoter sets (Figure 1A).

The pDK series includes 24 plasmids which carry an integrative module for 4 common genetic markers (*HIS3*, *URA3*, *ADE2*, and *TRP1*). Constitutive (*TEF1p*, *GPD1p*), daughter specific (*DSE4p*) and inducible (*CUP1p* and bidirectional *GAL1/10p*) promoters flank multiple cloning sites. We also include 4 bidirectional promoters (*GAL1p-GAL10p*, *TEF1p-GPD1p*, *TEF1p-CUP1p*, and *TEF1p-DSE4p*) which allow multiple integrations (Figure 1, Table 1). PCR is carried out with short primers specific to a selectable marker in order to integrate the module (see Supplemental Table 1). The pDK set can be used for multi-color imaging, metabolic engineering, or any set of experiments that require stable genomic integration of one or more genes.

RESULTS AND DISCUSSION

Vector overview

We designed 24 pDK plasmids for stable expression of multiple genes in the yeast *S. cerevisiae* (Figure 1, Table 1). Vectors can be used for stable integration into 4 common *S. cerevisiae* selective marker loci *HIS3*, *URA3*, *ADE2*, and *TRP1*. Plasmid variations include 6 promoters (Figure 1C): Constitutive *TEF1p*, inducible *CUP1p*, inducible bidirectional *GAL1-10p* in opposing orientation, constitutive bidirectional *TEF1-GPD1p*, constitutive-inducible bidirectional *TEF1-CUP1p*, constitutive-daughter-specific bidirectional *TEF1-DSE4p* (bidirectional promoter sets have two multiple cloning sites (MCS) in opposing orientation). The markers are split into two parts (see plasmid construction for details) and are flanking the region containing a promoter, MCS, and a terminator (Figure 1A). The fragment is amplified with primers specific to the marker (Supplemental Table 1). The orientation of the split markers allows integration of amplified region in a traditional homologous recombination driven way, resulting in doubling of the marker region (Figure 1B).

Integrative modules allow stable and efficient integration of multiple inserts

pDK vectors were examined for integration efficiency, correct integration, and stability of the insert. 24 integrative modules of pDK series were transformed into the W303 strain. pDK integrative modules have comparable integration efficiencies (Supplemental Figure 1A). Although double integration is possible we recommend sequential integration and using bidirectional promoters to expedite the work flow. We also compared the integration efficiency to other integration strategies: (1) pRS series – conventional linearized integrative plasmids [1], (2) EasyClone strategy – extended homology regions, excisable marker and restriction based integration [6], and (3) extended primers - 45bp homology region. In general, extended homology region based strategies provide higher efficiency of integration. pDK series integration is more effective than pRS series, and is comparable with EasyClone efficiency, which also has a reduced insert size but relies on restriction based integration.

To evaluate integration stability, a GFP reporter was cloned into the pDK series. The stability of the integration was tested by avoiding selection for more than 80 generations and then counting fluorescent cells (Supplemental Figure 1B). Modules are integrated with 95 - 100% accuracy into the marker region as verified by PCR (Supplemental

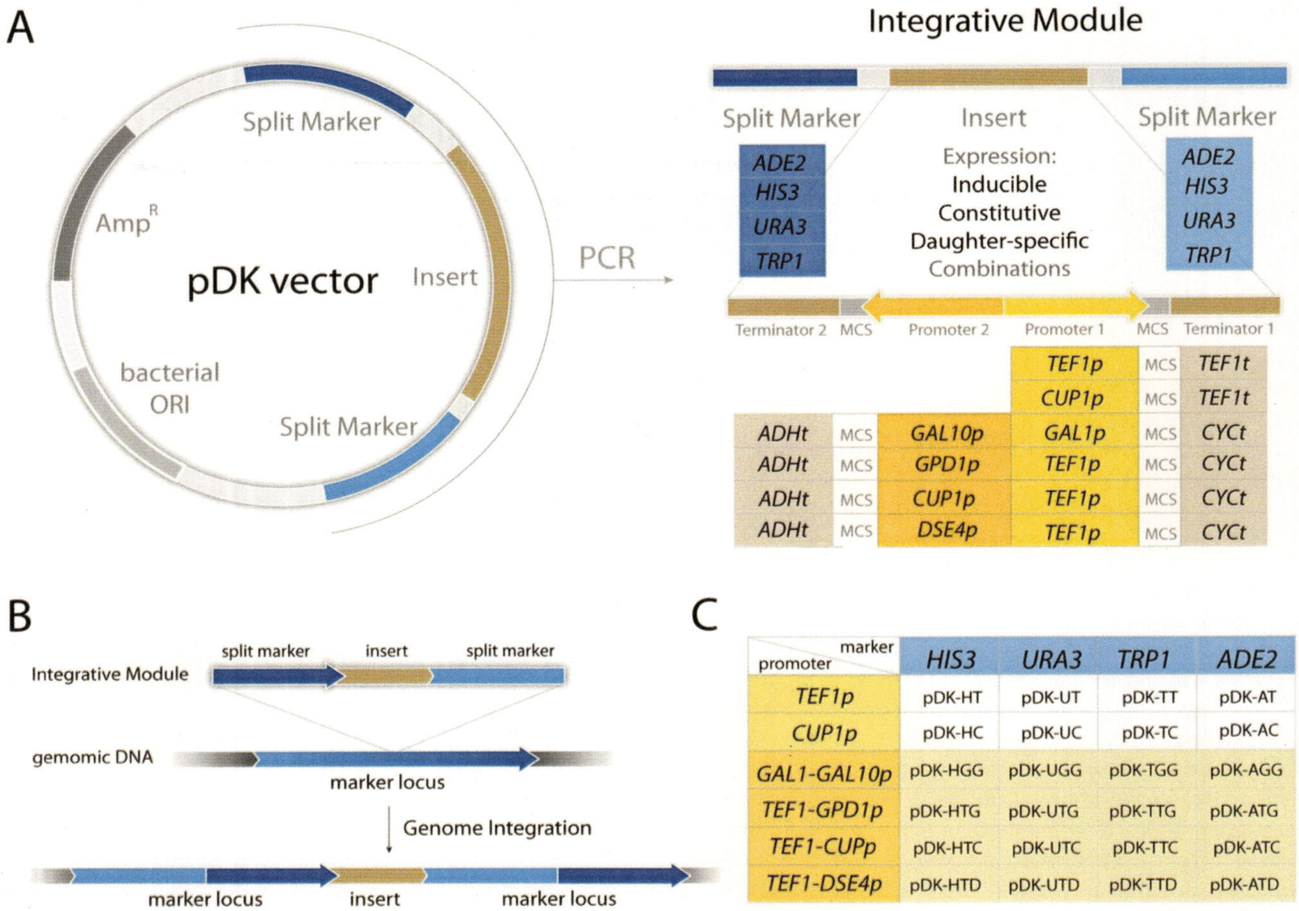

FIGURE 1: pDK vector series overview. (A) pDK vector with an integrative module flanked by a split marker. The customized insert is flanked by constitutive, inducible, or daughter-specific promoters. PCR is required for genetic integration. **(B)** Integration overview: the module is transformed into *S. cerevisiae* and inserted into the chromosome resulting in a marker locus duplication. **(C)** pDK vector set: constitutive, inducible, and four bi-directional promoter plasmids available for integration into four markers.

Figure 1B). Marker based integration is advantageous because it provides accurate targeting compared to the CRISPR/Cas9 strategy [15]. We also tested the stability of multiple integrations, since many of them carry same promoters and/or terminators, which can potentially lead to a loop-out of the fragments. Single and multiple integrations provide reliable homogenous expression levels under non-selective conditions (Supplemental Figure 1B, C).

Application of pDK plasmids to multi-color imaging reveals order of organelle inheritance

As proof of concept for multi-color imaging we constructed peroxisomal, nuclear, and actin [16] markers using the pDK vector series and used the strain for 3D time-lapse microscopy (4D imaging) (Figure 2) [17]. Nuclear localization signal fused to a far-red fluorophore was used to visualize the nucleus. Peroxisome localization signal - tripeptide SKL was fused to the mCherry C-terminus to visualize peroxisomes. We also used LifeAct fused to GFP to visualize actin in living yeast [16]. The vacuolar marker *VPH1* was endogenously tagged with mBFP or GFP. Peroxisomes are known to use actin cables for inheritance [18]. Inp2, an integral peroxisomal membrane protein, binds the Myo2 motor ensuring localization to the bud [19, 20]. Most of the organelles in yeast use actin cables for bud trafficking during division [21]. To determine whether organelle inheritance proceeds in a parallel or a serial way we examined the inheritance of the vacuole, peroxisomes, and the nucleus (Figure 2, Supplemental movie "wtdivision"). The nucleus itself is inherited in a microtubule dependent manner, however the astral microtubules are delivered on actin to the bud [22]. Multi-color imaging revealed that the first instance of peroxisome inheritance happens in parallel with the vacuole in the first 15 min of cell division (Figure 2B). Deletion of *INP2* abolishes peroxisome inheritance, which has no effect on the time of vacuole inheritance [19] (Figure 2C, Supplemental movie "inpdivision"). Interestingly, the nucleus is inherited prior to the end of cell division. The integrative vector series that we have developed allows for efficient integration of up to 8 markers, enabling us to image cellular compartments simultaneously.

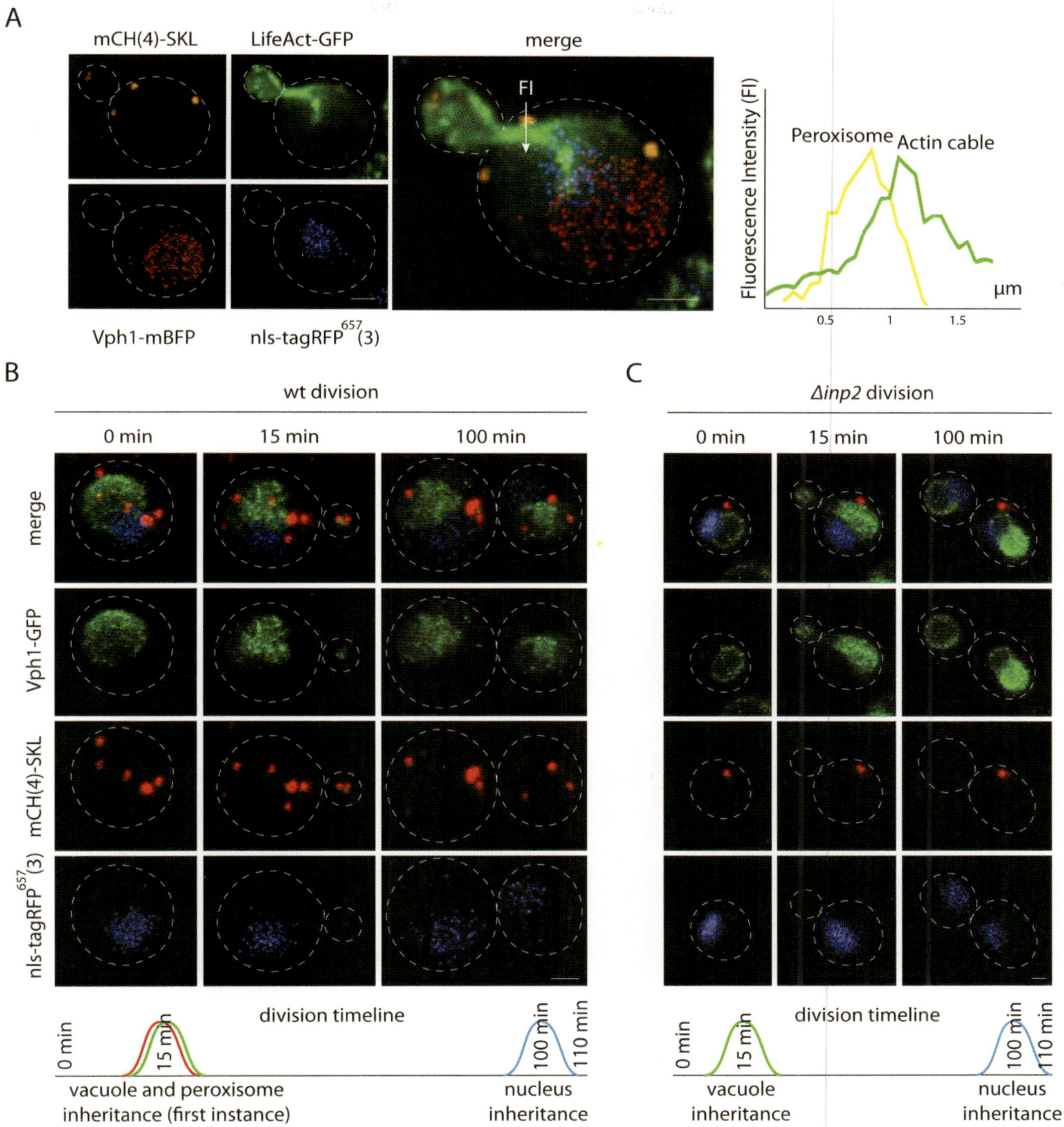

FIGURE 2: Proof of concept - Organelle inheritance order during division. **(A)** *In vivo* 4-color imaging of peroxisome (SKL signal C-terminally fused to 4 mCherry (mCH) on pDK-UT), actin (LifeAct fused to GFP on pDK-AT), nucleus (SV40 nls signal fused to 3 far-red fluorophores on pDK-HT), and vacuole (*VPH1* endogenously tagged with mBFP), the scale bar is 1 µm. The graph represents proximity of peroxisome to actin cable during division, Fluorescence Intensity (FI) was calculated along the x axes (arrow on the merge image) for red (peroxisome) and green (actin) channels. **(B)** The timeline of organelle inheritance in the wild type (wt) strain, scale bar is 1 µm, time points represent the average (n = 30) time of organelle inheritance from the start of division, strain carries SKL signal C-terminally fused to 4 mCherry (mCH) on pDK-UT module, SV40 nls signal fused to 3 far-red fluorophores on pDK-HT module, and *VPH1* endogenously tagged with GFP. **(C)** The timeline of the organelle inheritance in the *Δinp2* strain, scale bar is 1 µm.

Application of pDK plasmids to multi-copy integration

As proof of concept of exquisitely controlled inducible expression of a gene using multiple integrations we inserted GFP tagged VHL under inducible *CUP1* promoter in 4 marker loci. The von Hippel Landau tumor suppressor (VHL) has often been used as a model substrate to study protein aggregation in yeast [23]. Carefully controlling expression levels of integrated proteins is essential to many studies. In the case of VHL and similar proteins, increased protein concentration results in gradual protein self-association with subsequent decrease in the monomer fraction *in vitro*. This leads to the formation of protein inclusions or aggregates. To test if there is a correlation of misfolded protein concentration and inclusion formation *in vivo* we introduced 1-4 copies of GFP-VHL under *CUP1p* (Figure 3A, B) in yeast. We scored inclusion formation as a function of concentration and temperature. Aggregation strongly correlates with an increase in temperature, r(temperature) = 0.76 ($p < 0.05$), but has no significant correlation with VHL concentration, r(concentration) = 0.005 ($p = 0.98$). This could indicate a number of things. One interpretation of these data is that, regardless of the concentration, GFP-VHL forms a small percentage of overall protein content in inclusions (because of the abundance of misfolded and unstructured proteins in yeast). Another possibility is that without significant heat shock the quality control system is able to degrade a spectrum of misfolded VHL concentrations, but at higher temperatures it becomes uniformly inhibited. Additional experiments could make similar titrations of proteasome and chaperone activity. Together, these data are compelling as proof-of-concept for multi-gene integrations.

Bidirectional promoter sets

To improve the time of strain construction, we built versatile bidirectional promoter sets, allowing integration of two inserts under different promoters at the same time. The promoters are positioned in the opposite orientation (Figure 1A). Previously described *GAL1p/GAL10p* and 3 novel promoters: *TEF1p-GPD1p*, *TEF1p-CUP1p*, and *TEF1p-DSE4p*

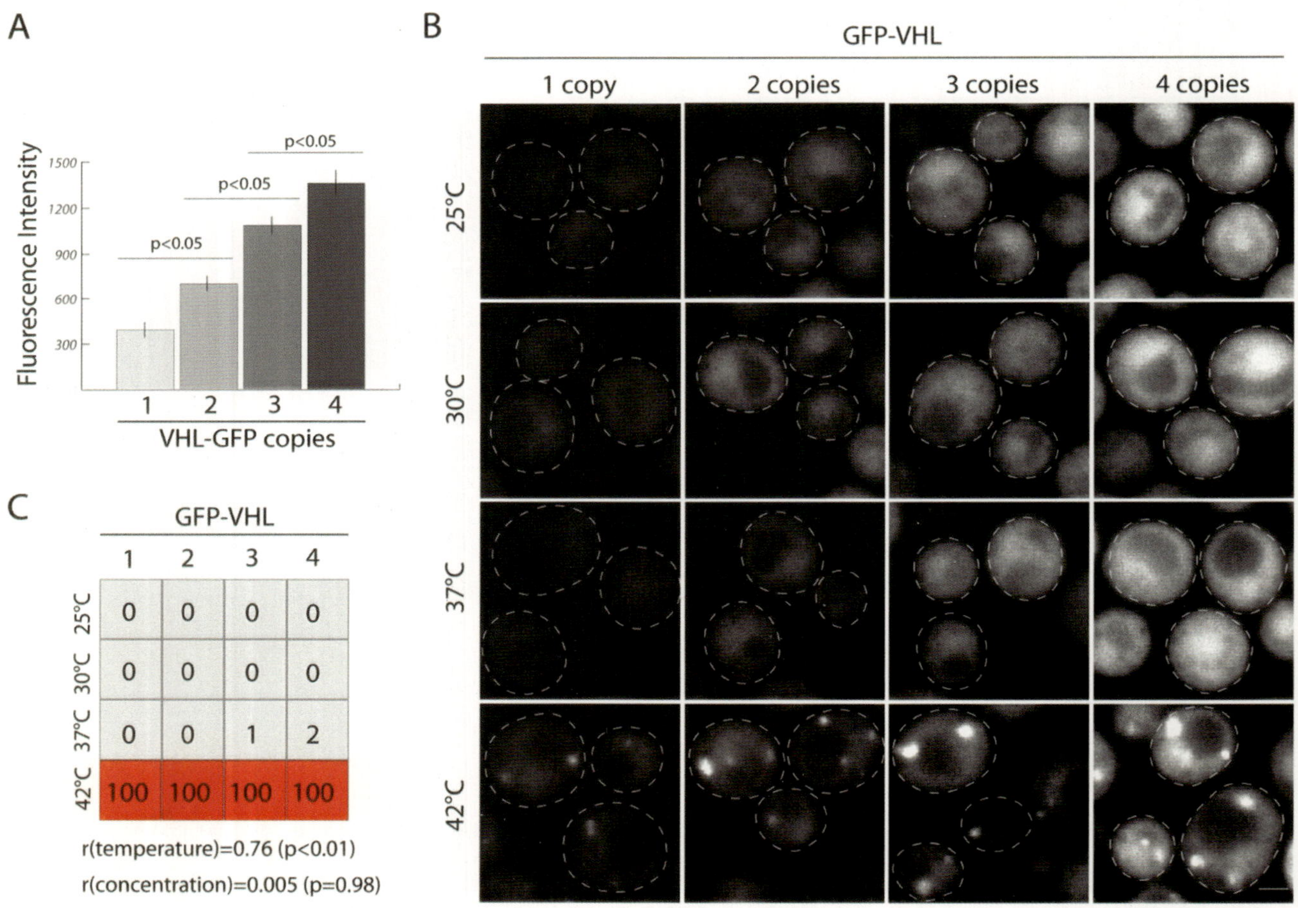

FIGURE 3: Proof of concept - Concentration does not correlate with inclusion formation. (A) Gradual increase in the amount of GFP-VHL according to integrated copies, Fluorescence Intensity was quantified in 30 single cells in the population, the average and standard errors are represented on the graph. **(B)** Confocal images of cells carrying 4 copies of GFP-VHL on pDK-AC, TC, UC, HC modules subjected to the range of temperatures for 1h, scale bar 1 µm. **(C)** Quantification of inclusion forming cells in the population according to the change in the temperature and the concentration, the numbers represent % of the cells with inclusions (n = 300), correlation coefficient of aggregation and temperature (r(temperature)) and aggregation and concentration (r(concentration)) is provided under the diagram.

were introduced into pDK series. To validate promoter sets we constructed GFP/mCherry reporters and visualized the cells during the log phase (Figure 4). Bidirectional promoters allow controlled inducible, semi-inducible, constitutive and daughter-specific expression (Figure 4A-D). Not only do they facilitate strain construction, bidirectional sets can also be used for cell sorting (daughter/mother cells), studying aging, or screening yeast genome collections [24, 25].

In summary, the pDK vector series allows for efficient multiple integrations and thus is a useful tool for multicolor imaging, metabolic engineering, controlled expression of genes of interest, and stable yeast strain production. We therefore hope that pDK vectors will be a useful tool for the yeast community.

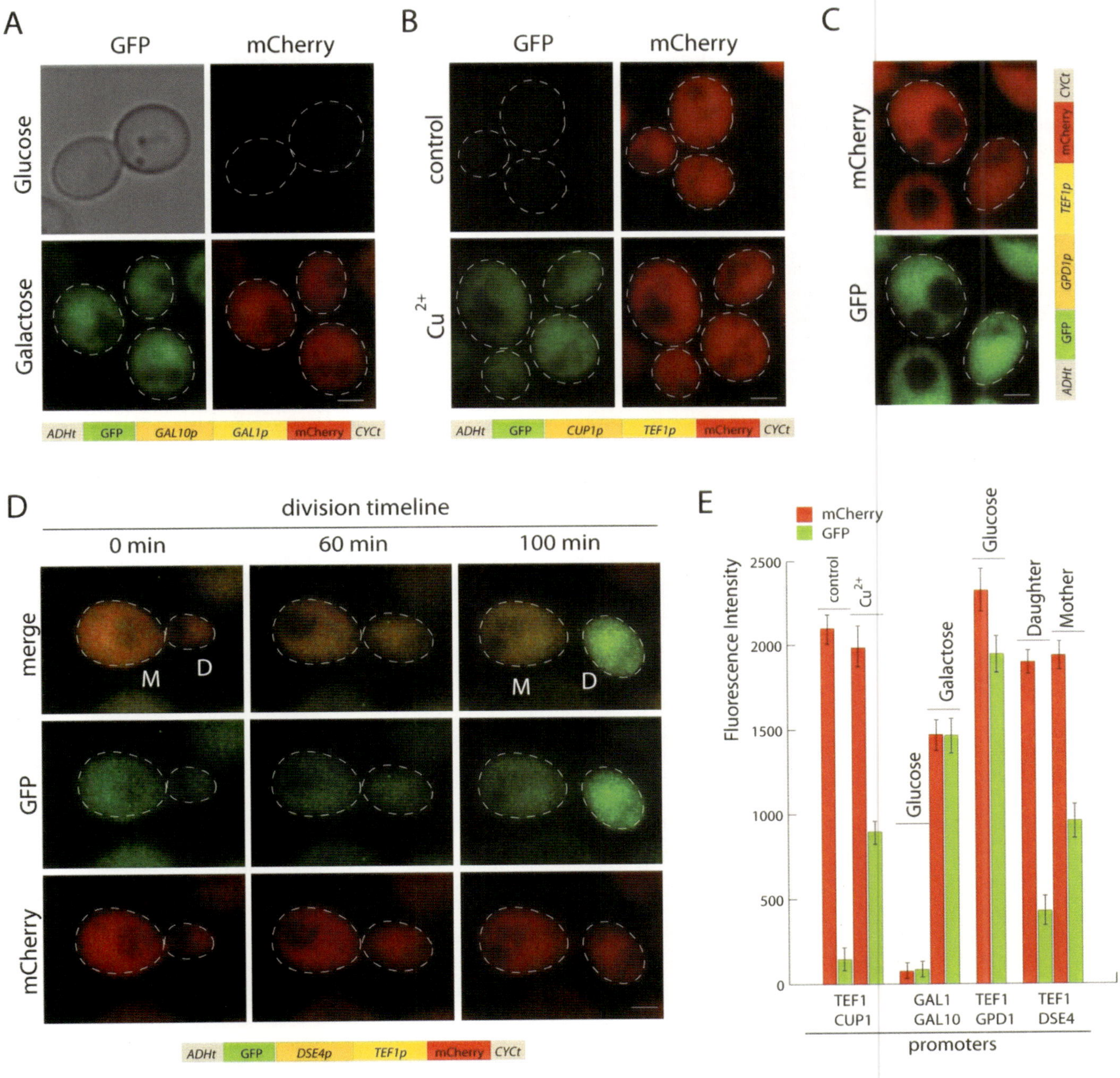

FIGURE 4: Bidirectional promoter reporters. (A) Galactose inducible bidirectional promoter, cells carrying pDK-HGG-GFP-mCherry were induced with galactose or grown on glucose for 6 hours, scale bar 1 µm. **(B)** Constitutive-inducible promoter, cells carrying pDK-HTC-GFP-mCherry were grown with and without (control) copper2+ for 4 hours, scale bar 1 µm. **(C)** Constitutive bidirectional promoter, cells carrying pDK-HTG-GFP-mCherry were grown on glucose containing medium to middle log phase, scale bar 1 µm. **(D)** Daughter-specific-constitutive bidirectional promoter, cells carrying pDK-HTD-GFP-mCherry module were grown on glucose containing medium on the microscope, frames are shown (0 min, 60 min, 100 min). **(E)** Comparison of fluorescence intensity of bidirectional reporters, the graph shows the average of fluorescence intensity of 20 cells and standard error.

MATERIALS AND METHODS

Strains and media

We used standard conditions for culturing yeast and bacterial cells [26]. Yeast were grown in the selective medium (1.7 g/l yeast nitrogen base without amino acids and ammonium sulfate (Difco Laboratories), 5 g/l ammonium sulphate, 0.77 g/l complete, 2 g/l amino acids supplement powder mix [27], 20 g/l glucose, and 20 g/l agar for the solid medium) or rich medium (20 g/l Peptone, 10 g/l yeast extract, 20 g/l glucose, and 20 g/l agar for the solid medium). Galactose induction was performed on selective medium supplemented with 20 g/l galactose instead of glucose for 6 hours. *CUP1p* induction was performed by addition of 50 μM Cupric Sulphate to the selective medium and growing cells for 4 hours. Plasmids were constructed using *Escherichia coli* strain DH5α. Yeast strain W303-1B (*MATα leu2-3,112 trp1-1 can1-100 ura3-1 ade2-1 his3-11,15*) was used for the experiments.

Plasmid construction

pUC19 plasmid [28] was used as a backbone for pDK vectors. Primers used in this study are summarized in supplemental tables: 1 - primers used for yeast integration, 2 - primers used for construction of plasmids. Markers loci (including promoters and terminators) were split in the following parts: *HIS3* (after 286bp of the ORF), *URA3* (after 420 bp of the ORF), *TRP1* (after 224 bp of the ORF), *ADE2* (after 462 bp of the ORF). Split fragments were amplified with: HIS3F1/R1 and HIS3F2/R2, URA3F1/R1 and URA3F2/R2, TRP1F1/R1 and TRP1F2/R2, ADE2F1/R1 and ADE2F2/R2, and cloned into pUC19's *Pci*I/*Hind*III and *Pfo*I/*Eco*RI, *Eco*RI/*Pfo*I and *Pci*I/*Hind*III, *Hind*III/*Pci*I and *Pfo*I/*Eco*RI, *Hind*III/*Pci*I and *Pfo*I/*Eco*RI respectively, resulting in the plasmids pDK-HH, pDK-UU, pDK-TT, and pDK-AA. Promoters and terminators (*TEF1p*, *CUP1p*, *TEF1t*) were amplified with: TEF1PF/R, CUP1PF/R, TEF1TF/R primers and cloned into pUC19's *Nde*I/*Eco*RI, *Nde*I/*Eco*RI, and *Sal*I/*Hind*III sites, respectively, resulting in pUC19-*TEF1p-TEF1t* and pUC19-*CUP1p-TEF1t* plasmids. *TEF1p-TEF1t*, *CUP1p-TEF1t* and *GAL1/10p-ADH1/CYC1t* modules were amplified from previously constructed pUC19-*TEF1p-TEF1t* and pUC19-*CUP1p-TEF1t*, and pESC-URA (Agilent) vectors using the primers: CUPTEFHHF, CUPTEFUUF, CUPTEFTTF, CUPTEFAAF and CUPTEFR for *TEF1p-TEF1* and, *CUP1p-TEF1t* modules, and GALHHF, GALUUF, GALTTF, GALAAF and GALR for *GAL1/10p-ADH1/CYC1t* module, and cloned using Gibson Assembly mix (NEB) into pDK-HH, pDK-UU, pDK-TT, and pDK-AA vectors linearized with *Eco*RI/*Sal*I restriction enzymes resulting in 12 pDK plasmids. Plasmid names contain the first letter of the marker followed by the first letter/s of the promoter/s (Table 1). *TEF1* promoter was amplified with TEFbF/R primers and cloned into pESC-URA *Age*I/*Bam*HI sites resulting in the pESC-TEFp plasmid. *CUP1*, *DSE4*, and *GPD1* promoters were amplified with CUPbF/R, DSEbF/R, and GPDbF/R primers and cloned into the pESC-TEFp vector into *Age*I/*Eco*RI sites resulting in pESC-TG, TC and TD plasmids carrying bidirectional promoters. Bidirectional promoters were subcloned using *Bam*HI/*Eco*RI into pDK-HG/UG/TG/AG, resulting in 12 bidirectional promoter sets (Table 1).

The nuclear cellular marker plasmid was constructed by cloning the SV40 nuclear localization signal fused to (tagRFP657)4 [29] into pDK-HT vector *Eco*RI/*Sma*I sites. The peroxisome marker was constructed by cloning the (mCherry)4-SKL sequence into pDK-UT plasmid. Actin was visualized via the LifeAct fragment fused to GFP [16], LifeAct-GFP was cloned into the pDK-AT vector. VHL plasmids were constructed by cloning the GFP-VHL sequence into pDK-HC, pDK-UC, pDK-AC, and pDK-TC vectors. pCfB2513 was a gift from Irina Borodina (Addgene plasmid # 67543).

For endogenous tagging we modified the pKT127 [30]

TABLE 2. Strains used in this study.

Strain	Genotype
BY4741	*MATa his3Δ0 leu2Δ0 met15Δ0 ura3Δ0*
W303α	*MATα leu2-3,112 trp1-1 can1-100 ura3-1 ade2-1 his3-11,15*
yDK - *inp2*, nuc, pex, vac	W303α *inp2Δ::KanMX his3-11,15::nls-(tagRFP657)$_3$::HIS3 ura3-1:(mCherry)$_3$-SKL::URA3 VPH1-GFP-HPH*
yDK - nuc, pex, vac	W303α *his3-11,15::nls-(tafRFP657)$_3$::HIS3 ura3-1:(mCherry)$_3$-SKL::URA3 VPH1-GFP-HPH*
yDK - nuc, pex, la, vac	W303α *his3-11,15::nls-(tafRFP657)$_3$::HIS3 ura3-1:(mCherry)$_3$-SKL::URA3 ade2-1::LifeAct-GFP::ADE2 VPH1-mBFP-HPH*
yDK - HHCTGV	W303α *his3-11,15::CUP1p-GFP-VHL-TEF1t::HIS3*
yDK – HHCTGV, UUCTGV	W303α *his3-11,15::CUP1p-GFP-VHL-TEF1t::HIS3 ura3-1::CUP1p-GFP-VHL-TEF1t::URA3*
yDK – HHCTGV, UUCTGV, TTCTGV	W303α *his3-11,15::CUP1p-GFP-VHL-TEF1t::HIS3 ura3-1::CUP1p-GFP-VHL-TEF1t::URA3 trp1-1::CUP1p-GFP-VHL-TEF1t::TRP1*
yDK – HHCTGV, UUCTGV, TTCTGV, AACTGV	W303α *his3-11,15::CUP1p-GFP-VHL-TEF1t::HIS3 ura3-1::CUP1p-GFP-VHL-TEF1t::URA3 trp1-1::CUP1p-GFP-VHL-TEF1t::TRP1 ade2-1::CUP1p-GFP-VHL-TEF1t::ADE2*

plasmid by inserting the mBFP sequence or GFP sequence into *PacI/AscI* sites and *HPH* marker sequence into *BglII/PmeI* sites. GFP reporters were constructing by cloning GFP into *SacI/XmaI* sites of pDK-HT/HC, and *Bam*HI/*XmaI* sites of pDK-HGG with primers GFPxR, GFPbF, GFPsF and subcloning the fragment containing the marker across different marker plasmids and bidirectional promoter ones. mCherry was amplified with primers CHeF and CHnR and cloned into the *EcoI/SpeI* sites of pDK-HGG-GFP, and subcloned to pDK-HTC-GFP, pDK-HTG-GFP, and pDK-HTD-GFP.

Strain construction

We introduced the *INP2* deletion using a PCR based deletion strategy [31] and primers delINPF/R. Gene deletions were verified by PCR. Strains with integrative modules were constructed by transforming yeast with a PCR fragment obtained from a corresponding plasmid with a set of primers listed in Table 1. For comparison experiments pRS plasmids were linearized in the marker locus prior to integration, pRS303 and pRS306 with *PstI* restriction enzyme, and pRS304 with *PmlI* enzyme.

VPH1 was tagged using modified pKT127 plasmid and eVPHF/R primers.

Protocol for yeast integration

The fragment for genomic integration is generated via PCR with primers listed in Table 2 using the following parameters (95°C- 5′, [95°C- 30″, 62°C- 30″ (increment 0.8°C per cycle), 72°C- X min (X = length of the fragment in kb)]25 cycles, 72°C-5′), and high fidelity polymerase generating blunt-end products, e.g. KAPA HiFi DNA Polymerase (KAPA Biosystems). A PCR protocol with fixed primer binding time can also be used. Up to 2 µg of PCR product is transformed using LiAc/PEG transformation [32] with modifications – 50 µl of DMSO is added prior to heat shock, heat shock time is reduced to 15 min at 42°C. 20 random clones carrying integrative modules were verified by PCR.

Stability of integration

Yeast strains were grown in rich medium in three replicates. The culture was diluted 1/100 every 24 hours for 10 days. The culture was analyzed by reporter fluorescence on the first and tenth day.

Confocal Microscopy

For imaging yeast cells were grown to mid-log phase and seeded on concanavalin A (Sigma) coated 4-well microscope plates (IBIDI). For induction galactose rich media (division experiments, Figure 2) or minimal media with 50 µm Cu^{2+} (Figure 3) were used. Copper induced cells were grown in 25°C to mid-log phase and then incubated for 1h at indicated temperatures (30, 37, 42°C). Confocal 3D images and movies were acquired using a dual point-scanning Nikon A1R-si microscope equipped with a PInano Piezo stage (MCL), using a 60 x Plan-Apo VC oil objective NA 1.40. Movies were acquired in resonant-scanning mode. Image processing was performed using NIS-Elements software.

Statistics

The experiments were repeated at least 3 times. Multiple correlation coefficients were calculated for 3 variables (aggregation, temperature, concentration), with subsequent regression analysis to determine p-values. Standard comparisons were performed using t-test.

ACKNOWLEDGEMENTS

We thank members of the Jerusalem Brain Community for their support. This work was supported by the European Research Council under the European Union's Seventh Framework Programme (FP/2007-2013)/ERC-StG2013 337713 DarkSide starting grant, as well as an Israel Science Foundation Grant ISF 843/11; a German Israel Foundation Grant GIFI-1201-242.13/2012; a Niedersachsen-Israel Research Program grant, an Abisch-Frenkel Foundation grant, and a joint Israel-Italy cooperation grant from the Israeli Ministry of Science, Technology, and Space. TA was funded by a Jerusalem Brain Community Doctoral Fellowship and by the Alexander Grass Center for Bioengineering.

CONFLICT OF INTEREST

The authors declare no conflicts of interest.

REFERENCES

1. Sikorski RS, Hieter P (**1989**). A System of Shuttle Vectors and Yeast Host Strains Designed for Efficient Manipulation of DNA in Saccharomyces-Cerevisiae. **Genetics** 122(1): 19-27.

2. Fang F, Salmon K, Shen MW, Aeling KA, Ito E, Irwin B, Tran UP, Hatfield GW, Da Silva NA, Sandmeyer S (**2011**). A vector set for systematic metabolic engineering in Saccharomyces cerevisiae. **Yeast** 28(2): 123-136.

3. Alberti S, Gitler AD, Lindquist S (**2007**). A suite of Gateway (R) cloning vectors for high-throughput genetic analysis in Saccharomyces cerevisiae. **Yeast** 24(10): 913-919.

4. Ronda C, Maury J, Jakociunas T, Jacobsen SAB, Germann SM, Harrison SJ, Borodina I, Keasling JD, Jensen MK, Nielsen AT (**2015**). CrEdit: CRISPR mediated multi-loci gene integration in Saccharomyces cerevisiae. **Microb Cell Fact** 14:97.

5. Agmon N, Mitchell LA, Cai YZ, Ikushima S, Chuang J, Zheng A, Choi WJ, Martin JA, Caravelli K, Stracquadanio G, Boeke JD (**2015**). Yeast Golden Gate (yGG) for the Efficient Assembly of S-cerevisiae Transcription Units. **Acs Synth Biol** 4(7): 853-859.

6. Stovicek V, Borja GM, Forster J, Borodina I (**2015**). EasyClone 2.0: expanded toolkit of integrative vectors for stable gene expression in industrial Saccharomyces cerevisiae strains. **J Ind Microbiol Biot** 42(11): 1519-1531.

7. Leite FCB, dos Anjos RSG, Basilio ACM, Leal GFC, Simoes DA, de Morais MA (**2013**). Construction of integrative plasmids suitable for genetic modification of industrial strains of Saccharomyces cerevisiae. **Plasmid** 69(1): 114-117.

8. Kung SH, Retchless AC, Kwan JY, Almeida RPP (**2013**). Effects of DNA Size on Transformation and Recombination Efficiencies in Xylella fastidiosa. **Appl Environ Microb** 79(5): 1712-1717.

9. Jensen NB, Strucko T, Kildegaard KR, David F, Maury J, Mortensen UH, Forster J, Nielsen J, Borodina I (**2014**). EasyClone: method for iterative chromosomal integration of multiple genes in Saccharomyces cerevisiae. **Fems Yeast Res** 14(2): 238-248.

10. Jahn M, Vorpahl C, Hubschmann T, Harms H, Muller S (**2016**). Copy number variability of expression plasmids determined by cell sorting and Droplet Digital PCR. **Microb Cell Fact** 15(1): 211.

11. Sikorski RS, Hieter P (**1989**). A system of shuttle vectors and yeast host strains designed for efficient manipulation of DNA in Saccharomyces cerevisiae. **Genetics** 122(1): 19-27.

12. Horwitz AA, Walter JM, Schubert MG, Kung SH, Hawkins K, Platt DM, Hernday AD, Mahatdejkul-Meadows T, Szeto W, Chandran SS, Newman JD (**2015**). Efficient Multiplexed Integration of Synergistic Alleles and Metabolic Pathways in Yeasts via CRISPR-Cas. **Cell Syst** 1(1): 88-96.

13. Shi SB, Liang YY, Zhang MZM, Ang EL, Zhao HM (**2016**). A highly efficient single-step, markerless strategy for multi-copy chromosomal integration of large biochemical pathways in Saccharomyces cerevisiae. **Metab Eng** 33:19-27.

14. Kuijpers NGA, Chroumpi S, Vos T, Solis-Escalante D, Bosman L, Pronk JT, Daran JM, Daran-Lapujade P (**2013**). One-step assembly and targeted integration of multigene constructs assisted by the I-SceI meganuclease in Saccharomyces cerevisiae. **Fems Yeast Res** 13(8): 769-781.

15. Taxis C, Knop M (**2006**). System of centromeric, episomal, and integrative vectors based on drug resistance markers for Saccharomyces cerevisiae. **Biotechniques** 40(1): 73-78.

16. Cho SW, Kim S, Kim Y, Kweon J, Kim HS, Bae S, Kim JS (**2013**). Analysis of off-target effects of CRISPR/Cas-derived RNA-guided endonucleases and nickases. **Genome Research** 24(1): 132-141.

17. Riedl J, Crevenna AH, Kessenbrock K, Yu JH, Neukirchen D, Bista M, Bradke F, Jenne D, Holak TA, Werb Z, Sixt M, Wedlich-Soldner R (**2008**). Lifeact: a versatile marker to visualize F-actin. **Nat Methods** 5(7): 605-607.

18. Spokoini R, Shamir M, Keness A, Kaganovich D (**2013**). 4D Imaging of Protein Aggregation in Live Cells. **Jove-J Vis Exp** (74).

19. Hoepfner D, van den Berg M, Philippsen P, Tabak HF, Hettema EH (**2001**). A role for Vps1p, actin, and the Myo2p motor in peroxisome abundance and inheritance in Saccharomyces cerevisiae. **J Cell Biol** 155(6): 979-990.

20. Fagarasanu A, Fagarasanu M, Eitzen GA, Aitchison JD, Rachubinski RA (**2006**). The peroxisomal membrane protein Inp2p is the peroxisome-specific receptor for the myosin V motor Myo2p of Saccharomyces cerevisiae. **Dev Cell** 10(5): 587-600.

21. Saraya R, Cepinska MN, Kiel JAKW, Veenhuis M, van der Klei IJ (**2010**). A conserved function for Inp2 in peroxisome inheritance. **Bba-Mol Cell Res** 1803(5): 617-622.

22. Knoblach B, Rachubinski RA (**2015**). Sharing the cell's bounty - organelle inheritance in yeast. **J Cell Sci** 128(4): 621-630.

23. Beach DL, Thibodeaux J, Maddox P, Yeh E, Bloom K (**2000**). The role of the proteins Kar9 and Myo2 in orienting the mitotic spindle of budding yeast. **Curr Biol** 10(23): 1497-1506.

24. Brock KP, Abraham AC, Amen T, Kaganovich D, England JL (**2015**). Structural Basis for Modulation of Quality Control Fate in a Marginally Stable Protein. **Structure** 23(7): 1169-1178.

25. Li A, Liu Z, Li Q, Yu L, Wang D, Deng X (**2008**). Construction and characterization of bidirectional expression vectors inSaccharomyces cerevisiae. **Fems Yeast Res** 8(1): 6-9.

26. Afonso B, Silver PA, Ajo-Franklin CM (**2010**). A synthetic circuit for selectively arresting daughter cells to create aging populations. **Nucleic Acids Research** 38(8): 2727-2735.

27. Sherman F (**2002**). Getting started with yeast. **Method Enzymol** 350:3-41.

28. Baryshnikova A, Costanzo M, Dixon S, Vizeacoumar FJ, Myers CL, Andrews B, Boone C (**2010**). Synthetic Genetic Array (SGA) Analysis in Saccharomyces cerevisiae and Schizosaccharomyces pombe. 470:145-179.

29. Yanischperron C, Vieira J, Messing J (**1985**). Improved M13 Phage Cloning Vectors and Host Strains - Nucleotide-Sequences of the M13mp18 and Puc19 Vectors. **Gene** 33(1): 103-119.

30. Morozova KS, Piatkevich KD, Gould TJ, Zhang JH, Bewersdorf J, Verkhusha VV (**2010**). Far-Red Fluorescent Protein Excitable with Red Lasers for Flow Cytometry and Superresolution STED Nanoscopy. **Biophys J** 99(2): L13-L15.

31. Sheff MA, Thorn KS (**2004**). Optimized cassettes for fluorescent protein tagging in Saccharomyces cerevisiae. **Yeast** 21(8): 661-670.

32. Baudin A, Ozierkalogeropoulos O, Denouel A, Lacroute F, Cullin C (**1993**). A Simple and Efficient Method for Direct Gene Deletion in Saccharomyces-Cerevisiae. **Nucleic Acids Res** 21(14): 3329-3330.

33. Gietz RD, Schiestl RH, Willems AR, Woods RA (**1995**). Studies on the Transformation of Intact Yeast-Cells by the Liac/S-DNA/Peg Procedure. **Yeast** 11(4): 355-360.

11

Thiol trapping and metabolic redistribution of sulfur metabolites enable cells to overcome cysteine overload

Anup Arunrao Deshpande[1,#], Muskan Bhatia[1,#], Sunil Laxman[2,*], Anand Kumar Bachhawat[1,*]

[1] Department of Biological Sciences, Indian Institute of Science Education and Research (IISER Mohali), S.A.S. Nagar, Punjab 140306, India.

[2] Institute for Stem Cell Biology and Regenerative Medicine (inStem), NCBS campus, Bangalore 560065, India.

[#] These authors contributed equally to the work.

* Corresponding Authors:

Anand Kumar Bachhawat, E-mail: anand@iisermohali.ac.in

Sunil Laxman, E-mail: sunil@instem.res.in

ABSTRACT Cysteine is an essential requirement in living organisms. However, due to its reactive thiol side chain, elevated levels of intracellular cysteine can be toxic and therefore need to be rapidly eliminated from the cellular milieu. In mammals and many other organisms, excess cysteine is believed to be primarily eliminated by the cysteine dioxygenase dependent oxidative degradation of cysteine, followed by the removal of the oxidative products. However, other mechanisms of tackling excess cysteine are also likely to exist, but have not thus far been explored. In this study, we use *Saccharomyces cerevisiae*, which naturally lacks a cysteine dioxygenase, to investigate mechanisms for tackling cysteine overload. Overexpressing the high affinity cysteine transporter, *YCT1*, enabled yeast cells to rapidly accumulate high levels of intracellular cysteine. Using targeted metabolite analysis, we observe that cysteine is initially rapidly interconverted to non-reactive cystine *in vivo*. A time course revealed that cells systematically convert excess cysteine to inert thiol forms; initially to cystine, and subsequently to cystathionine, S-Adenosyl-L-homocysteine (SAH) and S-Adenosyl L-methionine (SAM), in addition to eventually accumulating glutathione (GSH) and polyamines. Microarray based gene expression studies revealed the upregulation of arginine/ornithine biosynthesis a few hours after the cysteine overload, and suggest that the non-toxic, non-reactive thiol based metabolic products are eventually utilized for amino acid and polyamine biogenesis, thereby enabling cell growth. Thus, cells can handle potentially toxic amounts of cysteine by a combination of thiol trapping, metabolic redistribution to non-reactive thiols and subsequent consumption for anabolism.

***Keywords**: cysteine, toxicity, thiol, polyamine, biosynthesis, Saccharomyces cerevisiae.*

Abbreviations:
GSH - glutathione,
GSSG - oxidized glutathione,
SAM - S-Adenosyl L-methionine,
SAH - S-Adenosyl-L-homocysteine.

INTRODUCTION

Cysteine is essential for the survival of all cells, and is obtained either through *de novo* biosynthesis, or is transported into cells from the extracellular medium by amino acid transporters. In organisms such as bacteria, archaea, fungi and plants, cysteine biosynthesis occurs *de novo* from inorganic sulfate [1-4]. In mammals, cysteine biosynthesis from inorganic sulfate is absent, but cysteine is synthesized from methionine via the trans-sulfuration pathway, and is also obtained directly from nutritional sources [5, 6]. Cysteine transporters have been described across phyla [7-10]. Contrastingly, cystine transporters, though not as ubiquitous, are also prevalent in different organisms [11-15].

Due to its highly reactive thiol group, sudden, high concentrations of cysteine can be harmful to the cell. Indeed, cysteine toxicity has been reported in bacteria, yeasts and fungi [16-18]. In mammals, high cysteine levels are extremely toxic especially in neuronal cells [19, 20]. Compared to other amino acids, the intracellular cysteine concentrations in the cell appear to be maintained at much lower levels at steady state [21], and may therefore be a limiting amino acid for many processes.

Cysteine homeostasis, therefore, is maintained through tight regulatory control of biosynthesis and transport in microbial cells [1-4]. In mammalian cells, while the regulation of cysteine homeostasis is less studied, it is thought to be primarily maintained by the removal of excess cysteine through its degradation by cysteine dioxygenase [5]. Cysteine dioxygenase converts cysteine to cysteine-sulfinate, which in turn is utilized for the generation of hypotaurine/taurine, sulfite, sulfate, pyruvate and H_2S [5]. In the pathogenic yeast *C. albicans*, cysteine dioxygenase (*CDG1*) can degrade excess cysteine, and the sulfite generated is effluxed out through the sulfite efflux pump, *SSU1* [18]. However, despite the proposed central role of cysteine dioxygenase in cysteine homeostasis, a knockout of cysteine dioxygenase in mice surprisingly led to only a small increase in intracellular cysteine concentrations. Here, in the absence of cysteine dioxygenease, the non-oxidative desulfuration pathway was hypothesized to play a role in cysteine detoxification [22]. Collectively, these data suggest the existence of other mechanisms to handle cysteine overload, which remain uninvestigated.

The budding yeast, *S. cerevisiae,* naturally lacks a cysteine dioxygenase and thus appears to be an excellent organism to evaluate alternate modes of handling cysteine overload. Prior studies of cysteine and homocysteine toxicity in *S. cerevisiae* showed that yeast experienced an ER stress due to high intracellular concentrations of cysteine or homocysteine [17]. However, the mechanisms by which *S. cerevisiae* handled cysteine overload remain unknown. Furthermore, this and earlier models have required very high concentrations of extracellular cysteine in order to observe and investigate toxicity. This is partly because although *S. cerevisiae* has multiple transporters that can transport cysteine, these are all of low affinity and specificity. The primary high affinity cysteine transporter-*YCT1* is strongly repressed by cysteine in the medium [7]. Moreover, high levels of these amino acids in the medium can interfere with transport of other amino acids and other aspects of metabolism, and so the effects seen may be non-specific. We have therefore engineered and utilized a simple system where the specific cysteine transporter, *YCT1*, is constitutively expressed under a strong promoter in order to generate high cysteine levels in the cell, while adding relatively lower levels of the amino acid in the medium. Furthermore, we also employed a similarly expressed high affinity cystine transporter (*CgCYN1*) [12]. To understand cellular responses in yeast cells to high levels of intracellular cysteine we adopted a two-pronged approach. We used a targeted, quantitative metabolite analysis in cells exposed to cysteine/cystine at different time intervals where the metabolites in the sulfur assimilation and metabolic pathway were estimated. We combined this with genome-wide expression profiling of the cells exposed to high levels of either intracellular cysteine or cystine, to understand genetic responses in adaptation to cysteine overload. Our results show that cysteine and cystine can rapidly interconvert *in vivo*, and that a principal mechanism of removal of excess cysteine is through its conversion into more inert forms of thiols, effectively a form of 'thiol trapping'. Subsequently, cells can utilize these less-reactive thiols for multiple biosynthetic processes. These results reveal general ways of handling excess cysteine in the cell, and an elegant orchestration of metabolic processes to safely sequester and utilize cysteine.

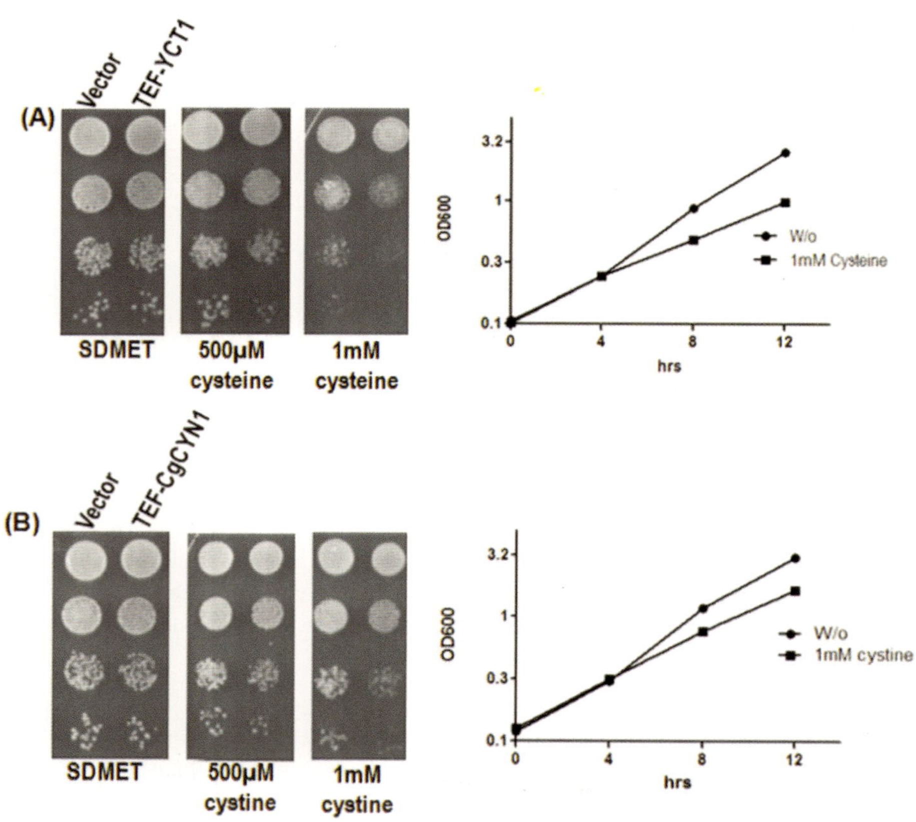

FIGURE 1: Effect of (A) cysteine and (B) cystine on *S. cerevisiae* cells expressing TEF-*YCT1* or TEF-*CgCYN1*. *S. cerevisiae* cysteine transporter *YCT1* and *C. glabrata* cystine transporter *CgCYN1* expressed under TEF promoter and corresponding vector (p416TEF) were transformed into *S. cerevisiae* ABC 1738. Transformants were evaluated for growth and toxicity by serial dilution on minimal media containing different cysteine and cystine concentrations and growth curve analysis was carried in SD (synthetic defined) medium where 1 mM cysteine or cystine was added after 3 hrs of incubation at 30°C.

RESULTS

Overexpression of the cysteine or cystine transporters leads to hypersensitivity of *S. cerevisiae* to cysteine or cystine

To determine if the constitutively expressed high affinity cysteine transporter (*YCT1*) or the heterologous *C. glabrata* cystine transporter (*CgCYN1*) in *S. cerevisiae* conferred sensitivity to these amino acids at lower media concentrations, *S. cerevisiae* cells were transformed with *YCT1* or *CgCYN1,* expressed under the strong constitutive TEF promoter, and analyzed on minimal SD media with different concentrations of cysteine or cystine.

The growth comparisons done on plates with increasing concentrations of cysteine or cystine suggested that cysteine or cystine were not toxic to the yeasts up to 0.5 mM concentration. However, beginning at 0.5 mM, we observed retarded growth and increased toxicity both on plates and in liquid broth (Fig. 1). Thus, toxicity to cysteine was observed at substantially lower concentrations than previous report [17]. We also confirmed these results in liquid broth with cysteine or cystine at concentrations of 0.5 and 1 mM (Fig. 1). These results suggest that these transporters allow the accumulation of these metabolites and thus did not require the high levels of cysteine in the medium to induce toxicity (as reported earlier) [17]. We have used this concentration (0.5 mM) where toxicity becomes observable in our subsequent experiments.

Metabolite analysis reveals a rapid inter-conversion of cysteine to cystine during cysteine overload

To understand how cells deal with an acute, large influx of either cysteine or cystine, we directly estimated the relative amounts of different metabolites in the sulfur related pathways over time. Beginning with 2.5 min after addition of cysteine, metabolites were extracted at 15, 60 and 300 min time intervals, and analyzed by targeted, quantitative LC-MS/MS. For comparative purposes, we looked at each individual metabolite under these different conditions.

In the *YCT1* expressing cells exposed to cysteine (Fig. 2A), an immediate, dramatic increase in intracellular cysteine amounts is observed. However, the level of cysteine drops after the first hour. Since the normal concentrations of cysteine in yeast cells grown in SD medium are in the range of 0.1-0.4 μM [21] under conditions where excess cysteine is pumped into the cells, the intracellular concentrations achieved were in the range of 60-240 μM. The intracellular cysteine concentrations eventually returned to initial, basal levels of the untreated conditions. Interestingly, in these same cells, we also observed high amounts of intracellular cystine (Fig. 2A). Since *YCT1* is known not to transport cystine [7], these data suggest that the cysteine transported inside the cells is converted to cystine.

CgCYN1 expressing cells when exposed to cystine expectedly showed a large increase in cystine levels (Fig. 2B). This amount was maximal in 1 hour, but by 5 hours the levels of cystine reduced to basal levels. This suggests that although initially cystine accumulation occurs, cystine is removed and converted to other compounds. These same cells also showed a significant increase in cysteine amounts (Fig. 2B) suggesting that much of the cystine that was pumped inside was being converted to cysteine and potentially utilized.

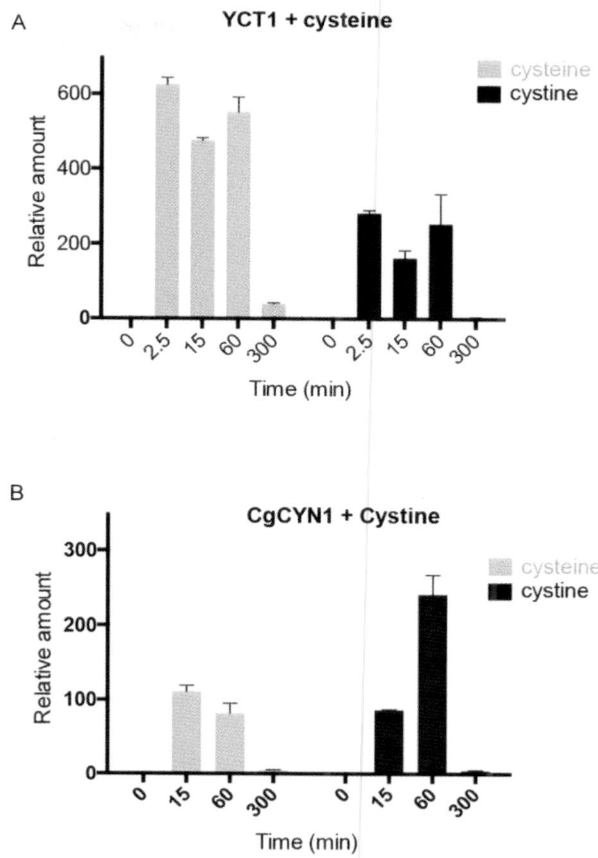

FIGURE 2: Relative intracellular amounts of cysteine and cystine as a function of time in cells overexpressing either cysteine transporter and challenged with cysteine or cystine transporter and challenged with cystine. (A) Relative levels of cysteine and cystine in *S. cerevisiae* cells expressing TEF-*YCT1* treated with 500 μM cysteine measured at different time points. **(B)** Relative levels of cysteine and cystine in *S. cerevisiae* cells expressing TEF-*CgCYN1* treated with 500 μM cystine measured at different time points.

Estimation of absolute concentrations of cystine in minimal media grown yeast cells reveal that intracellular cystine is in the 50-100 nanomolar range

In minimal media (SD) grown cells the levels of cysteine have been estimated to be in the range of 100-400 nM [21]. However, cystine levels have never been estimated. Since a ~200-fold increase in cystine was seen in some of the experiments we thought it was important to estimate intracellular cystine concentrations. We observed that the normal levels of cystine seen in a yeast cell grown in minimal media conditions was in the range of 50-100 nM, almost comparable to cysteine concentrations, and this increased up to ~10 μM concentration due to cysteine overload.

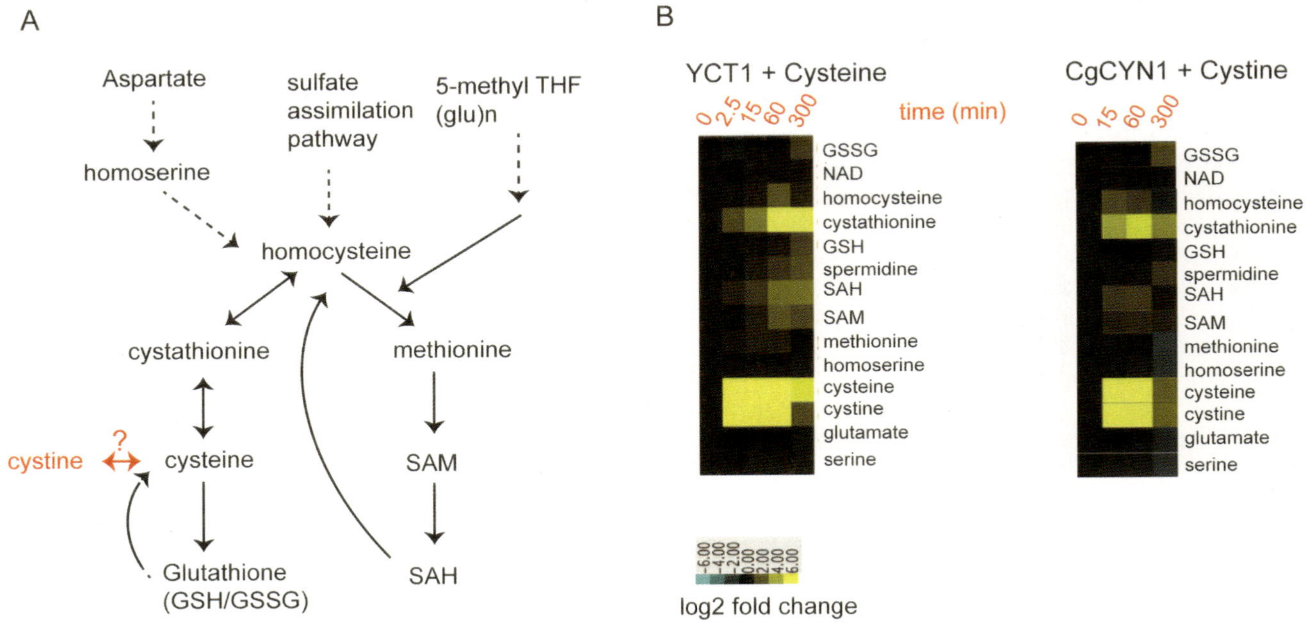

FIGURE 3. (A) A scheme of the sulfur amino acid metabolic pathway in yeast. **(B)** Heat maps illustrating a hierarchical clustering of changes in the indicated metabolites from cells expressing (i) TEF-*YCT1* treated with 500 µM cysteine and measured at 0, 2.5, 15, 60 and 300 min and (ii) TEF-*CgCYN1* treated with 500 µM cystine and measured at 0, 15, 60 and 300 min.

Relative amounts of homocysteine, cystathionine, SAM and SAH show a delayed increase compared to cysteine

The sulfur assimilation and amino acid biosynthesis pathway is a highly interconnected pathway, with homocysteine as a key intermediate metabolite (Fig. 3A). We therefore measured other key metabolites in the sulfur amino acid pathway, starting with homocysteine, using a targeted, quantitative LC-MS/MS based approach, in cells overexpressing *YCT1* and treated with a high dose (500 µM) of cysteine. We also used the same approach to measure these metabolites in cells overexpressing *CgCYN1* and treated with a high dose of cystine. Expectedly, there are substantial changes in the amounts of these metabolites, yet these metabolites show both distinct temporal profiles, as well as relative changes in their amounts (Fig. 3B). Importantly, the metabolites in the sulfur amino acid pathway increased in abundance over time in the order expected from their positions in the pathway (Figs. 3A and 3B), and the relative changes in amounts were entirely different for the different metabolites, ranging from several hundredfold for cysteine to much smaller changes for several other metabolites.

The intracellular amounts of homocysteine increased after cysteine treatment to ~5-fold above normal, after ~1 hour. Eventually, this decreases to ~2-fold above normal after 5 hours (Fig. 4). When we examined cystathionine under these conditions, we observed that cysteine exposure led to a 60-fold increase in cystathionine concentrations (Fig. 4). Furthermore, even after 5 hours, the levels of cystathionine remained significantly higher than the control cells.

SAM (S-Adenosyl L-methionine) levels also increased up to ~5-fold and remained high till ~5 hours (Fig. 4). Similarly, SAH (S-Adenosyl L-homocysteine) amounts increased up to 7-fold with maximum values achieved at ~5 hours (Fig. 4). Considering that the normal levels of SAM are in the 10-30 µM, and SAH are in the range of 100-400 µM in *S. cerevisiae* [21], these fold increases are quite substantial. In contrast, when we examined the relative levels of methionine in the cells accumulating cysteine, we observed that it led to only a ~2-fold change in methionine levels (Fig. 4), that returned to normal levels in about 5 hours. Methionine levels are normally in the range of 30-60 µM, and thus not much of the excess thiol was trapped in methionine. Indeed, there are other studies as well where sulfur amino acid metabolism change also indicates that cells maintain methionine concentrations within a fairly stable steady state range [21, 23].

Similar trends were observed when cystine was added to the cells expressing the *CgCYN1* transporter (Fig. 5), indicating an overall conserved metabolic output to acute addition of cysteine as well as cystine. However, the magnitude of increase in most metabolites was smaller. Interestingly, these data suggest how different metabolites in the sulfur amino acid pathways are maintained during basal homeostasis at entirely distinct concentration ranges, also reflecting why the relative change in amounts observed for each metabolite is likely to be entirely distinct, ranging from the ~100-fold changes for cysteine (normally maintained at very low absolute amounts) to single-fold changes for other metabolites present at much higher concentrations normally.

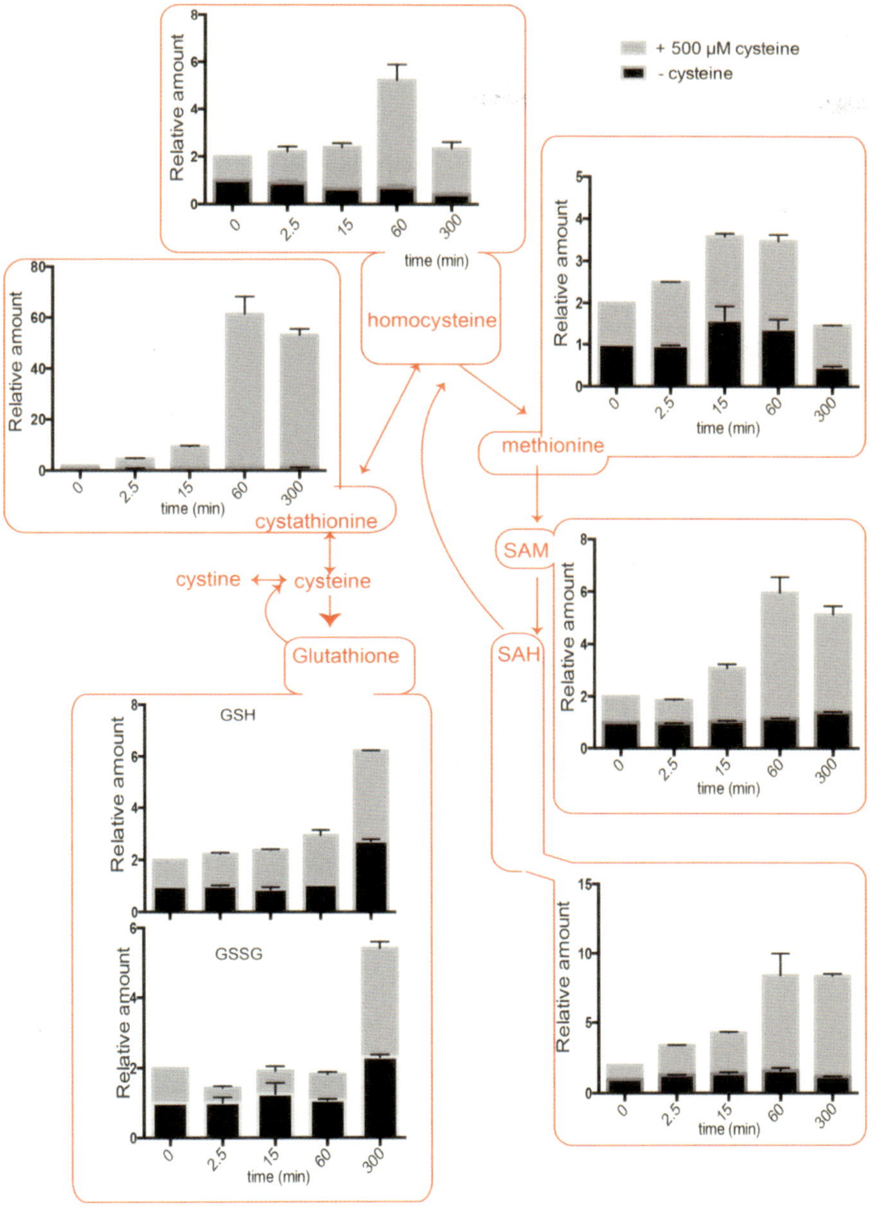

FIGURE 4. Positioned in the context of the sulfur amino acid pathway: Relative intracellular amounts of sulfur amino acid metabolites as a function of time in cells over-expressing the cysteine transporter *YCT1* and challenged with 500 μM cysteine.

Relative amounts of the naturally abundant GSH increase gradually to 3-fold higher than normal

We also observed a delayed increase in metabolites that could be considered long-term storehouses for sulfur containing amino acids. The total amounts of both GSH (glutathione) and GSSG (oxidized glutathione) did not change immediately after cysteine or cystine addition, but with time, in both the cysteine and cystine treated cells, the GSH and GSSG levels increased (Figs. 4 and 5). Although the fold increase was only about 3-fold, considering that GSH is found at very high concentrations between 1-10 mM in the cell, these changes are substantial.

Collectively, it is notable that different metabolites in the sulfur amino acid pathway show entirely different changes in abundance, both temporally as well as in the magnitude of change in relative abundance. These data reveal a sequential conversion and utilization of metabolites, as well as specific metabolites within the pathway that are maintained at more or less constant levels, upon encountering high cysteine (or cystine) levels in cells.

Genes up-regulated with TEF-YCT1 bearing cells exposed to cysteine

Our data revealed immediate, acute metabolic responses to cysteine overload, followed by a steady redistribution of the metabolites in the sulfur amino acid pathway. We therefore set out to understand global transcriptional changes in yeast to cysteine and cystine overload, to explain how cells eventually adapted to increased amounts of cysteine-derived metabolites. Using a genome-wide expression profiling approach, we identified genes that were differentially regulated in cells exposed to excess cysteine. For these experiments, *S. cerevisiae* cells bearing a deletion of *YCT1* and constitutively over-expressing the cysteine

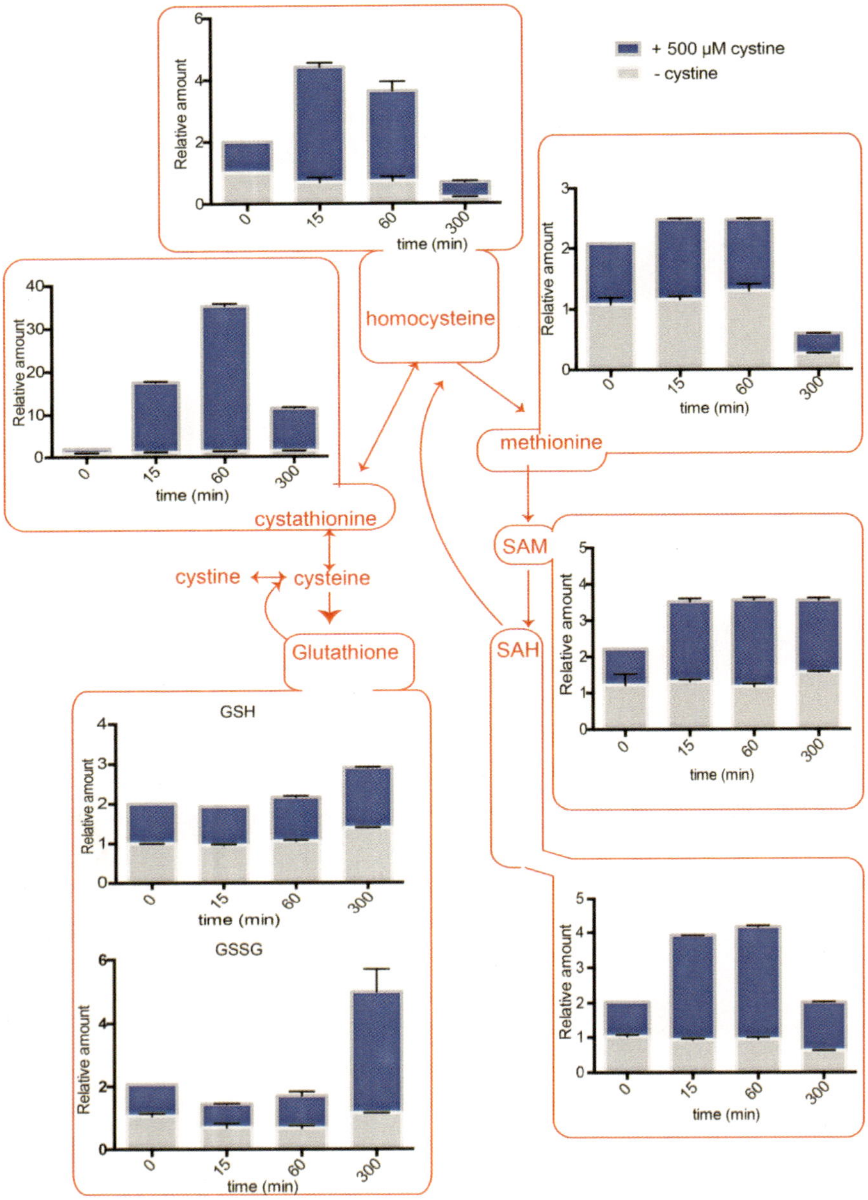

FIGURE 5. Positioned in the context of the sulfur amino acid pathway: Relative intracellular amounts of sulfur amino acid metabolites as a function of time in cells overexpressing the cystine transporter *CgCYN1* and challenged with 500 µM cystine.

transporter (TEF-YCT1) were grown in minimal medium and treated with 0.5 mM cysteine for 5 hours, and subsequently processed as described in Materials and Methods.

Interestingly, only a small subset of ~58 genes in a few functional clusters, was globally up or down regulated, revealing that a limited, highly specific response was sufficient for cells to adapt to cysteine overload (Table S2). To validate the microarray data a few of the upregulated genes in the TEF-YCT1 treated cells were also analyzed by qPCR (Figure S1). Strikingly, a gene ontology mapping of these genes to biological process revealed a very strong overrepresentation in cellular amino acid metabolic processes (Fig. 6A). Particularly among this cluster, genes from the ornithine/arginine biosynthetic pathway were highly up-regulated (Fig. 6B). This included the ornithine carbamoyl transferase, acetyl ornithine transferase, acetyl glutamate kinase and N-acetyl-γ-glutamyl-phosphate reductase and arginine succinate lyase (*ARG3, ARG8, ARG5,6,* and *ARG4*) (Fig. 6B). Additionally, other genes related to arginine metabolism including the vacuolar cationic amino acid transporters that principally transport arginine, and the mitochondrial ornithine transporter that exports ornithine from the mitochondria to the cytoplasm (*RTC2* and *ORT1*) were also induced (Fig. 6B), all suggesting that the adaptation response to acute cysteine overload eventually induced the ornithine/arginine biosynthetic pathway, which is typically considered an anabolic pathway.

This was surprising, but suggested a possibility that would metabolically make sense. A major consumer of the high amounts of SAM made during the redistribution of cysteine to other metabolites in the sulfur amino acid pathway (Figs. 3, 4, and 5) could be the polyamine pathway (Fig. 6C). Spermidine and spermine are two abundant, major metabolites that consume substantial amounts of SAM

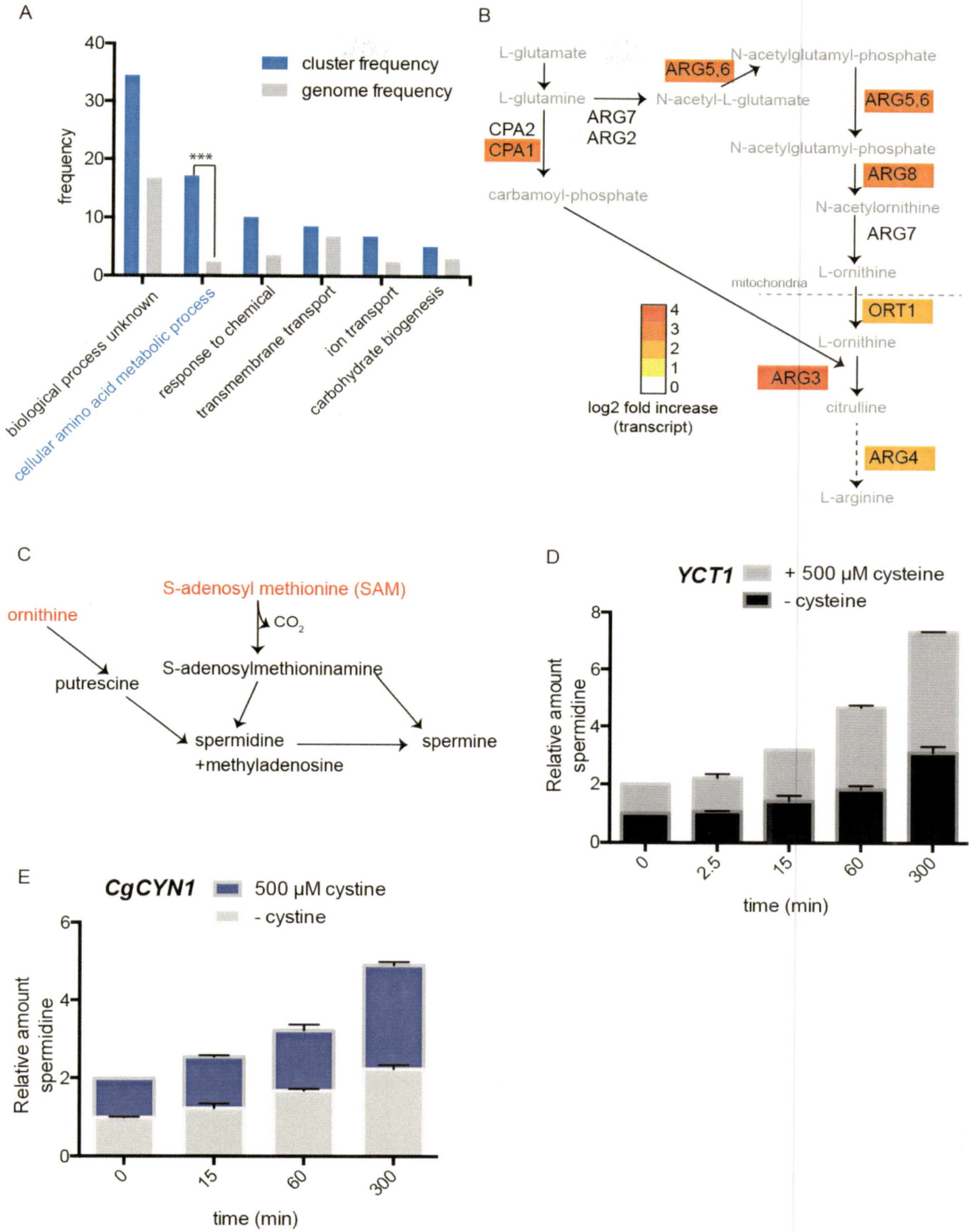

FIGURE 6. Gene expression changes due to cysteine or cystine addition to cells. (A) Gene ontology based analysis (by process), illustrating induced genes that are dramatically overrepresented (by process) upon cysteine addition. **(B)** A mapping of highly induced genes onto the ornithine/arginine biosynthesis pathway. The heat-map shows (log2) fold change in transcripts induced. **(C)** A schematic of the metabolic pathways used to make polyamines, spermidine and spermine, emphasizing the use of SAM (S-Adenosyl L-methionine) and ornithine in this process. **(D)** and **(E)** relative changes in spermidine amounts over time, upon the addition of cysteine (D) or cystine (E) to cells expressing *YCT1* or *CgCYN1* respectively.

(Fig. 6C). Importantly, this also requires ornithine (from the ornithine/arginine pathway, Fig. 6B). We therefore tested this hypothesis by directly measuring spermidine produced over time, post cysteine and cystine treatment. Spermidine showed a substantial increase in concentrations in the cysteine treated cells (Fig. 6D). Approximately 4-fold increase could be seen at the end of 5 hours. Considering the high levels of spermidine in the cell (2mM) [24], this signifies a very large increase in the absolute amounts of polyamines. Thus, it appears that one major destination (in addition to GSH) of the high amounts of cysteine are the polyamines, which would also require a simultaneous increase in the synthesis of ornithine as well as arginine, and together these might serve as anabolic molecules for subsequent growth.

Additionally, genes involved in iron and/or siderophore acquisition such as *FMP23*, *TIS11*, *FIT2*, *SIT1* and *ARN1* were found to be upregulated (Table S2). Relatedly, the enzyme trans-aconitate methyltransferase (*TMT1*) of the leucine biosynthetic pathway was also significantly upregulated, and this is known to be highly induced during iron limitation [25]. Apart from these, *ZRT1* (high-affinity zinc transporter of plasma membrane) gene was up regulated that is known to be involved in maintaining zinc homeostasis in the cell (Table S2). Finally, two genes of the sulfur metabolic pathway genes, *STR2* (cystathionine gamma-synthase) and *SSU1* (sulfite efflux pump), were found to be up regulated (Table S2). One possible explanation could be some non-oxidative degradation similar to what has been observed in mammalian cells by cystathionine beta synthase and thus the sulfite efflux pump might also be required under these conditions. Indeed, the sulfite efflux pump has been shown to be important for cysteine tolerance in *Candida albicans* [18]. In addition several genes with unknown function such as YJL160c and *SRL4* were upregulated. These are most likely to be associated with arginine biosynthesis, iron (or iron-sulfur) metabolism, or sulfur metabolism.

Genes upregulated with TEF-CgCYN1 bearing cells exposed to cysteine

In the case of the cystine overload experiments, we again observed an upregulation of the arginine biosynthetic pathway (Table S3, Fig. 7A). This included ornithine carbamoyl transferase (*ARG3*), acetyl glutamate kinase and N-acetyl-gamma-glutamyl phosphate reductase (*ARG5,6*), vacuolar cationic amino acid transporter (*RTC2*), acetyl-ornithine amino transferase (*ARG8*), small subunit of carbamoyl phosphate synthase (*CPA1*), mitochondrial ornithine transporter (*ORT1*), argininosuccinate lyase (*ARG4*) (Table S3). Similarly, we estimated changes in polyamine (spermidine) amounts after cystine treatment, and observed a substantial increase in spermidine post cystine treatment (Fig. 6E). These data matched the earlier data observed with cysteine treatment.

Similarly, genes involved in iron and/or siderophore acquisition and leucine biosynthesis were up-regulated as seen under cysteine treated cells. Mitochondrial succinate-fumarate transporter (*SFC1*-4.7-fold) was also found to be up regulated. Finally, two genes of the sulfur metabolism, cystathionine beta-lyase (*STR3*) and the sulfite efflux pump *SSU1*, and the zinc transporter *ZRT1* were upregulated. The genes up-regulated due to cystine overload strikingly overlapped with the genes up-regulated due to the overload of cysteine (Fig. 7A), however their fold induction was lower and hence a lower cut off has been used. This correlates well with the observed metabolite data, and the relative changes in abundance of these metabolites, all suggesting a conserved response to cysteine and cystine, with a relatively modest response for cystine.

Collectively, for both cysteine and cystine overload, there is a striking overlap of genes that are induced (Fig. 7A). Particularly, the induced genes are more related to amino acid metabolism that leads towards anabolism, particularly the ornithine/arginine biosynthesis pathway (Fig. 7A), leading to a final increase in polyamines that we have observed. Other changes include reducing sulfite assimilation, as well as dealing with iron and zinc limitations (all known to occur under conditions of cysteine toxicity), and not directly related to primary sulfur amino acid metabolism.

Genes down-regulated under conditions of cysteine or cystine overload in the cells

The down-regulated genes for the cysteine overload experiments (*YCT1*) (Table S4, Fig. 7B) included the genes from inorganic sulfur assimilation pathway such as the high affinity sulfate permease (*SUL1*), homoserine-O-acetyltransferase (*MET2*), ATP sulfurylase (*MET3*), sulfite reductase (*MET5*), adenylylsulfate kinase (*MET14*), and S-methyl methionine homocysteine methyltransferase (*MHT1*). All of these can be explained by a lack of the need to reduce and assimilate sulfur, which is no longer needed since cysteine provides a ready form of usable sulfur to cells. In case of the down-regulated gene list with cystine overload (*CgCYN1*) using the log2 cut-off of two, we did not observe any genes. However, on lowering the cut-off, some genes were indicated (Table S5). This list again included genes primarily from inorganic sulfur assimilation pathway and genes of the inorganic sulfur metabolic pathways such as *SUL1* (2.0-fold), *MET2* (2.1-fold), *MET3* (2.1-fold), *MET6* (1.8-fold), *MET10* (2.0-fold), as well as the glutathione transporter (*OPT1/HGT1*: 2.2-fold). Collectively, for both cysteine and cystine overload, there is a striking overlap of genes that are down regulated (Fig. 7B). All these data appear to be logical, since with an overload of cysteine, cells will no longer have to rely on challenging reductive biosynthesis of homocysteine from free sulfite.

Supplementation of arginine, zinc or iron does not reverse cysteine toxicity

Our metabolite analysis suggests that cells deal with cysteine toxicity by converting cysteine to less reactive metabolites, which lead to multiple anabolic pathways. This would require cells to be able to assimilate these metabolites for anabolism. Importantly, our gene expression analysis revealed a strong upregulation of ornithine/arginine metabolism ~5 hours after the addition of cysteine, when these

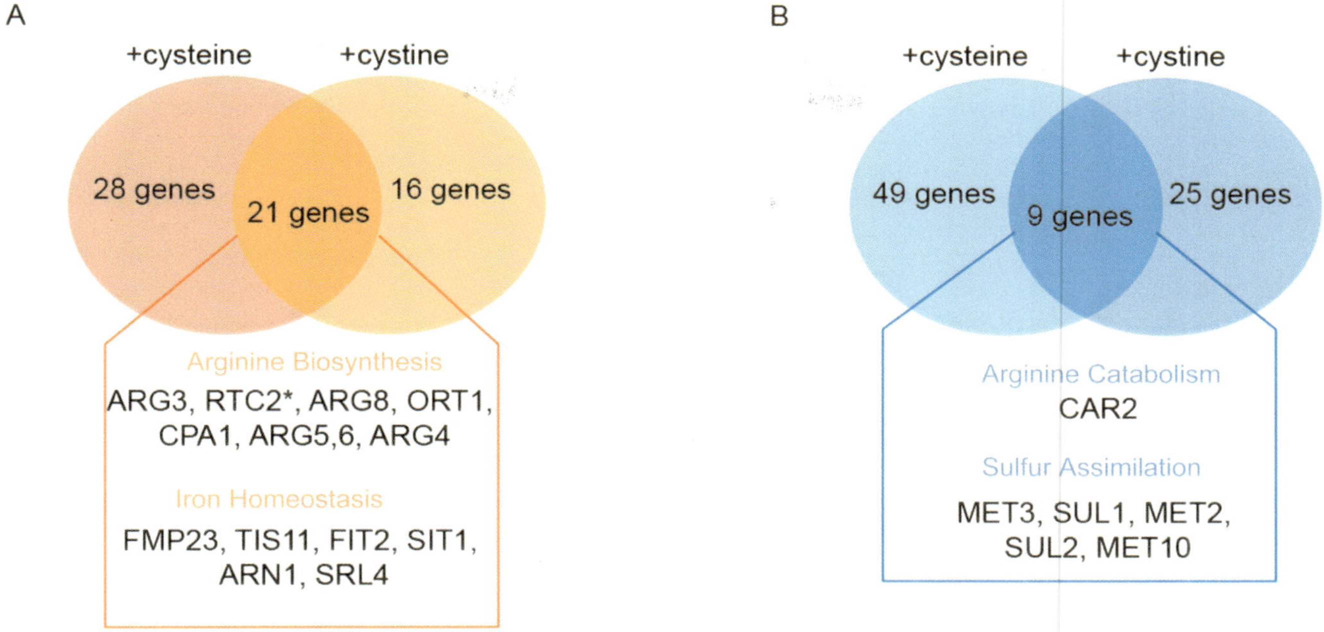

FIGURE 7. (A) Venn diagram showing a list of genes involved in specific pathways that are upregulated in *S. cerevisiae* cells expressing TEF-*YCT1* treated with cysteine (500μM), TEF-*CgCYN1* treated with cystine (500μM), where the genes commonly induced in both [$p<0.05$, fold change ≥ 2.0 (cysteine) and fold change ≥ 1.40 (cystine)] are emphasized. **(B)** Venn diagram showing a list of genes involved in specific pathways that are down regulated in *S. cerevisiae* cells expressing TEF-*YCT1* treated with cysteine (500μM), TEF-*CgCYN1* treated with cystine (500μM) and set of genes common to both [$p<0.05$, fold change ≥ 1.5(cysteine) and fold change ≥ 0.8 (cystine)].

other metabolites derived from cysteine metabolism were most abundant. This suggested that the most likely role of the ornithine/arginine metabolic pathway would be to synthesize and assimilate arginine and arginine derived metabolites, particularly polyamines, which require SAM metabolism. However, to eliminate the remote possibility that arginine limitation might determine cysteine toxicity responses, growth assays were carried out with strains over-expressing cysteine transporter under the TEF promoter in media supplemented with a range of arginine concentrations (0.01 - 2 mM). The addition of the amino acid expectedly enhanced the growth of the strains, but had no effect on the toxicity of cysteine that resulted from cysteine overload (not shown). The strains were similarly tested for reversal of cysteine toxicity by either zinc or iron since these pathways were also induced during cysteine or cystine overload. We added these metal ions in concentrations ranging from 0.01 to 0.5 mM. However, none of these metals or growth conditions could rescue the toxicity to any discernable level (data not shown).

Collectively, our data permit an interpretation suggesting that the excess cysteine, which is immediately channeled to a number of less reactive thiols, and subsequently to metabolites such as SAM and GSH, can be effectively finally utilized for anabolic processes typically derived from arginine metabolism. We explain these possibilities in the discussion section.

Investigation of deletion mutants in different pathways reveals no single enzyme is critical for survival under cysteine or cystine overload

Our data strongly suggests that the simultaneous occurrence of multiple metabolic processes allow cells to overcome cysteine overload. However, we wanted to determine if any one of the overexpressed gene clusters were particularly important for cells to handle cysteine toxicity. We utilized multiple strains from the Euroscarf collection (listed in Table S1) with deletions in the different pathways that were upregulated in response to excess cysteine or cystine. Particularly, these included the trans-sulfuration pathway (for cysteine to homocysteine conversion), and the arginine and the polyamine biosynthetic pathways. We also tested gene deletions of iron homeostasis and/or siderophore receptor synthesis that were upregulated.

Of particular interest were the single deletion mutants of the pathway from cysteine to homocysteine that included, *STR2* (cystathionine gamma synthase) and *STR3* (Cystathionine beta-lyase). However, none of these single mutants showed any enhanced toxicity as seen by plate assays. The *SSU1* sulfite efflux pump also did not show any enhanced toxicity suggesting that sulfite efflux was not the principal pathway being employed. We also evaluated the enzymes in polyamine biosynthesis that included *SPE1*, *SPE2*, *SPE3* and *SPE4*. No enhanced cysteine toxicity was observed. Finally, we tested mutants of iron/or siderophore acquisition that were upregulated with cysteine or cystine overload as well as several genes of unknown function. These genes included *FIT2, FMP23, SIT1, AFT1* and *AFT2, TMT1,* and the ORFs YGL117w and YEL057c. None of these mutants showed any enhanced toxicity in the cysteine toxicity experiments (data not shown). Collectively, these data strongly support our conclusion that no single gene or pathway is responsible for cells to handle cysteine overload, and multiple pathways allow the final assimilation of metabolites derived from cysteine.

DISCUSSION

In this manuscript, we first established a system of evaluating cysteine overload in yeast arising from high intracellular levels of cysteine without the need for addition of very high extracellular levels of the amino acid. This thereby enabled precise and systematic manipulations and thus is likely to better reflect the consequences of high intracellular cysteine accumulation. To sharpen our analysis, we additionally used the heterologously expressed *C. glabrata* cystine specific transporter (*CgCYN1*) in *S. cerevisiae,* which clarifies the metabolic consequences of the effects of cysteine influx. The effectiveness of the experimental setup was immediately apparent from the toxicity towards cysteine observed at lower media concentrations than previously reported. This was further validated when we observed that although relatively lower concentrations of these amino acids were added in the medium (10-fold lower than previous studies [17], we achieved high intracellular concentrations (over 600-fold higher than the steady-state, normal concentrations). Interestingly, the higher concentrations of cysteine also led to a large increase in intracellular cystine amounts within the cell, and vice versa. This provides a satisfying validation for the excellent correlation between the microarray experiment under both conditions, as well as the metabolite analysis carried out using similar conditions.

The observation that cysteine rapidly converts into cystine suggests that the reactive thiol of cysteine is quickly trapped in a less-reactive form. Cysteine was also subsequently distributed to other less-reactive thiols, such as cystathionine where a 60-fold increase in levels was seen. Over time, this was distributed among other thiols that included SAH, SAM, and also as GSH, all of which are present at concentrations several orders of magnitude greater than cysteine at normal steady state [21]. Thus, it appears that cells convert excess cysteine to cystine and other inert thiols as a mechanism to detoxify, and possibly store excess cysteine, an otherwise valuable nutrient to the cells. This happens through systematic, temporally separated changes in the amounts of different thiol metabolites, with the reactive thiol of cysteine sequestered away by 'thiol trapping'. It is possible that additional methods of detoxification that include sequestration into the vacuole of the oxidized cystine, might also exist in a manner similar to the sequestration of oxidized glutathione, GSSG, that plays an important role in glutathione redox homeostasis [26]. However, this is more difficult to test experimentally, due to challenges in isolating metabolites from cellular compartments.

We suggest that such 'thiol trapping' to sequester and utilize cysteine is general to all organisms, and not just to organisms like *S. cerevisiae*, lacking the oxidative degradative pathway of cysteine. Considering that sulfur assimilation is an energy consuming process, it seems likely that organisms would preferentially store and recover such forms of reduced sulfur, rather than degrade and dispose them from the cell. It is reasonable to therefore suggest that this might be a primordial strategy adopted by living cells, with degradation and consumption coming as a subsequent step. Consistent with this hypothesis, it is interesting to note that in the mouse cysteine dioxygenase knockouts, the GSH levels increased approximately two-fold [27, 28]. Other thiol intermediates such as Cys-GSH mixed disulfides which may also form were, however, not examined. Nevertheless, the elevated GSH levels alone suggest that 'thiol trapping' might be an effective mechanism in metazoans as well, although not explicitly recognized as such, till now.

The metabolite estimations were carried out at early, intermediate and late time points, providing a temporal sequence of early and later metabolic events. However, the gene expression arrays were performed at the later time point (5 hours), to provide an understanding of the eventual homeostatic adaptations for tackling high levels of cysteine. These microarray results revealed that when high levels of cysteine and cystine accumulate, only one major pathway seems to be primarily induced - the ornithine/arginine biosynthesis pathway. However, it must be pointed out that this is based on the 5 hour time point data. At earlier time points, it is possible that other pathways might also be induced, but this was not examined, since this was not the purpose of this study. The increase in the expression of the mitochondrial ornithine transporter *ORT1* and the vacuolar cationic amino acid transporter *RTC2* can also be explained by this conversion to arginine. Interestingly, *RTC2,* which has been shown to be upregulated by lysine limitation, preferentially transports arginine (as compared to lysine) [29]. The results from this study indicate that *RTC2* might also be under regulation by arginine limitation as well. Accounting for the upregulation of the siderophore receptors, many of these siderophores have in their core metabolites derived from ornithine (N-hydroxyl ornithine), which is itself derived from arginine metabolism [30, 31]. Although *S. cerevisiae* lacks the siderophore biosynthetic pathway [32], the up-regulated siderophore receptors could bind and take up these siderophores present in the extracellular medium and thus further help fulfill the cellular arginine/ornithine requirement.

Importantly, a simple metabolic process can explain why ornithine/arginine genes might be upregulated over time. SAM is intimately linked to arginine and polyamine biosynthesis [33, 34]. One explanation for the increased requirement of arginine/ornithine is that some excess cysteine can be channeled into the polyamine biosynthesis pathway to form spermine and/or spermidine with ornithine/arginine as a precursor for biosynthesis. The ornithine/arginine pathway is linked to SAM through 5'-methylthioadenosine (MTA) formation, a byproduct of polyamine biosynthesis in the methionine salvage pathway [35]. Importantly, the synthesis of spermidine is entirely SAM dependent, requiring SAM and decarboxylated SAM to convert putrescine (derived from ornithine/arginine) to spermidine [33, 34]. This simple feed-forward mechanism, converting reactive cysteine to SAM (observed in our data), and subsequently using this towards amino acid biosynthesis, arginine biosynthesis and consumption, polyamine biosynthesis and finally nucleotide biosynthesis will allow cells to restore growth after a time-period of reduced growth due to high cysteine concentrations, and thereby utilize a valuable, scarce resource. Polyamines themselves are both highly abundant and critical metabolites for growth, important for a variety of reasons [36-38] and so would be an effective final outcome when cysteine is converted into more usable forms of protected thiols. Indeed, our data showed a substantial increase in intracellular polyamine concentrations (as measured directly by spermidine amounts) in ~5 hours after cells encounter large concentrations of cysteine.

In the case of cysteine overload cells we observed, in addition to *SSU1*, the *STR2* gene being upregulated. *STR2* encodes for the gene cystathionine gamma synthase. Cysteine metabolism proceeds through *STR2*, the first enzyme that acts on cysteine to take it to cystathionine, and then subsequently to homocysteine. Surprisingly, though, the *str2Δ* did not lead to any significant enhancement of sensitivity to cysteine toxicity. *STR2* deletions are otherwise unable to utilize cysteine as a sulfur source. These data suggest that cysteine is not as toxic in yeast as has been presumed. Furthermore, this supports the conclusion that cysteine has multiple routes to detoxification. Particularly, our data suggest that conversion of cysteine to cystine, along with conversion to GSH may be an effective mechanism to remove excess cysteine.

An earlier report about homocysteine and cysteine toxicity showed that cells reduce toxicity by means of upregulation of genes involved in primary metabolism, such as glycolysis, folate metabolism, and serine biosynthesis. The study also suggested that the cysteine and homocysteine toxicity is due to elevated ER stress related genes [17]. In case of *E. coli,* growth inhibitory effects were apparently caused by inhibition of threonine deaminase by cysteine leading to isoleucine starvation and can be reversed by addition of isoleucine, leucine and valine [39]. In contrast, in neurons it is reported that the cysteine toxicity is a result of hydroxyl radicals generated during the auto-oxidation of cysteine generating cystine and is catalyzed by copper, iron and other transition metals and can be reversed by catalase and pyruvate [20]. In yeasts, in contrast, an oxidative stress was not observed, although an ER stress response was shown to occur, but this was done at much higher concentrations of externally added cysteine that could interfere with transport of other amino acids and thus could be a result of secondary consequences [17]. However, none of these mechanisms appeared important based on the studies described here. Instead, some of these can be explained by textbook metabolic feed-forward loops [40], which would be consistent with the actual metabolite data observed in this study. In particular, increased folate metabolism can be expected if there is an increase in SAM availability, and increased folate metabolism is central for nucleotide biosynthesis through SAM dependent one-carbon metabolism, which also requires ribose and deoxyribose sugars from the pentose phosphate pathway branching from glycolysis [40].

Finally, whether the mechanism by which the oxidation of cysteine to cystine occurs is purely non-enzymatic, or might involve the possibility of an active enzymatic regulation remains unclear. Enzymatic conversion has been shown to occur through copper or copper-dependent proteins [41]. Further, determination of *in vitro* cystine reduction activity by using purified thioredoxins, glutaredoxins, or screens leading to as yet unknown proteins might reveal proteins specifically involved in cystine reduction.

MATERIALS AND METHODS

Chemicals and Reagents

All chemicals used were obtained from commercial sources and were of analytical grade. Media components, fine chemicals and reagents were purchased from Sigma Aldrich (St. Louis, USA), HiMedia (Mumbai, India), Merck Millipore India Ltd (Mumbai, India), USB Corporation (Ohio, USA) or Difco, USA. Cysteine and cystine were also purchased from Sigma Aldrich (St. Louis, USA). RNA*later*® solution was procured from Thermo Fisher Scientific (Waltham, MA USA). Cysteine stock solutions that were prepared fresh were made by dissolving the required amount of cysteine in 1 ml of de-ionized water, which was then filter-sterilized using 0.2 μm filter membrane. Cystine stock solutions were prepared by dissolving the required amount of cystine in 1 ml of deionized water along with 25 μL of concentrated HCl, filter-sterilized using 0.2 μm filter membrane.

Strains and Plasmids

S. cerevisiae strain MATα *ura3Δ0 leu2Δ0 his3Δ1 lys2Δ0 yct1Δ::kanMX4* (ABC1738) was used as the parent host for all the transformations, and was obtained from Euroscarf (Y11543). This strain and all the other *S. cerevisiae* strains used in this study are listed in Table S1. The yeast strains were maintained on yeast extract (1%), peptone (2%) and dextrose (2%) (YPD) medium and grown at 28-30°C. The yeast transformants were selected and maintained on synthetic defined minimal medium containing yeast nitrogen base (0.17%), ammonium sulfate (0.5%) and dextrose (2%) supplemented with histidine, leucine, lysine, and uracil at 80 mg/l as per requirement.

TEF-YCT1 (ABE2196) contained a C-terminal HA-tagged YCT1 downstream of the TEF–promoter in p416TEF vector. TEF-CgCYN1 (ABE2847) contained a C-terminal HA-tagged CgCYN1 downstream of the TEF promoter in the centromeric vector p416TEF.

Cell growth and spotting experiments

For the growth assays, the *S. cerevisiae* strain (ABC1738) constitutively over-expressing TEF-YCT1 or TEF-CgCYN1 were grown overnight in SD minimal medium at 30°C at 220 rpm shaking, and inoculated in fresh medium to an OD_{600} of 0.1 and grown until the OD_{600} reaches 0.5 - 0.8. The cell suspensions (4 X 10^7 cells) were serially diluted to 1:10, 1:100 and 1:1000. 10 µl of these cell re-suspensions were spotted on minimal medium containing different concentrations of cysteine or cystine (0.01-1 mM). The plates were incubated at 30°C for approximately 2 days.

Total metabolite extraction and preparation

For metabolite analysis, the *S. cerevisiae* (ABC1738) transformants which were constitutively over-expressing cysteine transporter (*YCT1*) or cystine transporter (*CgCYN1*) were exposed to 500 µM cysteine or cystine. Samples were collected at different time intervals (0, 2.5, 15, 60, 300 min) and all samples except the zero time point were done in duplicate. Samples were rapidly quenched in a quenching solution, following which metabolite extraction was done. Metabolite quenching and extraction was performed as described previously [23, 42]. Metabolite extracts were dried down in a speed-vacuum (3-4 hours), and stored at -80°C until analyzed by mass spectrometer.

Targeted metabolite analysis

For detecting metabolites, a triple-quadrupole mass spectrometer (TSQ Vantage, Thermo Scientific) coupled to an HPLC was used, which allows one to detect each metabolite simultaneously using multiple reaction monitoring (MRM). Metabolites were separated chromatographically using a Synergi Fusion column (150 × 2.0 mm, 4 µm, Phenomenex), using a Shimadzu Prominence HPLC (high-performance liquid chromatography) autosampler coupled to the mass spectrometer. Extracted metabolites were measured using targeted LC-MS/MS methods, detailed method described previously [23, 42]. A library of common metabolites was constructed using standards, and metabolites were detected using a TSQ Vantage (Thermo Scientific) triple quadrupole–linear ion trap mass spectrometer for quantitative optimized detection of daughter ions upon collision-induced fragmentation of the parent ion. For each metabolite, parameters for quantitation of the two most abundant daughter ions (that is, two MRMs per metabolite) were included. To quantify metabolites, the area under each peak was quantitated by using the Xcaliber software, inspected for accuracy, and normalized against total ion count, after which relative amounts were quantified.

Estimation of intracellular metabolite concentrations

Measurement of absolute cystine concentrations in the yeast cells was done using transformants expressing *YCT1* which yielded highest fold change (200-fold) of cystine in the cells upon cysteine overload. LC-MS/MS system Agilent 1200 infinity series connected to a TSQ vantage (Triple quadrapole) Mass spectrometer which uses Electrospray Ionization (ESI) was used for detection of intracellular cystine. The samples were separated by the LC-MS/MS system using a synergy RP-fusion column. An MRM method was created for the targeted metabolite. Buffers used for positive mode analysis were: Buffer A 99.9% H_2O/0.1% formic acid, and Buffer B- 99.9% methanol/0.1% formic acid (T = 0 min, 0% B; T = 3 min, 5% B; T = 12 min, 80% B, T = 12.1 min, 90% B, T = 13 min, 0% B, T = 15 min, stop). Standards of known cystine concentrations were run under the same conditions and the area under the peak for each concentration was used to make a standard plot. The area under each peak was quantitated by using the Xcaliber software, inspected for accuracy, and the concentration of cystine extrapolated from the standard curve for a given volume of cells (≈10 ml or 9.38 OD). The intracellular concentration of cystine was estimated assuming the volume of a single yeast cell = 42 fl [43] and 1 OD = 2.0 X 10^7 cells.

Microarray analysis: growth of cells, RNA extraction and analysis

Growth of cells

The *S. cerevisiae* (ABC 1738) transformants constitutively over-expressing either cysteine transporter (TEF-YCT1) or cystine transporter (TEF-CgCYN1) were grown overnight in SD minimal medium, followed by secondary inoculation at 0.1 OD cells in same medium. After 3 hours of incubation at 30°C, we added either 500 µM cysteine or cystine, whereas respective control samples were grown without any further additions. After 5 hours of incubation of both test and respective control samples, cells were harvested, washed and re-suspended in RNA*later*® (10^8 cells). Four samples were in the cysteine analysis group (2 test samples and 2 control), and four samples were in the cystine treated group (2 test and 2 control).

RNA isolation

Total RNA was extracted using Qiagen's RNeasy minikit (Cat#74104) following the manufacturer's protocol. Total RNA integrity was assessed using RNA 6000 Nano Lab kit on the 2100 Bioanalyzer (Agilent, Palo Alto, CA) according to the manufacturer's protocol. Total RNA concentration was measured on the NanoDrop® ND-1000 UV-Vis Spectrophotometer (Nanodrop technologies, Rockland, USA). Total RNA with OD_{260}/OD_{280}>1.8 and $OD_{260}/OD_{230} \geq 1.1$ was used for microarray analysis.

cRNA synthesis, labeling and hybridization

The purified RNA samples were further processed for One-Color Microarray-Based Gene Expression Analysis. As per the manufacturer's protocol Agilent's Quick-Amp labeling Kit (Cat#5190-0442) was used for the cRNA synthesis and Cyanine-3 CTP labeling. Qiagen's RNeasy mini spin columns were used for purifying amplified cRNA samples. The quantity and specific activity of cRNA was determined by using NanoDrop ND-1000 Spectrophotometer version 3.2.1. Samples with specific activity >8 were used for hybridization. Hybridization solution was prepared using the Agilent's *in situ* Hybridzation kit (Cat#5190-0404). The resulting mixture was applied to the Yeast (V2) Gene Expression Microarray, 8x15K (Agilent Technologies AMADID: 16322), and hybridized at 65°C for 16 hours in the hybridization oven. After hybridization, slides were washed with Agilent Gene expression Wash Buffer-I for 1 min at room temperature followed by 1 min wash with Wash Buffer-II at 37°C. Slides were finally rinsed with acetonitrile for cleaning up and drying. These experiments were carried out at Genotypic Co., Bangalore, India.

Microarray scanning and data extraction

Hybridized arrays were scanned at 3μm resolution on an Agilent Microarray Scanner, Model G2565BA. The Images were manually verified and found to be devoid of uneven hybridization, streaks, blobs and other artifacts. Hybridization across the slides was good based on number of features that were positive and significantly above background. Data extraction from Images was done using Feature Extraction software of Agilent (V-11.5). Percentile shift Normalization was done using GeneSpring GX version 13. One-fold and above differentially regulated genes were filtered from the data and were considered for further analysis. With respect to control sets, fold change was calculated (logbase2) and genes showing fold change of ±0.6 were further considered for analysis. In order to identify the genes whose expression was affected in samples treated with cysteine, we used a cut off of log2 value of 2, which reflects the 4-fold change in expression of genes. We used lower cut off values for the samples treated with cystine, since the overall expression levels were lower. Only differentially regulated genes with a P≤0.05 were considered in the analysis. Microarray data has been submitted to the Gene Expression Omnibus (GEO) database (ACCID: GSE87794).

RNA isolation and RT-PCR

RNA isolation and cDNA synthesis was done as described previously [44]. The *S. cerevisiae* (ABC 1738) transformants constitutively over-expressing cysteine transporter (TEF-YCT1) were grown overnight in SD minimal medium, followed by secondary inoculation at 0.1 OD cells in same medium. After 3 hours of incubation at 30°C, we added 500 μM cysteine to the test samples, whereas control samples were grown without any further additions. After 5 hours of incubation of both test and respective control samples, cells were harvested. Total RNA was isolated by hot acid phenol method using 15 ml phase lock gel heavy tubes (5 prime), followed by DNase I treatment (Cat#M6101, Promega) for 15 min at 30°C. Zymo-Spin II column (Zymo Research, Cat#C1008-250) was used for clean-up of RNA. cDNA synthesis from 3 μg total RNA was done using reverse transcriptase (RT) and random-hexamer primers (GoScript™ Reverse Transcription System (Cat#A5000) Promega) at 42°C for 16 hours. Real time quantitative PCR (RT-qPCR) was carried out on LightCycler® 480 II System (Roche molecular Diagnostics) by using SYBR green dye-based reagents (Maxima Sybr Green QPCR Master Mix (Cat#K0251) Thermo Fisher). List of primers used for the analysis are listed in Table S8.

ACKNOWLEDGEMENTS

We acknowledge Hemanth Maddali in the SL lab, and the NCBS-inStem-CCAMP mass spectrometry facility for technical assistance. We thank Dr. Shantanu Sengupta, Institute of Genomics and Integrative Biology (IGIB), New Delhi for sharing unpublished results. AKB acknowledges grant in aid support (Project No: SB/SO/BB/017/2014) from the Department of Science and Technology, Government of India. AKB is a recipient of a JC Bose National Fellowship from the Department of Science and Technology, Government of India. SL is supported by a Wellcome Trust-DBT IA Intermediate fellowship (IA/I/14/2/501523), and DBT institutional support. AAD acknowledges a senior research fellowship from the University Grants Commission, Government of India, and in part from the Cystinosis Research Foundation, USA. MB acknowledges fellowship from the Indian Council of Medical Research, Government of India.

Author contributions: AAD did the initial growth experiments, microarray and RT-PCR related experiments and was involved in the analysis of the metabolite data and contributed to design, MB did the growth experiments, toxicity related experiments, participated in the metabolite estimations and contributed to the design, SL did the metabolite estimations and along with AKB supervised the project that included the experimental design, data interpretation and both SL and AKB wrote the manuscript.

CONFLICT OF INTEREST

The authors declare that they have no conflicts of interest with the contents of this article.

REFERENCES

1. Birke H, De Kok LJ, Wirtz M, Hell R (**2015**). The role of compartment-specific cysteine synthesis for sulfur homeostasis during H2S exposure in Arabidopsis. **Plant Cell Physiol** 56(2): 358-367.

2. Thomas D, Surdin-Kerjan Y (**1997**). Metabolism of sulfur amino acids in Saccharomyces cerevisiae. **Microbiol Mol Biol Rev** 61(4): 503-532.

3. Guédon E, Martin-Verstraete I (**2007**). Cysteine Metabolism and Its Regulation in Bacteria. In: Wendisch VF, editor Amino Acid Biosynthesis ~ Pathways, Regulation and Metabolic Engineering. Springer Berlin Heidelberg, Berlin, Heidelberg; pp 195-218.

4. Liu Y, Beer LL, Whitman WB (**2012**). Sulfur metabolism in archaea reveals novel processes. **Environ Microbiol** 14(10): 2632-2644.

5. Stipanuk MH, Dominy JE, Jr., Lee JI, Coloso RM (**2006**). Mammalian cysteine metabolism: new insights into regulation of cysteine metabolism. **J Nutr** 136(6 Suppl): 1652S-1659S.

6. Yin J, Ren W, Yang G, Duan J, Huang X, Fang R, Li C, Li T, Yin Y, Hou Y, Kim SW, Wu G (**2016**). L-Cysteine metabolism and its nutritional implications. **Mol Nutr Food Res** 60(1): 134-146.

7. Kaur J, Bachhawat AK (**2007**). Yct1p, a novel, high-affinity, cysteine-specific transporter from the yeast Saccharomyces cerevisiae. **Genetics** 176(2): 877-890.

8. Watts SD, Torres-Salazar D, Divito CB, Amara SG (**2014**). Cysteine transport through excitatory amino acid transporter 3 (EAAT3). **PLoS One** 9(10): e109245.

9. Lee CP, Wirtz M, Hell R (**2014**). Evidence for several cysteine transport mechanisms in the mitochondrial membranes of Arabidopsis thaliana. **Plant Cell Physiol** 55(1): 64-73.

10. Utsunomiya-Tate N, Endou H, Kanai Y (**1996**). Cloning and functional characterization of a system ASC-like Na+-dependent neutral amino acid transporter. **J Biol Chem** 271(25): 14883-14890.

11. Chonoles Imlay KR, Korshunov S, Imlay JA (**2015**). Physiological Roles and Adverse Effects of the Two Cystine Importers of Escherichia coli. **J Bacteriol** 197(23): 3629-3644.

12. Yadav AK, Bachhawat AK (**2011**). CgCYN1, a plasma membrane cystine-specific transporter of Candida glabrata with orthologues prevalent among pathogenic yeast and fungi. **J Biol Chem** 286(22): 19714-19723.

13. Nagamori S, Wiriyasermkul P, Guarch ME, Okuyama H, Nakagomi S, Tadagaki K, Nishinaka Y, Bodoy S, Takafuji K, Okuda S, Kurokawa J, Ohgaki R, Nunes V, Palacin M, Kanai Y (**2016**). Novel cystine transporter in renal proximal tubule identified as a missing partner of cystinuria-related plasma membrane protein rBAT/SLC3A1. **Proc Natl Acad Sci U S A** 113(3): 775-780.

14. Conrad M, Sato H (**2012**). The oxidative stress-inducible cystine/glutamate antiporter, system x (c) (-) : cystine supplier and beyond. **Amino Acids** 42(1): 231-246.

15. Baptist EW, Kredich NM (**1977**). Regulation of L-cystine transport in Salmonella typhimurium. **J Bacteriol** 131(1): 111-118.

16. Oguri T, Schneider B, Reitzer L (**2012**). Cysteine catabolism and cysteine desulfhydrase (CdsH/STM0458) in Salmonella enterica serovar typhimurium. **J Bacteriol** 194(16): 4366-4376.

17. Kumar A, John L, Alam MM, Gupta A, Sharma G, Pillai B, Sengupta S (**2006**). Homocysteine- and cysteine-mediated growth defect is not associated with induction of oxidative stress response genes in yeast. **Biochem J** 396(1): 61-69.

18. Hennicke F, Grumbt M, Lermann U, Ueberschaar N, Palige K, Bottcher B, Jacobsen ID, Staib C, Morschhauser J, Monod M, Hube B, Hertweck C, Staib P (**2013**). Factors supporting cysteine tolerance and sulfite production in Candida albicans. **Eukaryot Cell** 12(4): 604-613.

19. Olney JW, Zorumski C, Price MT, Labruyere J (**1990**). L-cysteine, a bicarbonate-sensitive endogenous excitotoxin. **Science** 248(4955): 596-599.

20. Wang XF, Cynader MS (**2001**). Pyruvate released by astrocytes protects neurons from copper-catalyzed cysteine neurotoxicity. **J Neurosci** 21(10): 3322-3331.

21. Laxman S, Sutter BM, Wu X, Kumar S, Guo X, Trudgian DC, Mirzaei H, Tu BP (**2013**). Sulfur amino acids regulate translational capacity and metabolic homeostasis through modulation of tRNA thiolation. **Cell** 154(2): 416-429.

22. Jurkowska H, Roman HB, Hirschberger LL, Sasakura K, Nagano T, Hanaoka K, Krijt J, Stipanuk MH (**2014**). Primary hepatocytes from mice lacking cysteine dioxygenase show increased cysteine concentrations and higher rates of metabolism of cysteine to hydrogen sulfide and thiosulfate. **Amino Acids** 46(5): 1353-1365.

23. Laxman S, Sutter BM, Shi L, Tu BP (**2014**). Npr2 inhibits TORC1 to prevent inappropriate utilization of glutamine for biosynthesis of nitrogen-containing metabolites. **Sci Signal** 7(356): ra120.

24. Chattopadhyay MK, Park MH, Tabor H (**2008**). Hypusine modification for growth is the major function of spermidine in Saccharomyces cerevisiae polyamine auxotrophs grown in limiting spermidine. **Proc Natl Acad Sci U S A** 105(18): 6554-6559.

25. Shakoury-Elizeh M, Tiedeman J, Rashford J, Ferea T, Demeter J, Garcia E, Rolfes R, Brown PO, Botstein D, Philpott CC (**2004**). Transcriptional remodeling in response to iron deprivation in Saccharomyces cerevisiae. **Mol Biol Cell** 15(3): 1233-1243.

26. Morgan B, Ezerina D, Amoako TN, Riemer J, Seedorf M, Dick TP (**2013**). Multiple glutathione disulfide removal pathways mediate cytosolic redox homeostasis. **Nat Chem Biol** 9(2): 119-125.

27. Roman HB, Hirschberger LL, Krijt J, Valli A, Kozich V, Stipanuk MH (**2013**). The cysteine dioxgenase knockout mouse: altered cysteine metabolism in nonhepatic tissues leads to excess H2S/HS(-) production and evidence of pancreatic and lung toxicity. **Antioxid Redox Signal** 19(12): 1321-1336.

28. Ueki I, Roman HB, Valli A, Fieselmann K, Lam J, Peters R, Hirschberger LL, Stipanuk MH (**2011**). Knockout of the murine cysteine dioxygenase gene results in severe impairment in ability to synthesize taurine and an increased catabolism of cysteine to hydrogen sulfide. **Am J Physiol Endocrinol Metab** 301(4): E668-684.

29. Jezegou A, Llinares E, Anne C, Kieffer-Jaquinod S, O'Regan S, Aupetit J, Chabli A, Sagne C, Debacker C, Chadefaux-Vekemans B, Journet A, Andre B, Gasnier B (**2012**). Heptahelical protein PQLC2 is a lysosomal cationic amino acid exporter underlying the action of cysteamine in cystinosis therapy. **Proc Natl Acad Sci U S A** 109(50): E3434-3443.

30. Philpott CC, Protchenko O, Kim YW, Boretsky Y, Shakoury-Elizeh M (**2002**). The response to iron deprivation in Saccharomyces cerevisiae: expression of siderophore-based systems of iron uptake. **Biochem Soc Trans** 30(4): 698-702.

31. Kosman DJ (**2003**). Molecular mechanisms of iron uptake in fungi. **Mol Microbiol** 47(5): 1185-1197.

32. Haas H (**2003**). Molecular genetics of fungal siderophore biosynthesis and uptake: the role of siderophores in iron uptake and storage. **Appl Microbiol Biotechnol** 62(4): 316-330.

33. Sauter M, Moffatt B, Saechao MC, Hell R, Wirtz M (**2013**). Methionine salvage and S-adenosylmethionine: essential links between sulfur, ethylene and polyamine biosynthesis. **Biochem J** 451(2): 145-154.

34. Valdes-Santiago L, Ruiz-Herrera J (**2013**). Stress and polyamine metabolism in fungi. **Front Chem** 1:42.

35. Pirkov I, Norbeck J, Gustafsson L, Albers E (**2008**). A complete inventory of all enzymes in the eukaryotic methionine salvage pathway. **FEBS J** 275(16): 4111-4120.

36. Pegg AE (**2009**). Mammalian polyamine metabolism and function. **IUBMB Life** 61(9): 880-894.

37. Kusano T, Berberich T, Tateda C, Takahashi Y (**2008**). Polyamines: essential factors for growth and survival. **Planta** 228(3): 367-381.

38. Thomas T, Thomas TJ (**2003**). Polyamine metabolism and cancer. **J Cell Mol Med** 7(2): 113-126.

39. Harris CL (**1981**). Cysteine and growth inhibition of Escherichia coli: threonine deaminase as the target enzyme. **J Bacteriol** 145(2): 1031-1035. PMID:7007336

40. Nelson D, Cox M (**2004**). Lehninger Principles of Biochemistry, Fourth Edition. W. H. Freeman.

41. Sengupta S, Wehbe C, Majors AK, Ketterer ME, DiBello PM, Jacobsen DW (**2001**). Relative roles of albumin and ceruloplasmin in the formation of homocystine, homocysteine-cysteine-mixed disulfide, and cystine in circulation. **J Biol Chem** 276(50): 46896-46904.

42. Tu BP, Mohler RE, Liu JC, Dombek KM, Young ET, Synovec RE, McKnight SL (**2007**). Cyclic changes in metabolic state during the life of a yeast cell. **Proc Natl Acad Sci U S A** 104(43): 16886-16891.

43. Jorgensen P, Nishikawa JL, Breitkreutz BJ, Tyers M (**2002**). Systematic identification of pathways that couple cell growth and division in yeast. **Science** 297(5580): 395-400.

44. Inada M, Pleiss JA (**2010**). Genome-wide approaches to monitor pre-mRNA splicing. **Methods Enzymol** 470:51-75.

12

Farnesol inhibits translation to limit growth and filamentation in *C. albicans* and *S. cerevisiae*

Nkechi E. Egbe[1,2], Tawni O. Dornelles[1], Caroline M. Paget[1], Lydia M. Castelli[1,3] and Mark P. Ashe[1,*]

[1] Division of Molecular and Cellular Function, School of Biological Sciences, Faculty of Biology, Medicine and Health, The University of Manchester, Manchester Academic Health Science Centre, Michael Smith Building, Oxford Rd., Manchester, M13 9PT, United Kingdom.

[2] Current address: Department of Biological Sciences, Nigerian Defence Academy, PMB 2109, Kaduna, Nigeria.

[3] Current address: Sheffield Institute for Translational Neuroscience, University of Sheffield, Sheffield, S10 2HQ, United Kingdom.

* Corresponding Author:

Mark P. Ashe, Division of Molecular and Cellular Function, School of Biological Sciences, Faculty of Biology, Medicine and Health, The University of Manchester, Manchester Academic Health Science Centre, Michael Smith Building, Oxford Rd., Manchester, M13 9PT, United Kingdom; E-mail: mark.p.ashe@manchester.ac.uk

ABSTRACT ***Candida albicans*** **is a polymorphic yeast where the capacity to switch between yeast and filamentous growth is critical for pathogenicity. Farnesol is a quorum-sensing sesquiterpene alcohol that, via regulation of specific signalling and transcription components, inhibits filamentous growth in *C. albicans.* Here we show that farnesol also inhibits translation at the initiation step in both *C. albicans* and *S. cerevisiae*. In contrast to fusel alcohols, that target the eukaryotic initiation factor 2B (eIF2B), farnesol affects the interaction of the mRNA with the small ribosomal subunit leading to reduced levels of the 48S preinitiation ribosomal complex in *S. cerevisiae*. Therefore, farnesol targets a different step in the translation pathway than fusel alcohols to elicit a completely opposite physiological outcome by negating filamentous growth.**

Keywords: *protein synthesis, translation control, farnesol, quorum sensing.*

Abbreviations:
QS - quorum sensing,
TC - ternary complex.

INTRODUCTION

The capacity to detect and respond to environmental change is essential for microorganism survival. This is especially true for opportunist pathogens like *Candida albicans*; where to initiate infection, the organism must adapt and persist in spite of host immune responses. Typically, *C. albicans* is a harmless commensal, yet in infected patients it causes various different conditions, from mucosal infections to life-threatening systemic infections [1].

Cell-cell signalling, particularly quorum sensing (QS), is a major focus of microbiological research. Farnesol is an acyclic sesquiterpene alcohol that represents the first QS molecule identified in eukaryotic microorganisms [2] where it causes a range of physiological effects [3]. In *C. albicans*, farnesol inhibits the yeast to hyphal switch [2] to prevent colonization of different niche environments [4, 5], it has antioxidant effects [6, 7] and it inhibits transporters [8]. In many species farnesol induces cellular death: for example in the fungal species, *Saccharomyces cerevisiae* [9], *Aspergillus nidulans* [10], *Penicillium expansum* [11], *Botrytis cinerea* [12] and even *C. albicans* under certain conditions [13]. Equally, farnesol triggers cell death in mammalian cells [14] and can have antibacterial properties [15, 16]. In fact, farnesol was first discovered as a constituent of plant essential oils with antimicrobial activities [17].

Cellular responses to stimuli act via signal transduction pathways to regulate gene expression. In *C. albicans,* farnesol targets pathways like the Ras-PKA pathway that, via the transcription factors Efg1p and Czf1p and the repressor Tup1p, regulates gene expression [18]. If a stimulus induces cellular stress, a transient inhibition of global protein synthesis is often observed, which further modulates the programme of gene expression to allow stress responsive gene expression programs to be initiated [19, 20]. Control of translation in this manner mostly occurs at the initiation stage in order to allow rapid and reversible management of gene expression.

Translation initiation is the assembly of an elongation competent 80S ribosome with an initiator methionyl-tRNA ($Met\text{-}tRNA_i^{Met}$) base paired via its anticodon loop to an mRNA Start codon [21]. Highly conserved controls allow eukaryotic cells to globally reduce translation [19, 20]: a prominent example involves eIF2α kinases, like Gcn2p in *S. cerevisiae* [22, 23]. Gcn2p activation after amino acid starvation causes phosphorylation of the α subunit of eukaryotic translation initiation factor 2 (eIF2) [24]. eIF2 is an essential GTP-binding protein that interacts with Met-tRNA$_i^{Met}$ to form a ternary complex (TC) that is competent for initiating translation [23]. Phosphorylated eIF2 competitively inhibits the eIF2B-mediated guanine nucleotide exchange reaction on eIF2, reducing TC levels and translation initiation [23]. However, specific mRNAs, such as yeast *GCN4,* continue to be translated under these conditions. *GCN4* encodes a transcription factor that regulates the expression of amino acid biosynthetic genes. This feedback regulatory circuit has proved a paradigm for studies on translation control [25]. Similarly, *C. albicans* expresses a single eIF2α kinase, Gcn2p, which phosphorylates eIF2α in response to various stresses [26, 27] and translational activation of *CaGCN4* also provides feedback regulation [28, 29].

As well as indirect attenuation of eIF2B activity via phosphorylation of eIF2α, cells can also modulate eIF2B activity more directly. In mammalian cells, phosphorylation of eIF2B has been identified as an important regulatory mechanism [30]. In addition, in both yeast and mammalian cells volatile anaesthetics appear to inhibit protein synthesis via eIF2B regulation [31, 32]. Moreover, in both *S. cerevisiae* and *C. albicans*, fusel alcohols, which are also characterised as quorum sensing molecules [33], have been shown to inhibit translation initiation in a mechanism that targets eIF2B but independently of the Gcn2p kinase or eIF2α phosphorylation [34-36].

Besides control via eIF2B, another regulated step in translation initiation is the mRNA selection phase [37]. eIF4E and Pab1p select mRNA via interaction with the 5′ cap and 3′ poly(A) tail, respectively. eIF4G can interact with both eIF4E and Pab1p to form a closed loop complex that, via interactions with eIF3, eIF5 and eIF1, can recruit the small ribosomal subunit to form a 48S preinitiation complex [21]. A variety of stress conditions have been shown to target these steps in the initiation pathway leading to transient reductions in translation to facilitate a switch to a new program of gene expression [19, 20].

In this study, we show that as well as hampering various filamentation pathways, farnesol inhibits protein synthesis. This inhibition of translation occurs at the initiation step and most likely impacts upon the assembly of the 48S preinitiation complex. Intriguingly, this means two different quorum sensing agents, farnesol and fusel alcohols that have conflicting effects on filamentous growth, both inhibit translation initiation but by different mechanisms.

RESULTS AND DISCUSSION

Farnesol inhibits growth and protein synthesis

Farnesol, a eukaryotic QS molecule inhibits filamentous growth in both *S. cerevisiae* and *C. albicans* [2], however, the concentration required varies according to the specific growth regime [38]. Under the growth conditions used here, we found concentrations in excess of 100 μM farnesol inhibited the growth of *C. albicans* and an isogenic *gcn2Δ* mutant (Fig. 1A). To study possible origins of the growth inhibition, the impact of farnesol on the rate of protein synthesis was monitored. The resulting [^{35}S]-methionine incorporation data show that farnesol

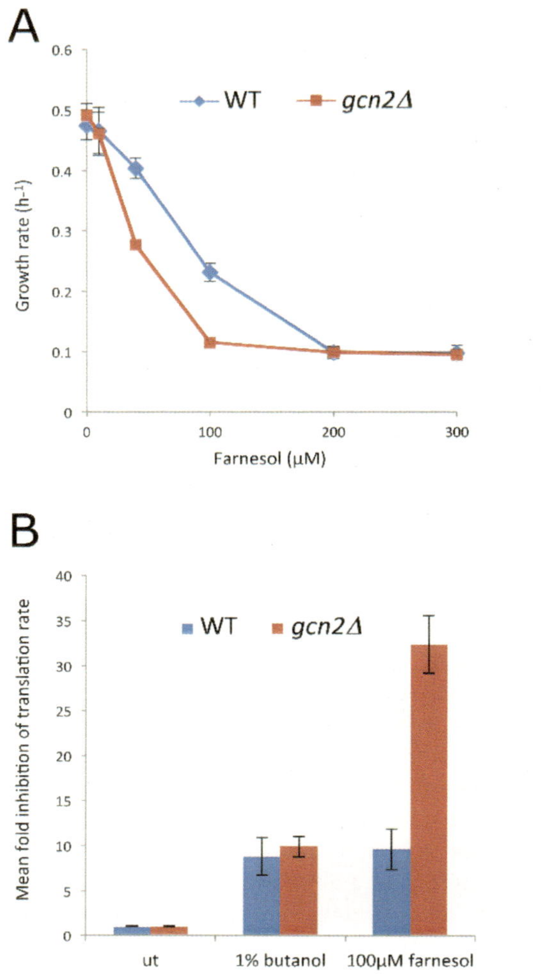

FIGURE 1: Farnesol inhibits growth and protein synthesis. (A) Growth rates for the CAI4 strain and the *gcn2Δ* mutant at various concentrations of farnesol as indicated. **(B)** Rates of radiolabelled methionine incorporation were measured for the CAI4 strain and the *gcn2Δ* mutant in untreated conditions (ut) or after 1% butanol or 100 μM farnesol treatment. These were used to calculate the fold inhibition for three biological replicates (error bars = SEM).

(300 μM) and butanol (2%) cause a 10-fold inhibition of protein synthesis in the CAI4 strain of *C. albicans* (Fig. 1B). Therefore, farnesol inhibits protein synthesis at very early stages after addition and this control could contribute to the growth inhibition observed.

To further investigate the stage of protein synthesis that is targeted by farnesol, polysome profiling was used, as this allows both the level of protein synthesis and the stage of regulation to be investigated [34-36]. Analysis of polysome distribution for the *C. albicans* CAI4 strain revealed that increasing concentrations of farnesol caused a change in the polysome profile (Fig. 2A). The 80S peak increased dramatically and the polysome peaks were reduced. This change in profile is characteristic of an inhibition of translation initiation [34] and has been observed for many stresses [19]. Similar results in terms of farnesol sensitivity were

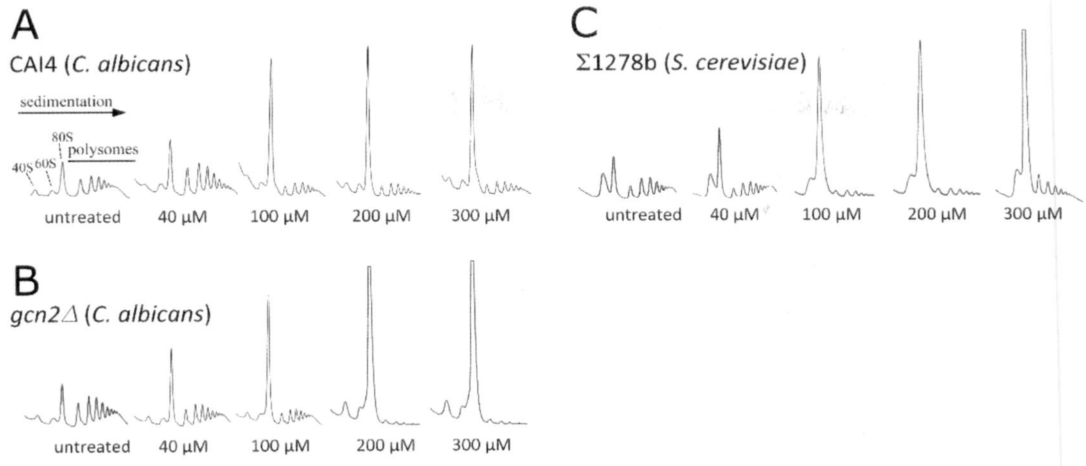

FIGURE 2: Translation initiation is inhibited by farnesol in a Gcn2p-independent manner in *C. albicans* and *S. cerevisiae*. Figure shows polysome analyses assessing the effect of farnesol on translation initiation in CAI4 **(A)** and *gcn2Δ* **(B)** strains of *C. albicans,* and the Σ1278b strain of *S. cerevisiae* **(C)**. Strains were grown in YPD and various concentrations of farnesol were added as indicated for 15 min prior to extract preparation. Extracts were sedimented on 15-50% sucrose gradients and the absorbance at 254 nm was continuously measured. The position of 40S, 60S and 80S peaks are labelled and the direction of sedimentation is noted.

obtained for the Σ1278b strain of *S. cerevisiae*, where similar concentrations elicited the response across the two yeast species (Fig. 2C). It has been noted previously that the level of free 60S is particularly high for the Σ1278b, although this does not appear to impact upon its growth or its sensitivity to translational stress [39]. The lowest farnesol concentration that caused a gross impact on polysome distribution for either *C. albicans* or *S. cerevisiae* was 100 μM (Fig. 2). This correlated well with the concentration that inhibited growth under the conditions used here (Fig. 1A) suggesting that the inhibition of translation initiation could be intrinsically connected to growth inhibition for farnesol.

The CAI4 strain of *C. albicans* used here is a commonly used lab strain that is auxotrophic in the uracil biosynthetic pathway by virtue of a homozygous deletion of the *URA3* gene [40]. This mutation has previously been shown to alter a number of aspects of *C. albicans* physiology including adhesion and virulence [41]. Therefore, a prototrophic strain of *C. albicans*, SC5314, was tested. Entirely analogous observations were made on the impact of farnesol on the growth (Fig. S1A) and translation (Fig. S1B) of this strain in response to similar concentrations of farnesol.

Previous studies evaluating QS have established that *trans,trans*-farnesol is produced by *Candida* as a QS molecule to inhibit filamentation [2, 42]. Intriguingly, the *trans,trans* form and mixed stereoisomer preparations both impact upon translation initiation equally (Fig. S1). Indeed 40 μM of each is sufficient to induce a mild inhibition of protein synthesis and 100 μM leads to a robust inhibition (Fig. 1B). Therefore, in order to explore the mechanism by which farnesol inhibits translation initiation, over the course of the rest of our studies 100 μM farnesol was used as this concentration elicits robust inhibition of both growth and translation. However, it should be noted that lower concentrations of farnesol (e.g. 40 μM) can lead to subtle alterations in the polysome profile (Fig. 2, 3 and S1). This level of sensitivity to farnesol correlates well with earlier studies using similar growth conditions [38].

GCN2 is not involved in the inhibition translation initiation by farnesol

In terms of the mechanism of translational regulation, like *S. cerevisiae*, *C. albicans* harbours a single eIF2α kinase gene, *GCN2*, which is involved in the regulation of translation initiation in response to various stresses [26, 27]. Indeed in *S. cerevisiae*, *gcn2Δ* mutants are incapable of inhibiting translation initiation in response to specific stress conditions [34], which prevents cells from mounting an appropriate stress response. Therefore, the role of Gcn2p in the farnesol-dependent inhibition of translation initiation was assessed using a *gcn2Δ* mutant strain of *C. albicans* [26]. Previous observations using this strain show that translation initiation remains uninhibited early after amino acid starvation [29]. In terms of the impact of farnesol on growth and translation, the *gcn2Δ* mutant strain is at least as sensitive as the wild type (Fig. 1A and B). In fact rather than the *gcn2Δ* mutant being resistant to farnesol in terms of translation, as might be expected if the Gcn2p kinase were involved in the control, the farnesol-treated *gcn2Δ* strain is even more inhibited than the wild type. Growth is inhibited at lower farnesol concentrations for the *gcn2Δ* mutant and methionine incorporation is inhibited up to 30-fold (Fig. 1A and B). Equally, a comparison of the polysome profiles shows that the *gcn2Δ* mutant exhibits somewhat greater sensitivity than the wild type mirroring the growth phenotypes (Fig. 2B). For instance, following treatment with 300 μM farnesol, greater polysome run-off is observed for the *gcn2Δ* mutant compared to the parent strain (*cf.* Fig. 2B with 2A). Overall, these results show that the

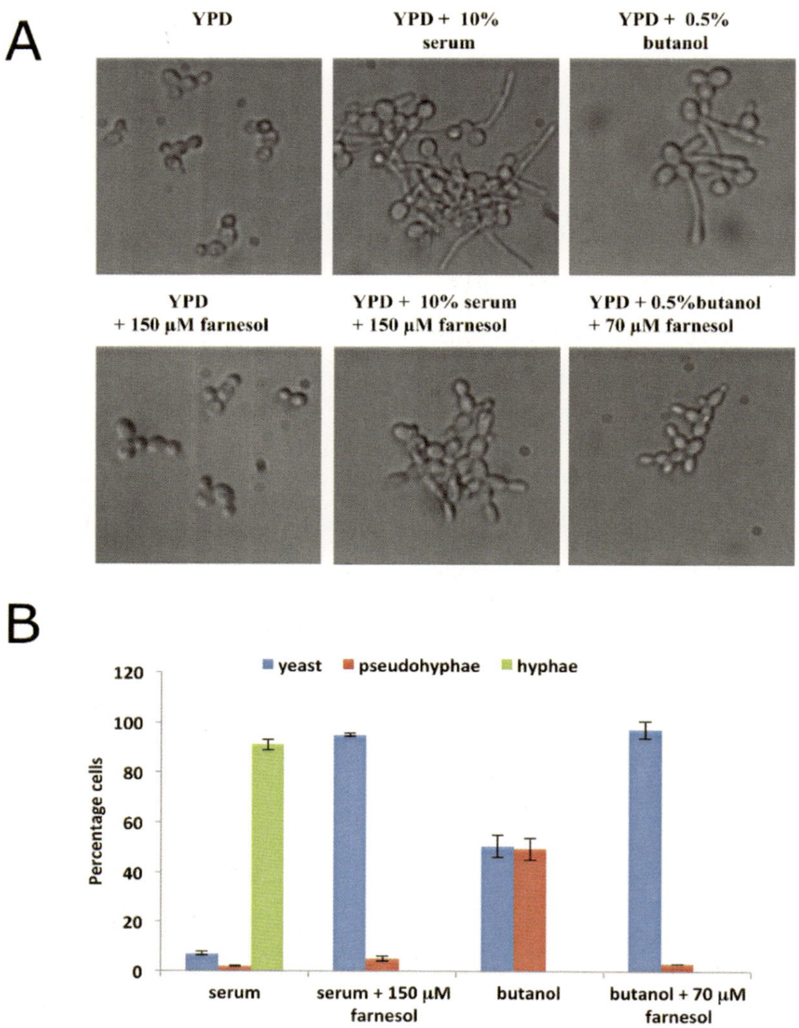

FIGURE 3: Farnesol represses morphological transition in *C. albicans*. (A) Overnight exponential cultures of the CAI4 strain were harvested, washed in water then incubated for 6 hours in different prewarmed media: YPD medium or YPD containing 150 μM farnesol; YPD containing 10% serum or YPD 10% serum with 150 μM farnesol; YPD containing 0.5% butanol or YPD 0.5% butanol with 70 μM farnesol. Cells were visualised via microscopy and photographed. **(B)** Cells from the above cultures were counted according to morphology using a cell counting chamber where pseudohyphae were scored if evidence of a restriction point between two cells was apparent, whereas hyphae were scored where elongated cells with no restriction points were viewed. Percentages are an average from three biological replicates (error bars = SEM).

Gcn2p kinase, that is a requirement for eIF2α phosphorylation and the subsequent regulation of translation initiation in response to a variety of stresses, is not required for the inhibition of translation initiation by farnesol in *C. albicans*; in fact the *gcn2Δ* mutant is more *sensitive* to treatment.

Fusel alcohols and other conditions that inhibit protein synthesis in *C. albicans* promote filamentous growth [36], whereas farnesol inhibits protein synthesis and *prevents* filamentation. An obvious query is whether the filamentation inducing signal generated by fusel alcohols can be overridden by farnesol or vice versa. Induction of filamentous growth is further complicated as different cues induce distinct forms of filamentation [43]. For instance, fusel alcohols are characterised as inducing pseudohyphal growth [44] where elongated, ellipsoid yeast cells remain attached to one another via constricted septation sites leading to growth of a colony in a branched pattern [45]. In contrast, serum addition elicits true hyphal growth [46], whereby cells are narrow, long, have parallel sides and no obvious constrictions points [45]. In the presence of serum alone over 90% hyphal growth was observed and the addition of 150 μM farnesol blocked the yeast to hyphae switch (Fig. 3A and B). In contrast, the fusel alcohol, butanol, induces a much less robust effect whereby roughly 50% of cells exhibit pseudohyphal morphology. Here just 70 μM farnesol was sufficient to block any filamentous growth. These results show that farnesol competes with both serum and butanol, but the concentration of farnesol required to effect competition varies according to the strength of the filamentation signal (Fig. 3A and B). Curiously, even though both fusel alcohols and farnesol target protein synthesis, they are in competition with respect to their physiological impact on filamentous growth. Thus previous observations suggesting that the inhibition of protein synthesis favours filamentation [36] cannot be generalised across all conditions.

Eukaryotic initiation factor 2B (eIF2B) is not regulated by farnesol

Fusel alcohols inhibit translation initiation in *S. cerevisiae* and *C. albicans* by targeting the guanine nucleotide exchange factor, eIF2B leading to reduced levels of the eIF2•GTP•Met-tRNA^{i} ternary complex [34-36].

Specific general control response reporters such as the *GCN4* reporter mRNA provide a sensitive indicator of changes in the ternary complex and are widely used to study translational regulation [25]. A key observation that pointed towards eIF2B as a target for fusel alcohols was the demonstration that these reporters of ternary complex levels are translationally up-regulated [34, 36]. In order to assess this response after farnesol treatment, strains carrying two renilla luciferase reporters were used: the first contains five copies of the general control response element (GCRE), while the second harbours the *GCN4* promoter and leader region upstream [28, 29]. Using the parent and *gcn2Δ* mutant strains bearing these reporters, the previous observation that 1% butanol elicits a non-Gcn2p dependent increase in the activity of *GCRE-Luc* and *GCN4-Luc* was confirmed (Fig. 4A and B). In stark contrast, farnesol elicits no significant increases of the *GCRE-Luc* or *GCN4-Luc* reporter expression (Fig. 4A and B) suggesting that farnesol does not alter ternary complex levels to activate the GCN response.

Previous studies on the localisation of eIF2B in organisms from *S. cerevisiae* to *Drosophila melanogaster* have defined a large cytoplasmic body called the eIF2B body (2B body) [35, 36, 47-49]. Exposure of either *S. cerevisiae* or *C. albicans* cells to fusel alcohols reduces the dynamics of this 2B body in a manner that correlates with the sensitivity/resistance of strains to alcohols [35, 36]. In order to ascertain whether farnesol also impacts upon 2B body dynamics, a *C. albicans* strain bearing GFP-tagged eIF2Bγ was used [36]. Epiflourescence time-lapse microscopy experiments were performed by acquiring images of untreated, butanol treated or farnesol treated cells over a 2 min period. Movement of the eIF2B body across the images was tracked and the total distance (μm) moved was calculated. Quantitation of the average displacement shows that 1% butanol causes total eIF2B body movement to drop by

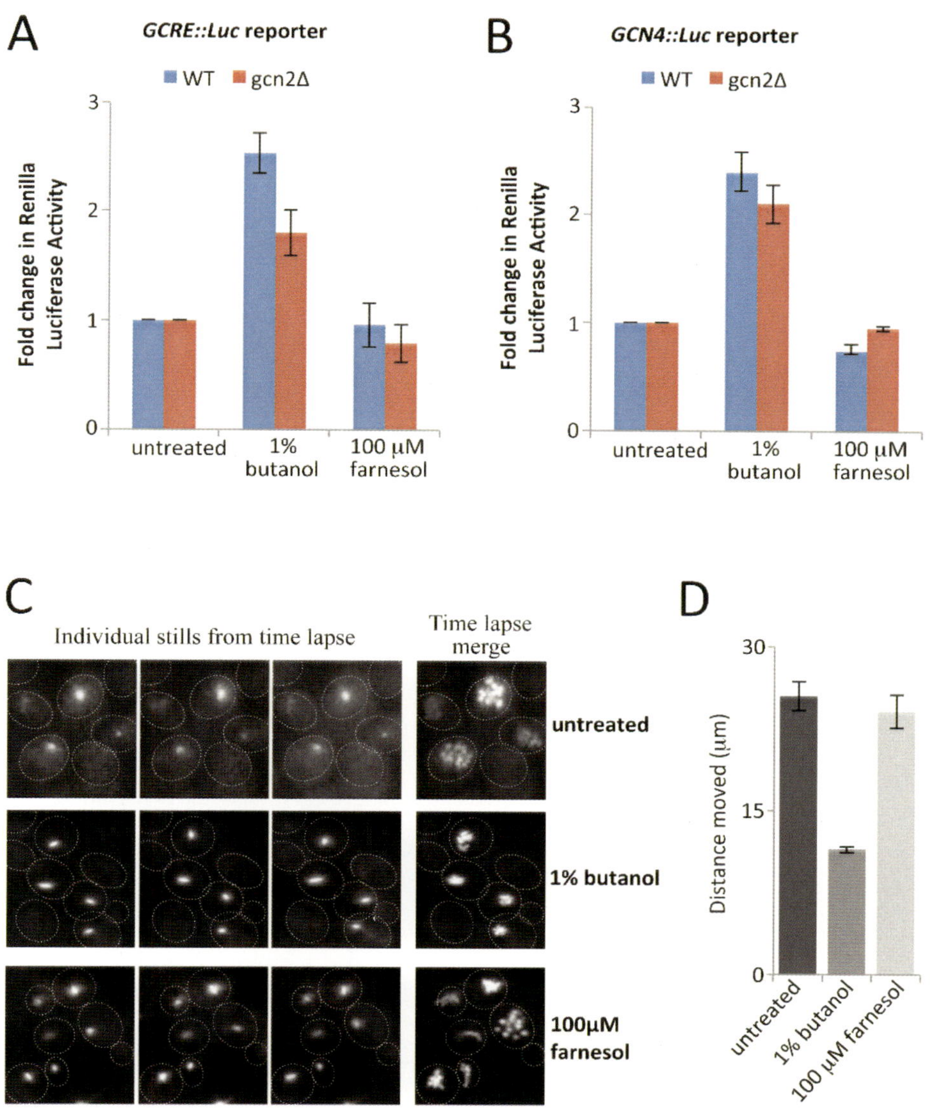

FIGURE 4: Farnesol does not induce *GCN4* expression or effect the dynamics of the eIF2B body. (A) and (B) CAI4 and *gcn2Δ* strains bearing the *GCRE-Luc* (A) and *GCN4-Luc* (B) reporters were treated with 1% (v/v) butanol or 100 μM farnesol for 2 h, extracts were prepared, then renilla luciferase activity was measured relative to untreated. Error bars = ± SEM. **(C)** Images from time-lapse microscopy studies using an eIF2Bγ-GFP expressing *C. albicans* strain. The strains were incubated in media with 1% butanol, or 100 μM farnesol, or they were left untreated (UT) for 15 min as indicated. Each row contains three stills from a series of 25 images over a period of 2 min, as well as a merged image of all 25 stills, which serves to depict the total extent of 2B body movement. **(D)** Bar chart depicting the mean distance moved in μ over a 2-min period from 24 time-lapse experiments. Error bars, ±1 SEM.

approximately 50%. In contrast, in farnesol treated cells the 2B body moves to the same extent as in the untreated cells (Fig. 4C and D). This observation again suggests that the regulatory mechanism by which farnesol inhibits translation initiation is distinct from that of fusel alcohols and is not dependent upon eIF2B regulation or the alteration of ternary complex levels.

Farnesol inhibits translation initiation by targeting 48S preinitiation complex formation

Translation initiation is controlled at other levels besides ternary complex formation and eIF2B. For instance, the interaction of mRNA with the 43S preinitiation complex, i.e. 48S preinitiation complex formation, can also be regulated [19, 20, 37]. This process relies upon interactions between proteins that bind the mRNA and proteins associated with the 40S ribosomal subunit complex. One way to assess the factors present with the 40S ribosomal subunit is to perform immunoblotting on fractions collected from across sucrose density polysomal gradients. Formaldehyde cross-linking prior to cell lysis stabilizes protein factors in such complexes during the subsequent sedimentation and fractionation steps [50, 51].

A limitation of such studies in *C. albicans* is that many antibodies against translation factors that are available for *S. cerevisiae* do not cross-react with *C. albicans* proteins (data not shown). Therefore, to further investigate the step in the translation pathway that is targeted by farnesol, investigations were undertaken in *S. cerevisiae*. The Σ1278b laboratory strain was selected, as like *C. albicans*, Σ1278b is diploid and can undergo morphogenetic switching to pseudohyphal growth [44]. In order to validate the

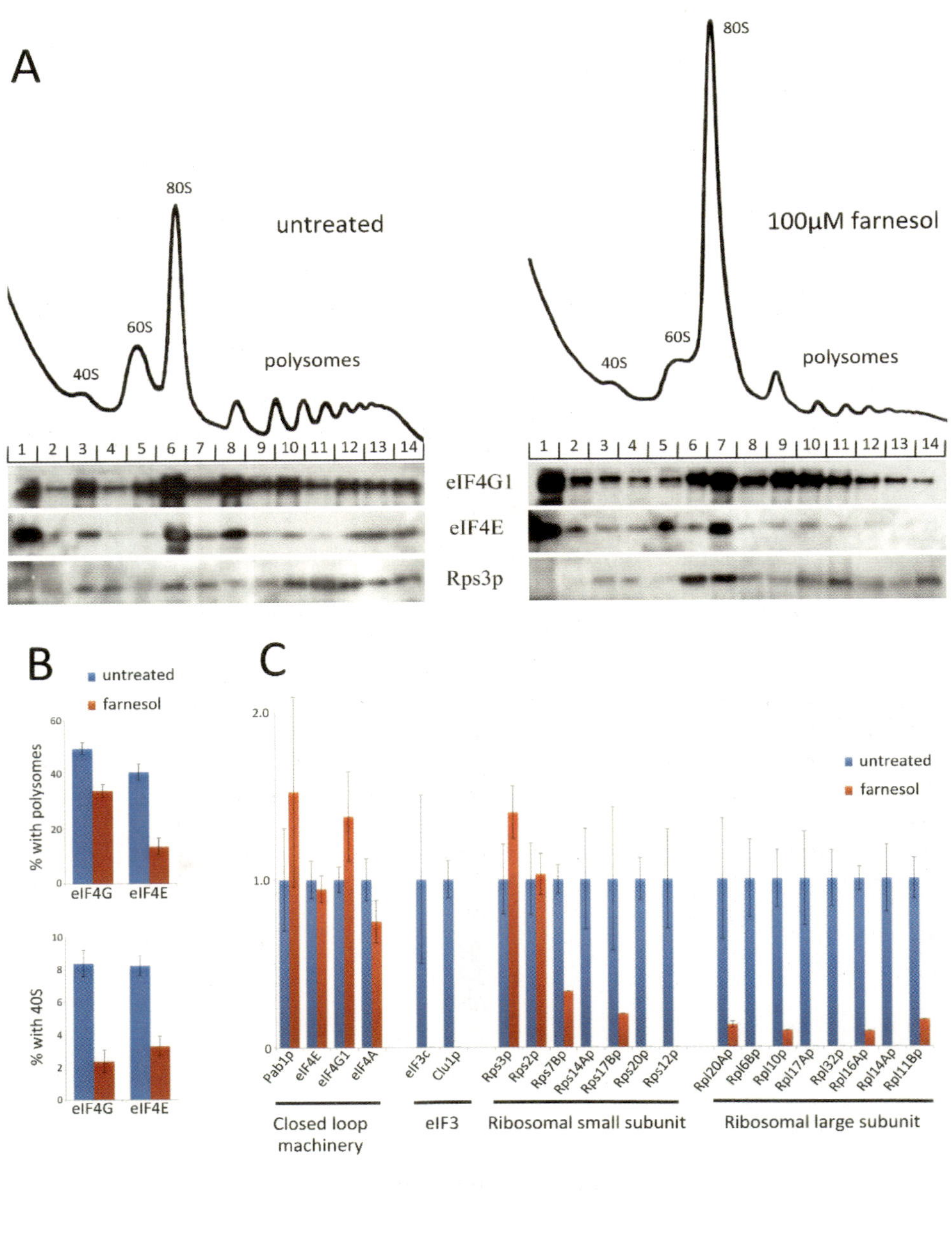

FIGURE 5: Farnesol affects the association of specific translation factors with the 48S pre-ribosomal complex. (A) Formaldehyde sucrose density gradient analysis on extracts from the MLY61 strain either treated with 100 μM farnesol for 15 min or untreated. Immunoblots on gradient fractions are shown below the traces and these were probed using antibodies against the indicated proteins. **(B)** Quantitation of the proportion of eIF4G and eIF4E present in the polysome fraction of the gradients before and after farnesol treatment. Quantitation of the proportion of eIF4G and eIF4E present in the 40S region of gradients before and after farnesol treatment. **(C)** Whole cell extracts and immunoprecipitation samples derived from the eIF4G1-TAP tagged strain treated with 100 μM farnesol or untreated were analysed via mass spectrometry. The plot shows the relative number of unique peptides that have been matched to the identified protein in the sample where the untreated was normalised to 1 to facilitate a direct comparison of the different translation factors.

use of this strain, the effects of farnesol on growth and translation were cross-compared. Farnesol inhibits growth and translation initiation over a similar concentration range for the two yeasts and for other lab strains of *S. cerevisiae*, such as BY4741 (Fig. 2C; data not shown), so it seems likely that the translational responses to butanol and farnesol are mechanistically conserved across these species.

Formaldehyde polysome analysis after farnesol treatment revealed an interesting effect in terms of the region of the gradient harbouring the 40S ribosomal subunit. This region not only contains free 40S ribosomal subunits but also the 48S ribosomal preinitiation complex where the 40S subunit is associated with mRNA and translation initiation factors i.e. an intermediate in the translation initiation process. A marker for this complex is the presence of translation initiation factors such as eIF4E and eIF4G that are specifically targeted to the mRNA rather than the 40S ribosomal subunit. The level of both eIF4G1 and eIF4E in the 40S region decreased dramatically after farnesol treatment (Fig. 5A, *cf.* fraction 3 untreated and fraction 3 treated). Quantitation confirmed that levels dropped from ~8-9% of total to 2-3% after farnesol treatment (Fig. 5B). Furthermore, both eIF4G and eIF4E are reduced in polysome regions and this likely reflects reduced levels of initiating ribosomes on mRNAs that are already being translated: although it should be noted that the scale of reduction is greater for eIF4E than eIF4G (Fig. 5B). This may relate to the fact that eIF4G can interact with RNA, Pab1p and other translation factors, whereas eIF4E is targeted to the mRNA cap. In sum, these data highlight the possibility that farnesol causes an alteration in protein-protein interactions that lie upstream of 48S complex formation.

In order to further investigate how farnesol treatment leads to eIF4G/eIF4E depletion from regions of the gradient, we undertook an immunopurification-mass spectrometry strategy using an eIF4G1-TAP tagged *S. cerevisiae* strain. In terms of the relative number of peptides observed in the immunopurified samples, peptides from known mRNA associated factors, such as eIF4G, eIF4E, eIF4A and Pab1p, were largely unaffected by farnesol (Fig. 5B). In contrast, peptides for other components of the translation machinery were reduced dramatically; including peptides for the ribosomal proteins, as well as subunits of eIF3, a translation initiation factor that is associated with the 40S ribosomal subunit. Overall, these data support a model where farnesol targets the formation of the 48S preinitiation complex to inhibit protein synthesis. The fact that farnesol targets a different step in translation to fusel alcohols, may mean that the impact of these agents and the stage targeted contributes to the opposite effects in terms of filamentous growth (Fig. 3). This is suggestive that translational control plays an important role in the physiological response of *C. albicans* to QS molecules.

DISCUSSION

A range of alcohols or their derivatives can act as signalling molecules across yeast species [2, 44, 52, 53]. The data presented here combined with that in our previous studies [36] show that both butanol and farnesol inhibit protein synthesis at the translation initiation stage in *C. albicans*. Both are metabolites of *C. albicans* that act as signalling molecules, yet have opposing effects on morphological transition [2, 36]. Previously, we have shown in both *S. cerevisiae* and *C. albicans* that short chain alcohols regulate protein synthesis by targeting the guanine nucleotide exchange factor eIF2B [34-36].

eIF2B regulation plays a critical role in reprograming gene expression as part of the response to stress across different eukaryotic cells [23]. For instance, eIF2 phosphorylation by eIF2α kinases, like Gcn2p in yeast, inhibits eIF2B in response to stresses such as amino acid starvation [24], purine starvation [54] and rapamycin treatment [55]. However, while these stresses target eIF2B in a Gcn2p-dependent manner, the mechanism by which short chain alcohols target eIF2B in both *S. cerevisiae* and *C. albicans* is Gcn2p-independent [34-36]. The *GCN4* reporter experiments and analysis of the eIF2B body in this study suggest that the longer chain sesquiterpene alcohol farnesol inhibits translation initiation in a mechanism that does not involve eIF2B regulation; either Gcn2p-dependent or independent.

Many studies have reported translational controls targeting steps upstream of 48S preinitiation complex formation. For instance, glucose starvation in *S. cerevisiae* causes a reorganisation of the closed loop mRNP translation complex, whereby eIF4A dissociates and the cosedimentation of eIF4E, eIF4G and Pab1p with ribosomal complexes is compromised [51]. The small non-coding BC RNAs in neuronal cells target the eIF4A helicase to inhibit 48S preinitiation complex formation on structured mRNAs [56]. Similarly, *Burkholderia* lethal factor 1, a toxin produced by *Burkholderia pseudomallei*, which causes the disease melioidosis, provokes a translational block via eIF4A [57]. 48S complex formation can also serve as the targeted step when *specific* mRNAs are translationally regulated. For instance, miRNAs have recently been shown to inhibit target mRNA translation by impacting upon eIF4A2 activity [58]. Therefore, a common translational regulatory mechanism that impacts upon the level of the 48S preinitiation complex is to target eIF4A activity.

In this study, we have investigated how farnesol effects different ribosomal complexes in *S. cerevisiae* using both formaldehyde-polysome analysis and immunoprecipitation followed by mass spectrometry. Both assays suggest that mRNA-associated translation factors (such as eIF4G, eIF4E and Pab1p) are associated less well with the ribosome and eIF3 following treatment with farnesol. Overall, the observed depletion of eIF4G and eIF4E from the 40S region of polysome gradients combined with the mass spectrometric analysis of eIF4G containing complexes lend support to a model where farnesol targets the formation of the 48S preinitiation complex to inhibit protein synthesis. This contrasts with the eIF2B dependent mechanism by which shorter chain alcohols target translation initiation.

Shorter chain alcohols and farnesol also differ in terms of their effects on morphological transitions in *C. albicans*. Short chain alcohols induce pseudohyphal growth in

C. albicans whereas farnesol inhibits this process [2, 44, 53]. Indeed we show that farnesol can impede the filamentation induced by a variety of triggers including short-chain alcohols. This is not without precedent, in *C. albicans* farnesol also blocks morphogenesis induced by the aromatic alcohol, tyrosol [59]. One intriguing question is how both the shorter chain alcohols and farnesol can target a key ubiquitous process like protein synthesis, yet elicit distinct outcomes in terms of filamentous growth. This question drives at the fundamental physiological rationale for translational regulation in response to changing external conditions. Does the regulation constitute a knee-jerk reaction allowing the preservation of cellular resources by inhibiting the expression of the vast majority of mRNAs, or does the regulation serve a different purpose allowing specific mRNAs to be altered in their translation? In terms of farnesol, we show that translation is mildly inhibited at 40 µM farnesol and robustly down-regulated at 100 µM farnesol. Various *Candida* strains produce farnesol up to a concentration of ~60 µM [60], which would appear to favour the option where translation of a specific subset of mRNAs is altered. Evidence from a number of systems including the induction of *GCN4* translation via amino acid starvation [25] would also favour this option. Under such a scenario, if two stresses impact upon different stages of translation initiation, they might alter the translation of different subsets of mRNA. We have previously observed evidence for such effects in *S. cerevisiae*, where fusel alcohols and amino acid starvation alter translational reprograming to allow continued translation of different cohorts of mRNAs [61]. With this in mind, we envisage that for farnesol and fusel alcohols, mRNAs encoding pro and anti-filamentation factors might be prominent in a set that are differentially regulated at the translational level. Such effects would also be integrated with well-defined transcriptional controls, especially for farnesol [18], to produce very different phenotypic outcomes.

MATERIALS AND METHODS

Media and growth conditions

The strains in Table 1 were grown and maintained as described previously [36]. Butanol and farnesol were routinely added for 15 min at the concentrations stated. Unless otherwise stated *trans,trans*-farnesol was used. Tolerance was assessed by adding butanol (0.5%, 1% and 2%) or farnesol (40 µM, 100 µM, 200 µM and 300 µM) to strains at OD_{600} 0.1 and then testing growth.

Morphogenesis assays

Exponential cultures were harvested, washed in water then re-inoculated into media with 0.5% butanol, 10% serum, 0.5% butanol - 70 µM farnesol or 10% serum - 150 µM farnesol. Filamentation was assessed microscopically as previously described [36].

Analysis of polysomes and other translation assays

Exponential strains were incubated with butanol/ farnesol for 15 min then treated with cycloheximide: 1 mg/ml (*C. albicans*) or 0.1 mg/ml (*S. cerevisiae*). Extracts were prepared then polysome analysis and fractionation were

TABLE 1. Strains used in this study.

***C. albicans* strains**		
Strain	**Genotype**	**Source**
CAI4	*ura3::λimm434/ura3::λimm434*	A. Brown
CAI8	*ura3::λimm434/ura3::λimm434 ade2::hisG/ade2::hisG*	A. Brown
gcn2Δ	*ura3::λimm434/ura3::λimm434 gcn2::hisG/gcn2::hisG*	A. Brown
SC5314	Prototroph	C. Grant
yMK2313	*ura3::λimm434/ura3::λimm434 GCD1-GFP::NAT/GCD1*	Ashe lab
CY2383	*ura3::λimm434/ura3::λimm434 ADE2::GCRE-rLUC/ade2::hisG*	C. Grant
CY2511	*ura3::λimm434/ura3::λimm434 ADE2::GCN4-rLUC/ade2::hisG*	C. Grant
CY2387	*ura3::λimm434/ura3::λimm434 gcn2::hisG/gcn2::hisG ADE2::GCRE-rLUC/ade2::hisG*	C. Grant
CY2492	*ura3::λimm434/ura3::λimm434 gcn2::hisG/gcn2::hisG ADE2::GCN4-rLUC/ade2::hisG*	C. Grant

***S. cerevisiae* strains**		
Strain	**Genotype**	**Source**
MLY61(Σ1278b)	*MATa/MATα ura3-52/ura3-52*	J. Heitman
yMK2197	*MATa his3Δ1 leu2Δ0 met15Δ0 ura3Δ0 HIS3*	Open biosystems
yMK2084	*MATa his3Δ1 leu2Δ0 met15Δ0 ura3Δ0 TIF4631-TAP::HIS3*	Open Biosystems

carried out as previous [36]. Formaldehyde polysome analysis was performed as described previously [50,51]. Immunoblots were probed with antibodies to yeast eIF4G, eIF4E and Rps3.

[^{35}S]-methionine incorporation assays were conducted by adding 60 ng/ml methionine, where 0.5 ng/ml was [^{35}S]-methionine (PerkinElmer), to exponential untreated or farnesol/ butanol treated cultures in synthetic complete dextrose (SCD) medium lacking methionine. Samples (1 ml) were taken at the indicated times and processed as described previously [36].

For the Luciferase reporter assays [29], lysates were prepared from exponential untreated or farnesol/ butanol treated cultures in RLUC buffer (0.5 M NaCl, 0.1 M K_2HPO_4, 1 mM Na_2EDTA, 0.6 mM sodium azide, 1 mM phenylmethylsulfonyl fluoride, 0.02% bovine serum albumin). 1.25 µM coelentrazine h (Promega) was added to the extracts to initiate the reaction, and activity was measured using a GloMax 20/20 luminometer (Promega). Luciferase activity (RLU) is expressed as relative luminescence per 10 s/mg protein.

For studies on the 2B-body [35, 36], real-time 2D deconvolved projections were generated via continuous z-sweep acquisition on a Delta Vision RT microscope (Applied Precision, Isaaquah, WA) with an Olympus 100× 1.40 NA DIC oil PlanApo objective (Melville, NY) and Roper CoolSnap HQ camera (Tucson, AZ) with Applied Precision Softworx 1.1 software for fast visualisation of all planes with minimal fluorescent bleaching. Images were acquired every 5 s over a 2 min period, and ImageJ (http://rsb.info.nih.gov/ij/; NIH) was used to track 2B body movement and calculate the mean total distance using at least 24 individual tracking experiments per condition.

Affinity Purification and mass spectrometry

For the eIF4G1-TAP purification, protein extracts were bound to IgG columns eluted with a TAP peptide, then samples were isolated from SDS PAGE gel slices [62]. Dried gel pieces containing the whole protein sample were digested using 100 ng trypsin and analysed by LC-MS/MS using an UltiMate® 3000 Rapid Separation liquid chromatography (Dionex Corporation, Sunnyvale, CA) coupled to a LTQ Velos Pro mass spectrometer (Thermo Fisher Scientific, Waltham, MA). Data were searched using Mascot (Matrix Science UK), against the Uniprot database with *S. cerevisiae* selected. Data were validated and further processed using Scaffold (Proteome Software, Portland, OR).

ACKNOWLEDGEMENTS

We thank A. Brown (Aberdeen University), C. Grant (University of Manchester (UoM)) and M. Pool (UoM) for strains and antibodies. We thank P. March and S. Mardsen for their help with the microscopy. The Bioimaging Facility microscopes used rely on grants by the Biotechnology and Biological Sciences Research Council (BBSRC), the Wellcome Trust and a UoM Strategic Fund. We are grateful to David Knight and Emma-Jane Keevil in the Biomolecular Analysis Core Facility, The University of Manchester for their help with mass spectrometry analysis. NEE was funded by the Tertiary education trust fund Nigeria, TD was funded by the Coordenação de Aperfeiçoamento de Pessoal de Nível Superior (CAPES)- science without borders program, CMP was funded by a BBSRC project grant (BB/K002767/1) and LMC was funded by a BBSRC LoLa grant (BB/G012571/1).

CONFLICT OF INTEREST

The authors declare no conflict of interest.

REFERENCES

1. Mayer FL, Wilson D, Hube B (**2013**). Candida albicans pathogenicity mechanisms. **Virulence** 4(2): 119-128.

2. Hornby JM, Jensen EC, Lisec AD, Tasto JJ, Jahnke B, Shoemaker R, Dussault P, Nickerson KW (**2001**). Quorum sensing in the dimorphic fungus Candida albicans is mediated by farnesol. **Appl Environmental Microbiol** 67(7): 2982-2992.

3. Langford ML, Atkin AL, Nickerson KW (**2009**). Cellular interactions of farnesol, a quorum-sensing molecule produced by Candida albicans. **Future Microbiol** 4(10): 1353-1362.

4. Ramage G, Saville SP, Wickes BL, Lopez-Ribot JL (**2002**). Inhibition of Candida albicans biofilm formation by farnesol, a quorum-sensing molecule. **Appl Environmental Microbiol** 68(11): 5459-5463.

5. Martins M, Henriques M, Azeredo J, Rocha SM, Coimbra MA, Oliveira R (**2007**). Morphogenesis control in Candida albicans and Candida dubliniensis through signaling molecules produced by planktonic and biofilm cells. **Eukaryot Cell** 6(12): 2429-2436.

6. Westwater C, Balish E, Schofield DA (**2005**). Candida albicans-conditioned medium protects yeast cells from oxidative stress: a possible link between quorum sensing and oxidative stress resistance. **Eukaryot Cell** 4(10): 1654-1661.

7. Deveau A, Piispanen AE, Jackson AA, Hogan DA (**2010**). Farnesol induces hydrogen peroxide resistance in Candida albicans yeast by inhibiting the Ras-cyclic AMP signaling pathway. **Eukaryot Cell** 9(4): 569-577.

8. Sharma M, Prasad R (**2011**). The quorum-sensing molecule farnesol is a modulator of drug efflux mediated by ABC multidrug transporters and synergizes with drugs in Candida albicans. **Antimicrob Agents Chemother** 55(10): 4834-4843.

9. Machida K, Tanaka T, Yano Y, Otani S, Taniguchi M (**1999**). Farnesol-induced growth inhibition in Saccharomyces cerevisiae by a cell cycle mechanism. **Microbiology** 145(Pt 2): 293-299.

10. Dinamarco TM, Goldman MH, Goldman GH (**2011**). Farnesol-induced cell death in the filamentous fungus Aspergillus nidulans. **Biochem Soc Trans** 39(5): 1544-1548.

11. Liu P, Luo L, Guo J, Liu H, Wang B, Deng B, Long CA, Cheng Y (**2010**). Farnesol induces apoptosis and oxidative stress in the fungal pathogen Penicillium expansum. **Mycologia** 102(2): 311-318.

12. Cotoras M, Castro P, Vivanco H, Melo R, Mendoza L (**2013**). Farnesol induces apoptosis-like phenotype in the phytopathogenic fungus Botrytis cinerea. **Mycologia** 105(1): 28-33.

13. Shirtliff ME, Krom BP, Meijering RA, Peters BM, Zhu J, Scheper MA, Harris ML, Jabra-Rizk MA (**2009**). Farnesol-induced apoptosis in Candida albicans. **Antimicrob Agents Chemother** 53(6): 2392-2401.

14. Joo JH, Jetten AM (**2010**). Molecular mechanisms involved in farnesol-induced apoptosis. **Cancer lett** 287(2): 123-135.

15. Gomes FI, Teixeira P, Azeredo J, Oliveira R (**2009**). Effect of farnesol on planktonic and biofilm cells of Staphylococcus epidermidis. **Curr Microbiol** 59(2): 118-122.

16. Jabra-Rizk MA, Meiller TF, James CE, Shirtliff ME (**2006**). Effect of farnesol on Staphylococcus aureus biofilm formation and antimicrobial susceptibility. **Antimicrob Agents Chemother** 50(4): 1463-1469.

17. Kuete V, Efferth T (**2013**). Molecular determinants of cancer cell sensitivity and resistance towards the sesquiterpene farnesol. **Pharmazie** 68(7): 608-615.

18. Langford ML, Hargarten JC, Patefield KD, Marta E, Blankenship JR, Fanning S, Nickerson KW, Atkin AL (**2013**). Candida albicans Czf1 and Efg1 coordinate the response to farnesol during quorum sensing, white-opaque thermal dimorphism, and cell death. **Eukaryot cell** 12(9): 1281-1292.

19. Simpson CE, Ashe MP (**2012**). Adaptation to stress in yeast: to translate or not? **Biochem Soc Trans** 40(4): 794-799.

20. Spriggs KA, Bushell M, Willis AE (**2010**). Translational regulation of gene expression during conditions of cell stress. **Mol Cell** 40(2): 228-237.

21. Hinnebusch AG (**2014**). The scanning mechanism of eukaryotic translation initiation. **Ann Rev Biochem** 83: 779-812.

22. Pavitt GD (**2005**). eIF2B, a mediator of general and gene-specific translational control. **Biochem Soc Trans** 33(Pt 6): 1487-1492.

23. Wek RC, Jiang HY, Anthony TG (**2006**). Coping with stress: eIF2 kinases and translational control. **Biochem Soc Trans** 34(Pt 1): 7-11.

24. Dever TE, Feng L, Wek RC, Cigan AM, Donahue TF, Hinnebusch AG (**1992**). Phosphorylation of initiation factor 2 alpha by protein kinase GCN2 mediates gene-specific translational control of GCN4 in yeast. **Cell** 68(3): 585-596.

25. Hinnebusch AG (**2005**). Translational regulation of GCN4 and the general amino acid control of yeast. **Annu Rev Microbiol** 59: 407-450.

26. Tournu H, Tripathi G, Bertram G, Macaskill S, Mavor A, Walker L, Odds FC, Gow NA, Brown AJ (**2005**). Global role of the protein kinase Gcn2 in the human pathogen Candida albicans. **Eukaryot cell** 4(10): 1687-1696.

27. Sundaram A, Grant CM (**2014**). Oxidant-specific regulation of protein synthesis in Candida albicans. **Fungal Genet Biol** 67: 15-23.

28. Tripathi G, Wiltshire C, Macaskill S, Tournu H, Budge S, Brown AJ (**2002**). Gcn4 co-ordinates morphogenetic and metabolic responses to amino acid starvation in Candida albicans. **EMBO J** 21(20): 5448-5456.

29. Sundaram A, Grant CM (**2014**). A single inhibitory upstream open reading frame (uORF) is sufficient to regulate Candida albicans GCN4 translation in response to amino acid starvation conditions. **RNA** 20(4): 559-567.

30. Wang X, Paulin FE, Campbell LE, Gomez E, O'Brien K, Morrice N, Proud CG (**2001**). Eukaryotic initiation factor 2B: identification of multiple phosphorylation sites in the epsilon-subunit and their functions in vivo. **EMBO J** 20(16): 4349-4359.

31. Palmer LK, Rannels SL, Kimball SR, Jefferson LS, Keil RL (**2006**). Inhibition of mammalian translation initiation by volatile anesthetics. **Am J Physiol Endocrinol Metab** 290(6): E1267-E1275.

32. Palmer LK, Shoemaker JL, Baptiste BA, Wolfe D, Keil RL (**2005**). Inhibition of translation initiation by volatile anesthetics involves nutrient-sensitive GCN-independent and -dependent processes in yeast. **Mol Biol Cell** 16(8): 3727-3739.

33. Hogan DA (**2006**). Quorum sensing: alcohols in a social situation. **Curr Biol** 16(12): R457-458.

34. Ashe MP, Slaven JW, De Long SK, Ibrahimo S, Sachs AB (**2001**). A novel eIF2B-dependent mechanism of translational control in yeast as a response to fusel alcohols. **EMBO J** 20(22): 6464-6474.

35. Taylor EJ, Campbell SG, Griffiths CD, Reid PJ, Slaven JW, Harrison RJ, Sims PF, Pavitt GD, Delneri D, Ashe MP (**2010**). Fusel alcohols regulate translation initiation by inhibiting eIF2B to reduce ternary complex in a mechanism that may involve altering the integrity and dynamics of the eIF2B body. **Mol Biol Cell** 21(13): 2202-2216.

36. Egbe NE, Paget CM, Wang H, Ashe MP (**2015**). Alcohols inhibit translation to regulate morphogenesis in C. albicans. **Fungal Genet Biol** 77: 50-60.

37. Richter JD, Sonenberg N (**2005**). Regulation of cap-dependent translation by eIF4E inhibitory proteins. **Nature** 433(7025): 477-480.

38. Langford ML, Hasim S, Nickerson KW, Atkin AL (**2010**). Activity and toxicity of farnesol towards Candida albicans are dependent on growth conditions. **Antimicrob Agents Chemother** 54(2): 940-942.

39. Ibrahimo S, Holmes LE, Ashe MP (**2006**). Regulation of translation initiation by the yeast eIF4E binding proteins is required for the pseudohyphal response. **Yeast** 23(14-15): 1075-1088.

40. Fonzi WA, Irwin MY (**1993**). Isogenic strain construction and gene mapping in Candida albicans. **Genetics** 134(3): 717-728.

41. Staab JF, Sundstrom P (**2003**). URA3 as a selectable marker for disruption and virulence assessment of Candida albicans genes. **Trends Microbiol** 11(2): 69-73.

42. Shchepin R, Hornby JM, Burger E, Niessen T, Dussault P, Nickerson KW (**2003**). Quorum sensing in Candida albicans: probing farnesol's mode of action with 40 natural and synthetic farnesol analogs. **Chem Biol** 10(8): 743-750.

43. Berman J (**2006**). Morphogenesis and cell cycle progression in Candida albicans. **Curr Opin Microbiol** 9(6): 595-601.

44. Lorenz MC, Cutler NS, Heitman J (**2000**). Characterization of alcohol-induced filamentous growth in Saccharomyces cerevisiae. **Mol Biol Cell** 11(1): 183-199.

45. Sudbery P, Gow N, Berman J (**2004**). The distinct morphogenic states of Candida albicans. **Trends Microbiol** 12(7): 317-324.

46. Ogletree FF, Abdelal AT, Ahearn DG (**1978**). Germ-tube formation by atypical strains of Candida albicans. **Antonie van Leeuwenhoek** 44(1): 15-24.

47. Campbell SG, Hoyle NP, Ashe MP (**2005**). Dynamic cycling of eIF2 through a large eIF2B-containing cytoplasmic body: implications for translation control. **J Cell Biol** 170(6): 925-934.

48. Browne CM, Samir P, Fites JS, Villarreal SA, Link AJ (**2013**). The yeast eukaryotic translation initiation factor 2B translation initiation complex interacts with the fatty acid synthesis enzyme YBR159W and endoplasmic reticulum membranes. **Mol Cell Biol** 33(5): 1041-1056.

49. Noree C, Sato BK, Broyer RM, Wilhelm JE (**2010**). Identification of novel filament-forming proteins in Saccharomyces cerevisiae and Drosophila melanogaster. **J Cell Biol** 190(4): 541-551.

50. Nielsen KH, Szamecz B, Valasek L, Jivotovskaya A, Shin BS, Hinnebusch AG (**2004**). Functions of eIF3 downstream of 48S assembly impact AUG recognition and GCN4 translational control. **EMBO J** 23(5): 1166-1177.

51. Castelli LM, Lui J, Campbell SG, Rowe W, Zeef LAH, Holmes LEA, Hoyle NP, Bone J, Selley JN, Sims PFG, Ashe MP (**2011**). Glucose depletion inhibits translation initiation via eIF4A loss and subsequent 48S preinitiation complex accumulation, while the pentose phosphate pathway is coordinately up-regulated. **Mol Biol Cell** 22(18): 3379-3393.

52. Chen H, Fink GR (**2006**). Feedback control of morphogenesis in fungi by aromatic alcohols. **Genes Dev** 20(9): 1150-1161.

53. Dickinson JR (**1996**). 'Fusel' alcohols induce hyphal-like extensions and pseudohyphal formation in yeasts. **Microbiology** 142(Pt 6): 1391-1397.

54. Rolfes RJ, Hinnebusch AG (**1993**). Translation of the yeast transcriptional activator GCN4 is stimulated by purine limitation: implications for activation of the protein kinase GCN2. **Mol Cell Biol** 13(8): 5099-5111.

55. Kubota H, Obata T, Ota K, Sasaki T, Ito T (**2003**). Rapamycin-induced translational derepression of GCN4 mRNA involves a novel mechanism for activation of the eIF2 alpha kinase GCN2. **J Biol Chem** 278(23): 20457-20460.

56. Eom T, Berardi V, Zhong J, Risuleo G, Tiedge H (**2011**). Dual nature of translational control by regulatory BC RNAs. **Mol Cell Biol** 31(22): 4538-4549.

57. Cruz-Migoni A, Hautbergue GM, Artymiuk PJ, Baker PJ, Bokori-Brown M, Chang CT, Dickman MJ, Essex-Lopresti A, Harding SV, Mahadi NM, Marshall LE, Mobbs GW, Mohamed R, Nathan S, Ngugi SA, Ong C, Ooi WF, Partridge LJ, Phillips HL, Raih MF, Ruzheinikov S, Sarkar-Tyson M, Sedelnikova SE, Smither SJ, Tan P, Titball RW, Wilson SA, Rice DW (**2011**). A Burkholderia pseudomallei toxin inhibits helicase activity of translation factor eIF4A. **Science** 334(6057): 821-824.

58. Meijer HA, Kong YW, Lu WT, Wilczynska A, Spriggs RV, Robinson SW, Godfrey JD, Willis AE, Bushell M (**2013**). Translational repression and eIF4A2 activity are critical for microRNA-mediated gene regulation. **Science** 340(6128): 82-85.

59. Ghosh S, Kebaara BW, Atkin AL, Nickerson KW (**2008**). Regulation of aromatic alcohol production in Candida albicans. **Appl Environmental Microbiol** 74(23): 7211-7218.

60. Weber K, Sohr R, Schulz B, Fleischhacker M, Ruhnke M. (**2008**). Secretion of E,E-farnesol and biofilm formation in eight different Candida species. **Antimicrob Agents Chemother** 52(5): 1859-61.

61. Smirnova JB, Selley JN, Sanchez-Cabo F, Carroll K, Eddy AA, McCarthy JE, Hubbard SJ, Pavitt GD, Grant CM, Ashe MP (**2005**). Global gene expression profiling reveals widespread yet distinctive translational responses to different eukaryotic translation initiation factor 2B-targeting stress pathways. **Mol Cell Biol** 25(21): 9340-9349.

62. Castelli LM, Talavera D, Kershaw CJ, Mohammad-Qureshi SS, Costello JL, Rowe W, Sims PF, Grant CM, Hubbard SJ, Ashe MP, Pavitt GD (**2015**). The 4E-BP Caf20p Mediates Both eIF4E-Dependent and Independent Repression of Translation. **PLoS Genet** 11(5): e1005233.

13

Mutational analysis of fructose-1,6-bis-phosphatase FBP1 indicates partially independent functions in gluconeogenesis and sensitivity to genotoxic stress

Ali Ghanem, Ana Kitanovic, Jinda Holzwarth, Stefan Wölfl*
Institute of Pharmacy and Molecular Biotechnology, Heidelberg University, Heidelberg, Germany.
* Corresponding Author:
Stefan Wölfl, E-mail: wolfl@uni-hd.de

ABSTRACT Fructose-1,6-bisphosphatase (*FBP1)* is a key enzyme in the evolutionary conserved pathway of gluconeogenesis. We had shown in an earlier study that *FBP1* is involved in the response and sensitivity to methyl-methanesulfonate (MMS)-induced DNA damage in yeast. In the work presented here we performed an alanine screen mutational analysis of several evolutionary conserved amino acid residues of *FBP1*, which were selected based on conserved residues and structural studies of mammalian and yeast homologues of *FBP1*. Mutants were examined for enzymatic activity, and yeast cells expressing these mutants were tested for growth on non-fermentable and MMS-containing media. The results obtained support predicted vital roles of several residues for enzymatic activity and led to the identification of residues indispensable for the MMS-sensitizing effect. Despite an overlap between these two properties, careful analysis revealed two mutations, Asn75 and His324, which decouple the enzymatic activity and the MMS-sensitizing effect, indicating two distinctive biological activities linked in this key gluconeogenesis enzyme.

***Keywords**: FBP1, S. cerevisiae, MMS, genotoxicity, DNA-damage, gluconeogenesis.*

Abbreviations:
MMS - methyl-methanesulfonate,
wt – wild type.

INTRODUCTION

Fructose 1,6 bisphosphatase (*FBP1*) has been known as the key-enzyme for gluconeogenesis, it irreversibly mediates the splitting of fructose 1,6 bis-phosphate into fructose 6 phosphate and inorganic phosphate. While gluconeogenesis shares most of its enzymes with glycolysis, Fbp1p is the only distinctive gluconeogenesis enzyme in the upper part of the gluconeogenesis pathway, through which the flux of this pathway can be regulated.

Gluconeogenesis has been thoroughly investigated under normal physiological conditions and in the adaptation to different nutrient conditions, and its importance in securing vital glucose homeostasis has been understood. However, the role of gluconeogenesis regulation and its effects in other conditions as for example during the development of tumours have long been overlooked [1].

The preference towards aerobic glycolysis as a major energy supply source is a common metabolic hallmark of both cancerous cells and many unicellular organisms including the yeast *S. cerevisiae*, both being aerobic fermenters with the major fermentation product being lactate in cancerous cells and ethanol in *S. cerevisiae* [2-4].

The induced glycolysis in cancers known as 'The Warburg Effect', its underpinnings and advantages for tumours form the cornerstone of our comprehension of cancer metabolism; nevertheless several additional aspects of tumour metabolism have remained in the shadow for decades [5]. In the last few years, a decisive role of FBP1-repression in the initiation of several tumour types has been described; metabolically, the loss of FBP1 increased glycolytic flux, and decreased respiration [6-8]; morphologically, FBP1-loss has been demonstrated to be essential for the epithelial to mesenchymal transition (EMT), an essential morphological reprogramming step of epithelial tumor cells required for the dissemination of tumor cells from primary tumors, subsequently leading to the formation of metastasis [7].

Aerobic fermentation is also a hallmark of the yeast *S. cerevisiae* when grown in glucose rich medium and tightly controlled by nutrient sensing mechanisms [9, 10]. Others and we have shown that there is a close link between metabolism and sensitivity to DNA damage in yeast cells [11, 12].

We had shown that expression of *FBP1* is induced in response to treatment with the genotoxic agent methyl-methanesulfonate (MMS) and that sensitivity of the yeast *S. cerevisiae* to MMS treatment is correlated with *FBP1* expression [13]. High levels of Fbp1p obtained by overexpression increased sensitivity, while lack of *FBP1* reduced sensitivity to MMS, which suggests a regulatory role for this enzyme in DNA damage response and in the resulting oxidative stress conditions [13]. This multi-faceted impact in addition to the high evolutionary conservation of several regulatory mechanisms and their related functions between the yeast and mammalian homologues, make yeast an attractive model-system to further probe and understand the multiple effects of *FBP1*. Here we present and discuss the results from a mutational screening of several evolutionary conserved amino acid residues of Fbp1p, predicted to be important for catalysis and/or allosteric regulation of this enzyme. We assessed the impact of these mutations on enzymatic activity, and further investigated their effect for growth on non-fermentable medium and sensitivity of yeast cells harbouring these mutant enzymes towards MMS treatment. In our study, we identified several indispensable residues for catalysis and MMS sensitization. The results obtained range from complete inactivation of the enzymatic activity of Fbp1p to partially deactivating mutations. Moreover, two mutations have been found, Asn75 and His325, which decouple catalysis from MMS-sensitization.

RESULTS

Selection of amino acid residues for mutational analysis

We prepared a small library of mutated yeast *fbp1* expression cassettes, each with a single point mutation, changing one evolutionary-conserved residue. Global alignment of the amino acid sequence of Fbp1p of *Saccharomyces cerevisiae* against that of human *Homo sapiens*, pig *Sus scrofa*, mouse *Mus musculus* and another yeast *Candida albicans*

```
CLUSTAL O(1.2.1) multiple sequence alignment

Human           -----------madqapfdtdvntltrfvmeegrk--argtgeltqllnslctavkaiss   47
Pig             -----------mtdqaafdtnivtltrfvmeegrk--argtgemtqllnslctavkaist   47
Mouse           -----------manhapfetdistltrfvmeqgrk--aqgtgeltqllnslctaikaiss   47
C.albicans      ----msgpvnsvskqmnvdtdiitltrfilqeqqtvaptatgelslllnalqfafkfiah   56
S.cerevisiae    mptlvngprrd--stegfdtdiitlprfiiehqkq-fknatgdftlvlnalqfafkfvsh   57
                             .   .:*:: ** **:::. :     .**::: :**:*  *.* ::
                                     A
Human           avrkagiahlygiagstnvtgdqvkkldvlsndlvmnmlkssfatcvlvseedkhaiive   107
Pig             avrkagiahlygiagstnvtgdqvkkldvlsndlvinvlkssfatcvlvseedknaiive   107
Mouse           avrqagiaqlygiagstnvtgdqvkkldilsndlvinmlkssyatcvlvseentnaiiie   107
C.albicans      nirraelvnligvsgsanstgdvqkkldvigdeifinamrssnnvkvlvseeqed-livf   115
S.cerevisiae    tirraelvnlvglagasnftgdqqkkldvlgdeifinamrasgiikvlvseeqed-livf   116
                 :*:* :..* *::*::* ***  ****::.:::.:* :::*    ******: . :*:
                               B
Human           pekrgkyvvcfdpldgssnidclvsvgtifgiyrkkstdepsekdalqpgrnlvaagyal   167
Pig             pekrgkyvvcfdpldgssnidclvsigtifgiyrknstdepsekdalqpgrnlvaagyal   167
Mouse           pekrgkyvvcfdpldgssnidclvsigtifgiyrkkstdepsekdalqpgrdlvaagyal   167
C.albicans      pg-ggtyavctdpidgssnidagvsvgtifgvyklqegstggisdvlrpgkemvaagytm   174
S.cerevisiae    ptntgsyavccdpidgssnldagvsvgtiasifrllpdssgtindvlrcgkemvaacyam   176
                *   *.*.** **:*****:*. **:*** .:::    .    .*.*: *:::*** *::

Human           ygsatmlvlamdcgvncfmldpaigefilvdkdvkikkkgkiyslnegyardfdpavtey   227
Pig             ygsatmlvlamvngvncfmldpaigefilvdrdvkikkkgsiysinegyakefdpaitey   227
Mouse           ygsatmlvlamdcgvncfmldpsigefimvdrdvkmkkkgniyslnegyakdfdpainey   227
C.albicans      ygasahlalttghgvnlftldtqlgefilthpnlklpdtkniyslnegysnkfpeyvqdy   234
S.cerevisiae    ygssthlvltlgdgvdgftldtnlgefilthpnlrippqkaiysinegntlywnetirtf   236
                **::: *.*:   **: * **  :****:.. ::::     ***:*** :  :   :  :
                                                      C
Human           iqrkkfppd--nsapygaryvgsmvadvhrtlvyggiflypankkspngklrllyecnpm   285
Pig             iqrkkfppd--nsapygaryvgsmvadvhrtlvyggifmypankkspkgklrllyecnpm   285
Mouse           lqrkkfppd--gsapygaryvgsmvadihrtlvyggiflypankkspsgklrllyecnpi   285
C.albicans      lkdi-------kkegyslryigsmvadvhrtllyggifaypt------lklrvlyecfpm   281
S.cerevisiae    iekvkqpqadnnnkpfsaryvgsmvadvhrtflygglfaypcdkkspngklrllyeafpm   296
                ::          .  :. **:******:***::***:* **        ***:***. *:
                             C
Human           ayvmekaggmattg-keavldviptdihqrapvilgspddvleflkvyekhsaq   338
Pig             ayvmekagglattg-keavldivptdihqrapiilgspedvtelleiyqkhaak   338
Mouse           ayvmekagglattg-dkdildivpteihqkapvvmgssedvqefleiyrkhkak   338
C.albicans      allmeqaggsavtikgerildilpkgihdkefycvgiqg---------------   320
S.cerevisiae    aflmeqaggkavndrgerildlvpshihdkssiwlgssgeidkfldhigksq--   348
                * :**:*** *..   : :**::*. **::    :*
```

FIGURE 1: Multiple global-alignment of *FBP1* amino-acid sequences of *Saccharomyces cerevisiae* and several species using the online-available ClustalW2 multiple sequence alignment tool, provided by EMBL-EBI. Identical residues are marked with (|) ; similarities (:) and totally different ones with (.). Residues included in the mutational analysis are highlighted. Three regions (a, b and c) of sequence homology are indicated: **(A)** representing the mostly conserved part of the regulatory loop, **(B)** the conserved metal binding site, and **(C)** the largest homology domain on the c-terminus containing the active site.

shows around 46% identity (Fig. 1) [14, 15]. This allows a large choice of conserved residues to be manipulated. Therefore, we employed data from structural and functional studies performed on Fbp1p from yeast and mammalian homologues to identify residues that are more likely to have a key role in catalysis and/or allosteric regulation.

A general overview of the alignment indicates a lack of homology in the N terminal region that extends for further 9 amino acids in the yeast Fbp1p compared to the mammalian homologues [15]. This part has been correlated to the glucose-mediated irreversible deactivation of yeast Fbp1p. The Pro1 residue has been implicated in targeting the enzyme to ubiquitin-mediated proteasome degradation in presence of glucose [4, 16]. The residue Ser11 is linked to cAMP-mediated phosphorylation leading to loss of activity following glucose reintroduction to the medium [17]. Structural studies on porcine Fbp1p identified residues 52 to 72, as homologues to residues 63 to 83 in yeast, which form a loop that plays a crucial role in determining the tertiary structure of the tetramer, depending on the presence or absence of AMP and thus mediating AMP allosteric inhibition on the enzyme [18].

Two amino acids of interest in this loop are conserved between yeast and mammalian Fbp1p. The first is Asp68 in mammalian Fbp1p, Asp79 in the yeast homologue, which orients towards the active site, and facilitates the transfer of a proton from the phosphate group 1 of the substrate F1,6P2 to the OH group of the product F6P. The second amino acid is the Asn64, Asn75 in yeast, which in addition to two other amino acids Asp 74 and Glu 98, Asp 85 and Glu 109 in yeast, contributes to the formation of a water-binding site. The water molecule bound to this site indirectly facilitates the nucleophilic attack of an adjacent water molecule on the phosphorus atom of the phosphate group on the position 1 of the substrate which consequently leads to the hydrolysis of the phosphate group and its splitting from the substrate [18].

The stride of residues Asp128 to Ser133 in yeast Fbp1p, or Asp118 to Ser 123 in mammalian Fbp1p, form an evolutionary conserved metal-binding motif. This sequence is largely conserved through pro and eukaryotic evolution; moreover it is encountered in various phosphatases that similarly require divalent cations as allosteric activators. In case of both mammalian and yeast Fbp1p the sequence is comprised of the following residues Asp-Pro-Ile-Asp-Gly-Ser [19]. Phe 194 is a conserved residue in both pro and eukaryotes, it lies within a conserved stretch of amino acids. Due to this, in addition to its high bulkiness, we presumed that this residue could be of significance for either the enzyme's folding or its protein/protein interactions.

The largest region of sequence homology in Fbp1p is on the carboxylic terminus and spans the residues Ala254 to Ser333 of the yeast homologue, and Ala243 to Ser321 of the mammalian enzyme, shows 66% sequence identity between the yeast and mammalian homologues. This region has been linked to the catalytic activity of the enzyme; a notion that has been deployed in recent studies on the human Fbp1p as well [8, 20]. This c-terminal catalytic sequence contains two evolutionary conserved histidine residues; His265 and His324, both are very likely to comprise a part of the enzyme's catalytic domain.

TABLE 1. Summary of the analyzed mutations and utilized mutagenesis primers.

Mutations (Altered residue)	Mutagenesis primers	Abbreviation of the mutated FBP1 form
Pro1→Ala1	fw: GGATTCTAGAACTAGTATGGCAAC TCTAGTAAA TGGACC rev: GGTCCATTTACTAGAGTTGCCATACTAGTTCTAGAATCC	Pro1
Asn75→Ala75	fw: GTTAGCAGGCGCTTCCGCCTTCACTGGTGACCAGC rev: GCTGGTCACCAGTGAAGGCGGAAGCGCCTGCTAAC	Asn75
Asp79→Ala79	fw: GCTTCCAACTTCACTGGTGCCCAGCAAAAGAAGTTGGAC rev: GTCCAACTTCTTTTGCTGGGCACCAGTGAAGTTGGAAGC	Asp79
Glu109→Ala109	fw: GGTCCTTGTATCTGAAGCACAGGAAGATTTGATCG rev: CGATCAAGTCTTCCTGTGCTTCAGATACAAGGACC	Glu109
Asp131→Ala131	fw: GTGTGTTGTGATCCTATTGCTGGCTCCTCAAATTTGGAC rev: GTCCAAATTTGAGGAGCCACGAATAGGATCACAACACAC	Asp131
Ser133→Ala133	fw: GTGATCCTATTGATGGCGCCTCAAATTTGGACGCCGG rev: CCGGCGTCCAAATTTGAGGCGCCATCAATAGGATCAC	Ser133
Phe194→Ala194	fw: GGTGATGGAGTTGATGGGGCTACCTTAGACACAAACTG rev: CAAGTTTGTGTCTAAGGTAGCCCCATCAACTCCATCACC	Phe194
His65→Ala265	fw: CCATGGTTGCTGATGTTGCCAGGACGTTTCTTTACGG rev: CCGTAAAGAAACGTCCTGGCAACATCAGCAACCATGG	His265
His324→Ala324	fw: GATTTGGTGCCAAGTCATACCGCTGACAAATCTTCTATT rev: AATAGAAGATTTGTCAGCGATATGACTGGCACCAAATC	His325

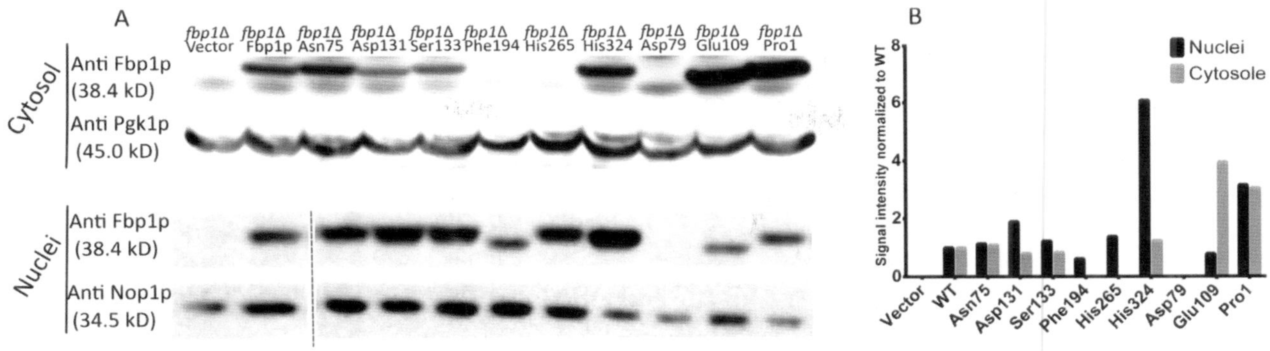

FIGURE 2: Immuno-detection of the over-expressed Fbp1p mutants. (A) Western-blot showing the levels of the over-expressed mutants in both cytosolic and nuclear fractions. Over-expression was verified in absence of endogenous Fbp1p expression in *fbp1Δ* strains. PGK1 and NOP1 were utilized as cytosolic and nuclear loading controls, respectively. **(B)** Quantification of the western-blot signals, each band was normalized to the intensity of its loading control, resulting values where then normalized to those of the wt.

Based on the aforementioned data, we screened 9 different mutated versions of yeast Fbp1p each having one of the formerly presented key-residues substituted with an alanine residue through a point mutation, alanine mutagenesis screening, as shown in Table 1.

In the following parts of this manuscript, mutated Fbp1p forms will be referred to using the number and three-letter symbol of the residue that had been replaced with alanine, these abbreviations are listed in Table 1.

Overexpression levels of Fbp1p mutants

Using immuno-blotting, 6 out of the 9 mutated forms of Fbp1p were successfully detected in the cytosolic fraction of the *fbp1Δ* yeast strains in which the mutated Fbp1p forms were over-expressed (Fig. 2A). No cytosolic bands were detected for the Asp79, Phe194 and His 265 mutants. More mutants were successfully detected in nuclear fraction lysates of the different strains with mutated Fbp1p, with only the Asp79 mutant remaining undetected (Fig. 2A). Quantification of band intensities showed that the Asp131, Ser133 and His324 mutants were highly enriched in the nucleus compared to the wild-type (wt) form, and, interestingly, Phe194 and His265 mutants were fully retained in the nucleus (Fig. 2B).

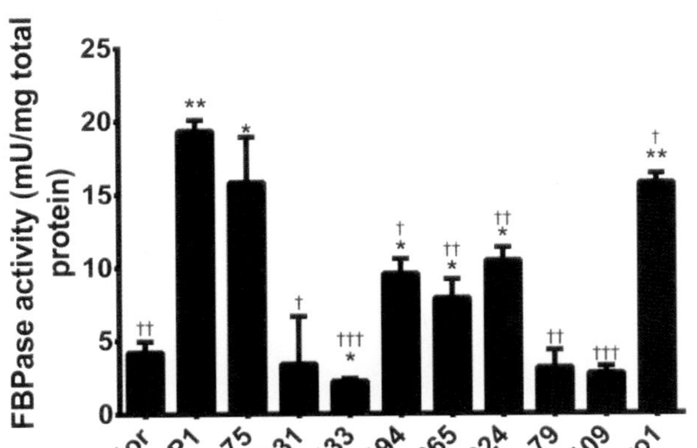

FIGURE 3: Enzymatic activity of the utilized *FBP1* mutants. FBPase catalytic activity was measured in native lysates of wt BY4741 strains with over-expression of wt and mutated *fbp1* cassettes. Measurements were performed in triplets, and error bars represent SD. Statistical significance was determined using student t-test for each of the mutants compared to both cells with control vector and those with over-expression of the wt Fbp1p. (*) Indicates statistically relevant difference compared to empty vector and (†) compared to the wt Fbp1p.

Catalytic activity of the Fbp1p mutants

In order to screen for the impact of amino-acid substitutions on the catalytic activity of Fbp1p, we over-expressed the mutated cassettes of *fbp1* as well as wt *fbp1*, in BY4741 wt haploid yeast strain. Subsequently we measured the fructose 1,6 bis-phosphatase (FBPase) enzymatic activity in the native protein lysates of the strains over-expressing the mutated forms of Fbp1p and statistically compared them to both cells over-expressing wt Fbp1p and those containing the control vector.

The assay shows that the mutants roughly cluster into 3 groups in respect of their exhibited enzymatic activity. Pro1 and Asn75 showed close-to-wild-type, only slightly reduced activity, nevertheless significantly different; Phe194, His265 and His324 diminished the activity to around 50% compared to the wt; while Asp79, Glu109, Asp131 and Ser133 completely abrogated the activity stemming from over-expression showing a similar activity to extracts of cells transformed with empty vector (Fig. 3).

Rescue of growth on non-fermentable medium (Drop-test on SDEG)

In order to assess the mutants' potential of rescuing the gluconeogenic pathway in cells lacking *FBP1* expression. The mutant *FBP1* forms were introduced into the *FBP1* knockout strain BY4741-*Δfbp1*. Drop tests on SDEG-medium, containing glycerol and ethanol as exclusive carbon sources, were then performed with the *fbp1Δ* strains over-expressing the Fbp1p mutant (Fig. 4A). Since the enzymatic activity of Fbp1p is indispensable for gluconeogenesis, the determining factor for the growth of each strain on the non-fermentable SDEG medium is the catalytic activity of the Fbp1p mutant expressed.

Four mutants, Asp79, Glu109, Asp131 and Ser133, completely failed to rescue growth on SDEG. The other screened mutants, Pro1, Asn75, Phe194, His265 and His 324, clearly rescued growth on SDEG-agar exhibiting various levels of growth (Fig. 4A).

Sensitivity to low doses of MMS

Since we had shown previously that exposure to low doses of MMS triggers *fbp1* transcription with the toxic outcome strongly depending on the presence of endogenous Fbp1p, we analyzed the effects of mutations in key residues of Fbp1p on its ability to mediate increased sensitivity to MMS. For this purpose, expression vectors with mutated *fbp1* cassettes were introduced into both wt and Fbp1p-deficient *fbp1*Δ BY4741 cells. Subsequently, the resulting strains were spotted on SD-Ura agar containing MMS concentrations ranging from 0.01% to 0.0175%. These drop tests showed that the catalytically inactive mutants Asp79, Glu109, Asp131, Ser133 and His324 contributed no additional sensitivity towards MMS, and had growth levels comparable to wt cells harbouring control vector (Fig. 4B to I). Three catalytically active mutants Pro1, Phe194 and His265 conferred increased sensitivity toward MMS, when over-expressed in wt or Fbp1p-deficient *fbp1*Δ cells, these

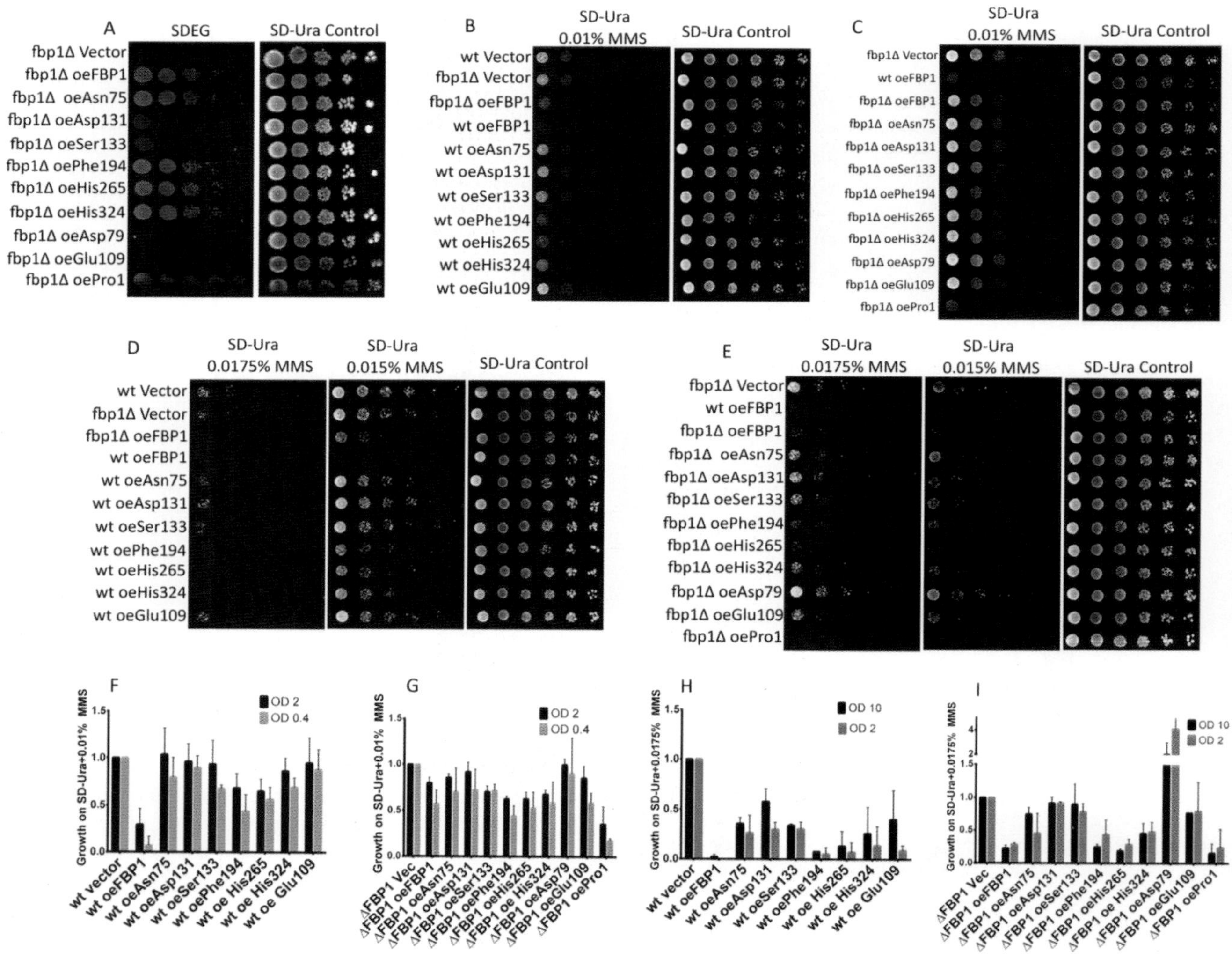

FIGURE 4: Drop-tests on Non-fermentable (SDEG) and MMS-containing agar-media. (A) Drop-test on SDEG non-fermentable medium. Wt and mutated *FBP1* cassettes were introduced into BY4741 *fbp1*Δ cells lacking endogenous *FBP1*, the resulting strains were grown in liquid SD-Ura medium, diluted and spotted on SDEG, same suspensions were spotted on SD-Ura agar as a control. Agar-plates were incubated for 48 h at 30°. **(B and C)** Drop tests on SD-Ura agar with 0.01% MMS. Mutants and wt cassettes were introduced into both wt **(B)** and Δ*fbp1* **(C)** BY4741 cells. The resulting strains were grown in SD-Ura and then diluted (1:5) and spotted on freshly prepared SD-Ura containing 0.01%, and on SD-Ura as non-treated control. Agar plates were incubated for 48 h at 30°C. **(D and E)** Drop tests on SD-Ura agar containing 0.015% and 0.0175% MMS. Plates were incubated for 72 h at 30°C. **(F to I)** Quantifications of the spot-intensities of the drop tests on SD-Ura 0.01% MMS second and third spots (OD_{600}: 2 and 0.4 respectively) and 0.0175% first and second spots (OD_{600}: 10 and 2 respectively). Featured results are taken from three biologically independent replicates, (error-bars: SD; N = 3).

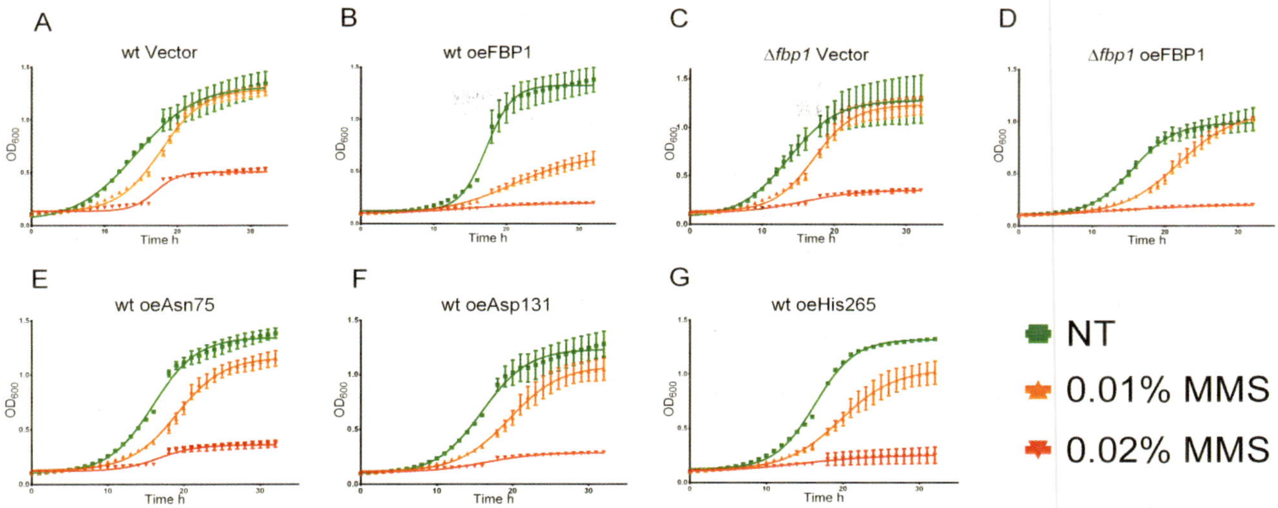

FIGURE 5: Growth curves upon MMS treatment. Figures A to G feature the growth curves of the non-treated and MMS-treated (0.01% and 0.02%) yeast strains in liquid SD-Ura medium. Curves feature the increasing OD_{600} values over-time at 30°C with intermittent shaking, error bars represent SD, N = 3. **(A)** wt Vector, **(B)** wt oe *FBP1*, **(C)** *fbp1Δ* Vector, **(D)** *fbp1Δ* oe Fbp1p, **(E)** wt oeAsn75, **(F)** wt oe Ser133, **(G)** wt oe His265.

mutants impeded growth on MMS compared to the corresponding strains with control vector (Fig. 4B to I). Interestingly, one catalytically active mutant, Asn75, contributed much less additional sensitivity towards MMS compared to wt and to other less catalytically active mutants (Fig. 2B to I). Similar trends were observed in both wt and *fbp1Δ* cells, with all wt-based strains showing less over-all growth with MMS concentrations of 0.01% and 0.0175%. These results consist with our previously reported finding that *fbp1* deletion confers resistance towards low doses of MMS [13]. Quantifications of spot intensities of the drop-tests confirmed the eye-observable changes in MMS sensitivity with different Fbp1p mutations (Fig. 4F, G, H and I).

Over-production of Fbp1p sensitises cells to MMS in liquid medium

To better understand the cytotoxic effects of MMS, we also evaluated its impact on yeast proliferation in liquid media. Cultures were pre-incubated for 3 h and then treated with 0.01% and 0.02% MMS in 96 well plates, subsequently OD_{600} was measured over-time using a plate reader. While treatment with 0.02% MMS resulted in severe inhibition of growth of all tested strains, the impact of 0.01% on proliferation in SD-Ura was clearly influenced by Fbp1p activity (Fig 5A to G). Over-expression of Fbp1p generally reduced growth, however a clear contribution of the endogenous Fbp1p was also observed. Over-expression of Fbp1p in wt cells sensitized them to the lower MMS concentration of 0.01% leading to less over-all growth (Fig. 5A and B). In contrast, the effect of Fbp1p over-expression in cells lacking endogenous Fbp1p was very mild and limited to a temporary inhibition of growth (Fig. 5C and D). Three mutants were tested in comparison to over-expression of wt Fbp1p in the wt cell background. Over-expression of the catalytically active mutants Asn75 and His265 led to a sensitizing effect at 0.01% MMS similar to the effect seen with the wt Fbp1p (Fig. 5E and G), while the inactive mutant Asp131 had no noticeable effect (Fig. 5F).

The effect of Fbp1p over-production on cell survival depends on its genomic context of *fbp1*

We then sought to assess the impact of over-expression of Fbp1p or its mutants on cell survival upon MMS treatment using flow cytometry (FACS) analysis of PI-stained cells. After 12 h of MMS treatment in liquid SD-Ura cultures, all strains had a clear and significant increase in the amount of PI positive cells. Over-expression of wt Fbp1p in wt cells possessing endogenous Fbp1p led to a significant increase in the PI positive fraction upon MMS treatment with both 0.02% and 0.03% concentrations (Fig. 6A), whereas no significant increase in the PI-positive fraction was observed upon over-expression of Fbp1p in cells lacking endogenous Fbp1p (Fig. 6B). We also evaluated the effects of two catalytically active mutants, His265 and Asn75, and the totally inactive mutant Asp131 both in wt and *fbp1Δ* cells. Interestingly, non-of the tested mutants led to any significant difference in the PI-positive fraction after 12 h of MMS treatment (Fig. 6A and B).

Please note, that over-expression of Fbp1p was associated with a significant decrease in the PI-positive fractions of non-treated cells, which was also observed with tested mutants (Fig. 6A and B).

Over-production of Fbp1p hampers colony-forming capacity upon MMS treatment

To further investigate cell survival upon MMS treatment, we tested the impact of wt and mutated forms of Fbp1p on the colony forming capacity of cells following MMS treat-

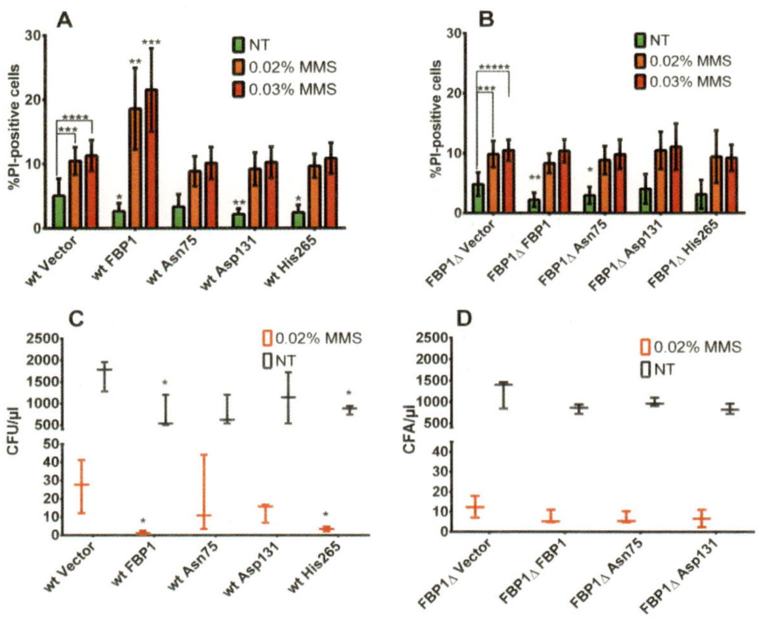

FIGURE 6: Cell survival and colony formation upon MMS treatment. (A and B) FACS analysis of PI staining following 12 h of MMS treatment. Bars represent the PI-positive percentage fractions of cells as measured using FACS. Results feature three biologically independent replicates with three technical replicates each. Error bars show the standard deviation. Statistical significance was determined using unpaired student t-test assuming equal variance. (*: $p < 0.05$; **: $p < 0.01$; ***: $p < 0.001$; ****: $p < 0.0001$). The brackets over the wt Vector and *fbp1Δ* Vector strains the control vector strains show the significance comparing treated and non-treated of the same strain. While significance stars over the mutants represent significance when comparing the results form these strains with those of the control vector. **(C and D)** colony forming on SD-Ura agar following 12 h of MMS treatment in liquid SD-Ura. Results are from 3 independent experiments. Error bars show the standard deviation. Asterisks indicate statistically significant differences compared to the control vector strains. Significance was assessed using student t-test (*: $p < 0.05$).

ment. Generally, 12 h of 0.02% MMS treatment lead to a two to three orders of magnitude decrease of the colony forming units CFU/µl (Fig. 6C and D). In context of wt cells with endogenous Fbp1p expression, this decrease was strongest when wt Fbp1p or the active mutant His265 were over-expressed (Fig. 6C). In contrast, over-expression of wt Fbp1p in cells lacking the endogenous gene (*fbp1Δ*) had no further inhibitory effect on colony forming capacity (Fig. 6D). It is also worth noting that in the wt strain, over-expression of Fbp1p or its active mutant His265 lead to a significant decrease in CFU/µl in the non-treated samples as well, compared to non-treated cells harbouring the control vector.

DISCUSSION

This work aimed at exploring possible key evolutionary-conserved residues in Fbp1p linked to its apparent multiple-roles in both gluconeogenesis and response to MMS toxicity. In this study, we managed to identify several key residues essential for the enzymatic activity and the role in response to MMS-inflicted damage. The full activity exhibited by the Pro1 mutant completely fits with its reported role as a ubiquitination-mediating residue [16]. Replacing Pro1 should abrogate ubiquitination and lead to increased stability, which is clearly reflected in the increased intensity of the western-blot band of this mutant [4]. Additionally, when ectopically expressed in *fbp1Δ*, the Pro1 mutant showed the highest MMS-sensitisation in the drop tests, even higher than wt Fbp1p, which could be explained by the higher stability of this mutant through evasion of proteasomal degradation. Replacing the residue Glu109 completely abolished the enzymatic activity, along with the ability to grow on non-fermentable carbon sources, however the Asn75 mutant almost maintained full activity compared to the wt, and also rescued growth on SDEG medium. This suggests that Glu109 is significantly participating in binding the catalytically essential water molecule while Asn75 seems to lack an important role in the formation of the water-binding pocket. The complete lack of enzymatic activity when Asp131 or Ser133 are replaced grants validity to their structurally presumed participation in the divalent-cation binding pocket and to the importance of the divalent cation for the activity of this enzyme.

Substituting Phe194 leads to a partial loss in enzymatic activity. This could be due to a direct influence on the catalytic activity. However, this lack of activity could also result from a change in molecular interaction and cellular localization, which was predominantly nuclear for this mutant.

Histidine residues in proximity to active sites on enzymes are usually presumed to contribute to the catalysis due to the amphiphilic nature of their side-chain imidazole ring [32]. In our study, altering either His265 or His324 reduced the FBPase catalytic activity by more than one fold; this supports the importance of these residues in facilitating the catalysis at the active site near the c-terminus.

Nevertheless, the activity is not entirely dependent on these residues. In both cases, the enzyme retains sufficient activity, and mediates the ability of growing on SDEG medium in transformed strains.

For most mutants, the observed sensitivity to MMS correlated to the measured FBPase catalytic activity, and the capacity to grow on SDEG non-fermentable medium. However, two mutants clearly defied this pattern. Asn75 showed full enzymatic activity, and also permitted growth on SDEG medium. However, Asn75 showed much less inhibition of growth on SD-Ura with MMS compared to the wt enzyme and other catalytically active mutants. This effect was verified repeatedly in drop-tests using both wt and *fbp1Δ* strains. Similarly, the His324 mutant, exhibiting al-

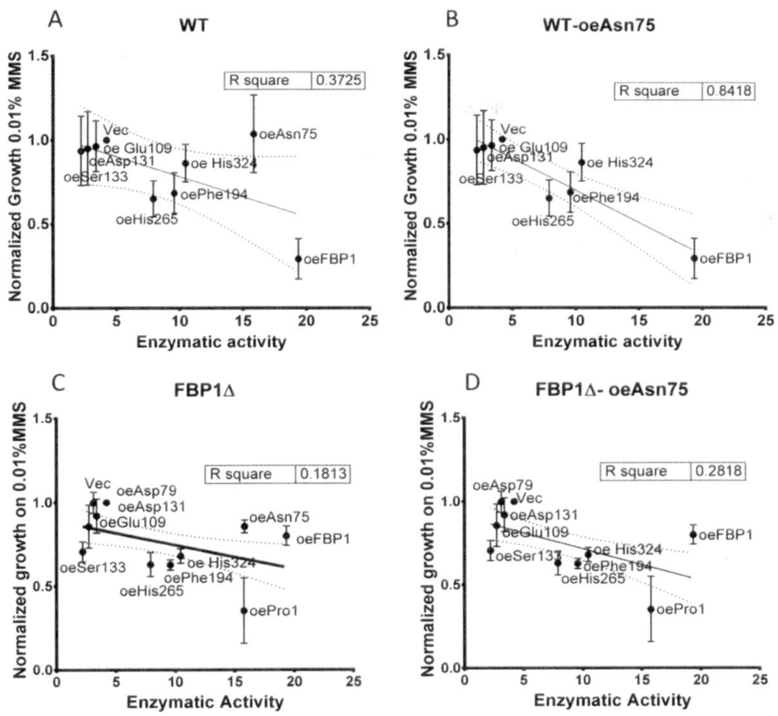

FIGURE 7: Linear correlation between enzymatic activity and growth on SD-Ura agar with 0.01% MMS in drop-tests. (A) wt strains; **(B)** wt strains excluding Asn75; **(C)** *fbp1Δ* strains; **(D)** *fbp1Δ* strains excluding Asn75. MMS growth values are taken from the spot-intensity quantification of three independent drop tests on SD-Ura agar with 0.01% MMS. N = 3; error bars represent SD.

most 50% of the original catalytic activity, did not significantly increase sensitivity towards MMS in the *fbp1Δ* strain. The decoupling of MMS-sensitisation from the FBPase activity in case of the Asn75 is further highlighted by investigating the linear correlation between the measured FBPase activity of the mutants and their quantified growth on 0.01% MMS in drop-tests (Fig. 7). In case of wt cells with intact endogenous Fbp1p, Asn75 lies far from the correlation line out of the confidence interval, showing less negative correlation between its catalytic activity and the growth of its hosting strain on MMS in drop-tests (Fig. 7A). Furthermore, the exclusion of the Asn75 from the analysis, leads to substantial improvement in the linear correlation marked by an increase of R-square value from ≈0.37 to ≈0.84 (Fig 7A and B). Thus, the Asn75 and also possibly the His324 decouple the catalytic activity of the enzyme from its MMS-sensitizing role.

Substituting the residue His265 reduced enzymatic activity to less than half of wt Fbp1p, whereas this mutant conveyed even higher sensitivity towards MMS compared to the wt enzyme. In case of this His265 mutant it must be noted that it was enriched in the nuclear fraction and could not be detected in cytosolic extracts by western-blot, which may lead to lower activity in the enzymatic activity assay and may contribute to its increased MMS-sensitizing effect.

Taken together, our observations that for some mutants the catalytic activity of Fbp1p is decoupled from its additional role mediating sensitivity to MMS treatment suggest that these effects are mediated by two at least partially independent sites in the enzyme, one for enzymatic activity and one underlying the MMS-response. This clearly contradicts the alternative model, in which the MMS-sensitizing role of *FBP1* would be directly linked to its catalytic activity and its role in gluconeogenesis.

Hypothesising on more speculative explanations that potentially could clarify the role of *FBP1* in response to MMS treatment, we searched for previously reported interactions of *FBP1* in the literature. A physical interaction with TORC1 was reported in studies analyzing Fbp1p degradation upon glucose reintroduction following long-term glucose starvation [21, 22]. High throughput screenings also showed a genetic interaction with TOR2 [23]. TORC1 is a nutrient-sensing protein-kinase complex that responds to nutrient abundance by driving anabolic pathways required for cell growth and proliferation in addition to promoting cell cycle progression [24, 25]. Under glucose starvation, Fbp1p binds to TORC1, with the latter being essential for glucose-mediated Fbp1p degradation [21, 22]. These findings suggest a potential scenario in which the Fbp1p expressed in response to MMS treatment also binds to and inhibits TORC1 leading to a more significant inhibition of cell proliferation. Such Fbp1p-mediated loss of TORC1 function would consist with the increased viability upon MMS treatment in absence of Fbp1p, it could also explain the observed uncoupling of the enzymatic activity from MMS-sensitization. Other reported genetic interactions suggest an involvement of *FBP1* in cell cycle regulation, these include: CDH1 [23], a cell-cycle regulated mitotic exit protein, GDH1 [26] glycogen phosphorylase postulated to co-ordinate metabolism to cell-cycle, TAF1 [27] involved in G1/S progression of cell cycle and ADK1 [26], involved in purine metabolism and response to DNA-replication stress. The aforementioned interactions make it very intriguing to functionally probe these suggested interactions and their effects on cell cycle progression in context of DNA damage,

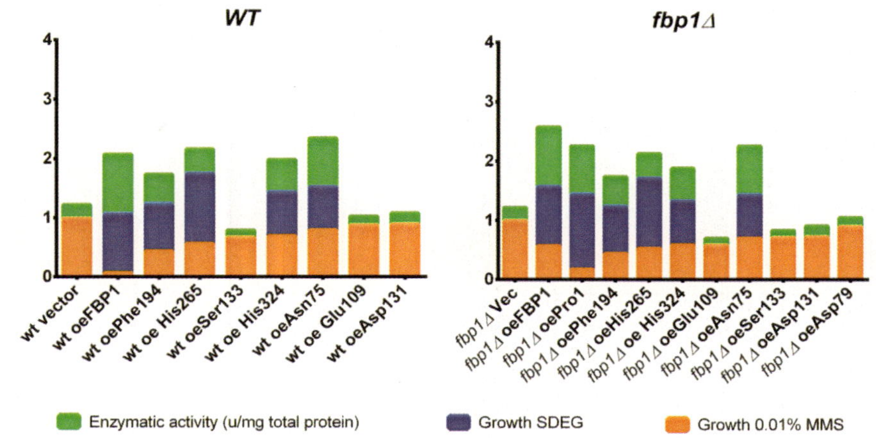

FIGURE 8: Summary figure. This figure summarizes the observed effects of the investigated mutations on FBPase catalysis, growth on non-fermentable SDEG and growth on SD-Ura agar containing 0.01% MMS. Featured are the normalized values corresponding to enzymatic activity measured in the wt-strain background, image quantification of drop-tests on SDEG of the *fbp1Δ*-derived strains and averages of three independent replicates of drop-tests on SD-Ura agar with 0.01% MMS with both wt and *fbp1Δ* strains. Enzymatic activity and growth on SDEG were normalized to the strain over-expressing wt Fbp1p, while growth on SD-Ura agar with 0.01% MMS was normalized to the strain harbouring empty vector.

an interesting open question we intend to pursue in the near future.

Our findings consistently showed that the outcome of *FBP1* over-expression is very dependent on its genetic context and the presence of endogenously expressed Fbp1p. This is in line with earlier reports that MMS treatment induces the transcription of *FBP1* within the first hour of treatment [13]. It is also already known that the lack of *fbp1* on a genomic level confers more resistance against MMS treatment [13]. In this work we found that the impact of the ectopically expressed Fbp1p on cell survival upon MMS treatment largely depends on the presence of the endogenously expressed enzyme. Whereas over-expression of Fbp1p inhibits growth on SD-Agar MMS in both wt and *fbp1Δ* genotypes, the presence of endogenous Fbp1p seems to be the decisive factor when it comes to the impact of ectopically expressed Fbp1p on cell survival and sustained damage upon MMS treatment, this is made clear by the drastically different outcomes of PI-staining and colony formation of MMS-treated cells with Fbp1p over-expression when using the wt or *fbp1Δ* cells. Additionally our observations support the notion that *FBP1* delays aging in yeast as evident by the significant decrease in PI-positive fractions of non-treated cells with Fbp1p over-expression compared to the vector control.

A summarizing figure (Fig. 8) depicts the observed effects of the mutants on three studied phenomena: catalytic FBPase activity, growth on non-fermentable carbon sources (SDEG) and growth on SD-Ura agar with 0.01% MMS in drop tests. The bars in the graph are sorted by decreasing the MMS-sensitivity (increasing values of growth on MMS). The decoupling of MMS-sensitisation from catalytic activity, discussed above, is again clearly visible in this figure. While the Asn75 mutant, catalytically active and conferring growth on SDEG, does not significantly sensitize cells to MMS in both wt and *fbp1Δ* genetic backgrounds, in case of His324, which is also catalytically active and confers growth on SDEG, the decoupling is most clearly visible in the wt strain.

In this work we successfully identified several conserved key residues for the catalytic function and/or the increased sensitivity to MMS and provide a clear separation of both activities. Our observation in yeast fits well with a recent report on FBP1 protein interactions in human tumours, which also seem to be independent of its catalytic activity [8, 28].

MATERIAL AND METHODS

Utilized yeast strains, media and growth conditions

All yeast strains were cultured and utilized in their haploidic-a form. Strains with the genetic background BY4741 (either wt or *Δfbp1*) were used to over-express the mutated versions of Fbp1p and screen for the effects of the mutations on enzymatic activity, growth on non-fermentable SDEG medium and sensitivity towards MMS treatment.

Transformed strains were maintained on synthetic minimal medium without uracil (SD-Ura) agar, containing 6.7 g/l yeast nitrogen base, 20 g/l glucose 100 mg/l hestidine, lucien and lysine. Prior to treatments starting cultures were made by inoculating a single colony into 2 ml of SD-Ura liquid medium and incubating at 30°C with shaking overnight.

Plasmid vectors

Wt and mutated cassettes of FBP1 were over-expressed under the yeast GPD promoter using the high-copy number pRS426 plasmid vector containing *Ura3* as a selection marker in yeast.

Preparation of *fbp1* mutant plasmids

The plasmid pRS426fbp1containing the wt fbp1 cassette was obtained from previous work performed in our lab [13].

Mutants of *fbp1* were prepared using the PCR-based site-directed mutagenesis method using primers containing the desired point mutations, in combination with a commercially available kit "QuickChange® XL Site-directed mutagenesis kit" (Stratagen). Mutagenesis-PCR reactions were carried out according to the kit's instructions and 9 mutants of *fbp1* were prepared, each containing a single point mutation leading to the conversion of the codon of the targeted residue into an alanine codon.

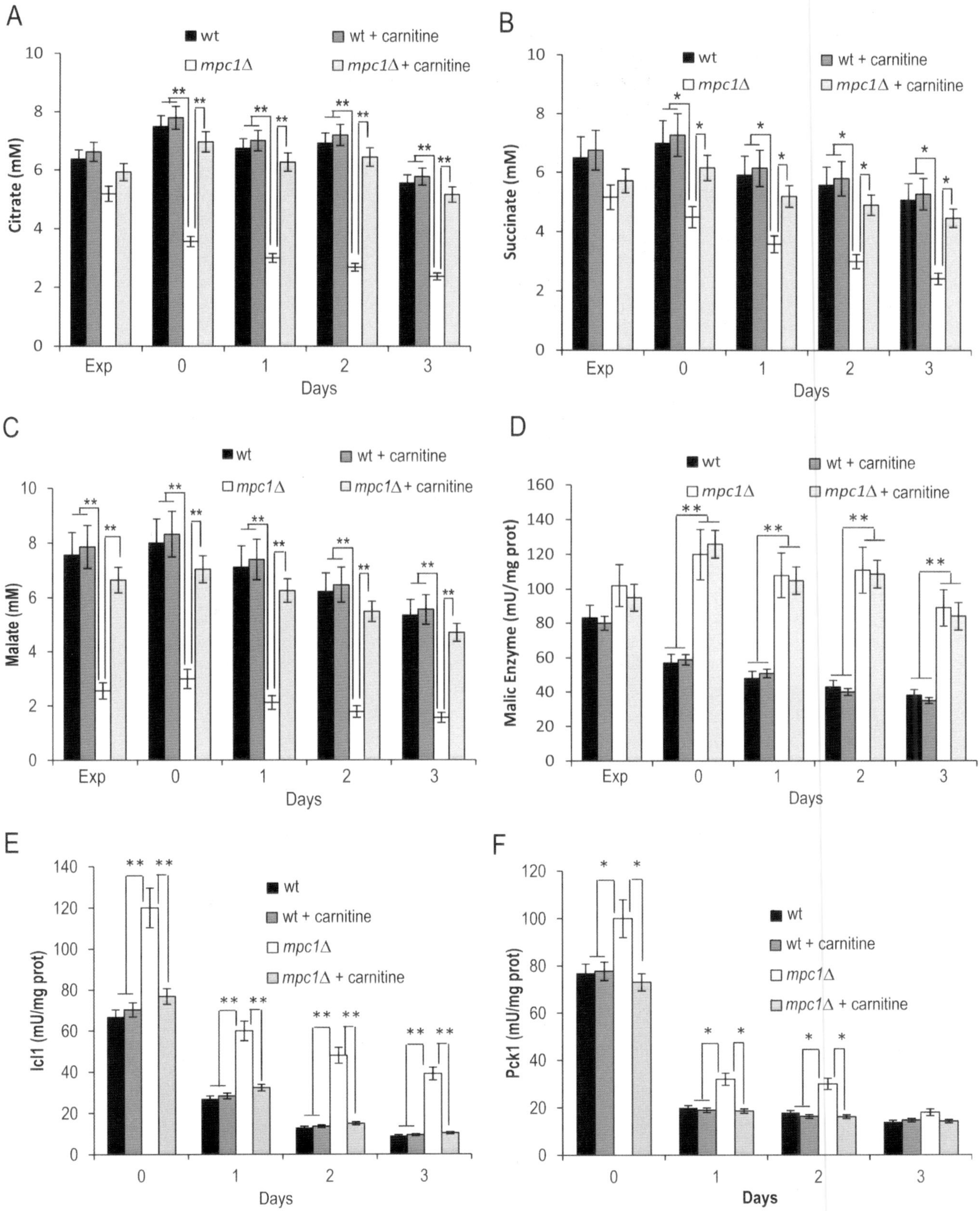

FIGURE 7: Carnitine increases the levels of the TCA cycle intermediates in the *mpc1Δ* mutant. Wt and *mpc1Δ* cells were grown in minimal medium/2% glucose supplemented with carnitine (10 mg/L) and, at the indicated time points, the concentrations of citrate **(A)**, succinate **(B)** and malate **(C)** were measured together with Mae1 **(D)**, Icl1 **(E)** and Pck1 **(F)** enzymatic activities. Day 0, diauxic shift. Data refer to mean values determined in three independent experiments. SD is indicated. * P $\leq$ 0.05 and ** P $\leq$ 0.01.

[40]. As shown in Fig. 6A, a typical tubular network was observed for wt cells whereas for the mutant fluorescent punctiform structures appeared at Day 2 after the diauxic shift. These structures are indicative of mitochondrial fragmentation and are linked to an elevated activity of the mitochondrial fission machinery [41]. In addition to an altered morphology, the mitochondria of the mutant displayed a time-dependent reduction in membrane potential and, at Day 4, did not accumulate $DiOC_6$ (Fig.6A). Mitochondrial dysfunctions are intrinsically related to reactive oxygen species (ROS) of which superoxide anion is one of the most potentially harmful. This radical derives mainly from leakage of electrons from the respiratory chain and, among others, can target mitochondria with detrimental effects [42, 43]. Chronologically aging *mpc1Δ* cells had a higher ROS content, measured as the superoxide-driven conversion of non-fluorescent dihydroethidium (DHE) into fluorescent ethidium (Eth), compared with that of the wt cells (Fig. 6B). Notably, culturing *mpc1Δ* cells in a carnitine-supplemented medium was sufficient to avoid this phenomenon (Fig. 6C). Moreover, oxygen consumption measurements indicated that in these cells cellular respiration increased (Fig. 6D). In *S.cerevisiae,* carnitine is involved in a process referred to as the carnitine shuttle which allows the transport of acetyl-CoA to the mitochondria. This transport system which is non-functional unless carnitine is supplied with the medium [44, 45], involves the transfer of the acetyl moiety of acetyl-CoA to carnitine and the subsequent transport of the acetylcarnitine to the mitochondria. Here, a mitochondrial carnitine acetyltransferase catalyses the reverse reaction generating carnitine and acetyl-CoA which enters the TCA cycle [44, 46]. As shown in Fig. 7A-C, the supplemental carnitine did not significantly affect the levels of citrate, succinate and malate in the wt whilst this was not the case for the *mpc1Δ* mutant where the levels of all three intermediates increased and were restored to wt-like ones. No effect was observed on the malic enzyme activity which at/after the post diauxic shift in the *mpc1Δ* cells was still the double of that of the wt (Fig. 7D) suggesting that the presence of carnitine does not abolish the Mae1-dependent flux towards mitochondrial pyruvate generation. Concomitantly, in the *mpc1Δ* cells the enzymatic activities of Icl1 and Pck1 were reduced to the physiological levels measured in the wt (Fig. 7D and E). Thus, all this suggests that in the *mpc1Δ* mutant the activation of the carnitine shuttle can properly feed the TCA cycle by supplying acetyl-CoA to the mitochondria. Hence, the compensative metabolite feeding from the cytosol provided by the glyoxylate shunt seems to be no longer required. Moreover, following carnitine supplementation, during the post-diauxic phase no effect was observed on the ethanol consumption in both the wt and the mutant strains (Fig. 8A). Similarly, the acetate utilization in the wt was not affected, while in the *mpc1Δ* mutant its utilization was promoted (Fig. 8B). This indicates that in the latter the activation of the carnitine shuttle and the consequent acetyl-CoA transport to the mitochondria can result in an enhancement in the flux downstream from the acetate activation allowing acetate utilization. Finally, in the *mpc1Δ* mutant these metabolic changes matched the almost completely

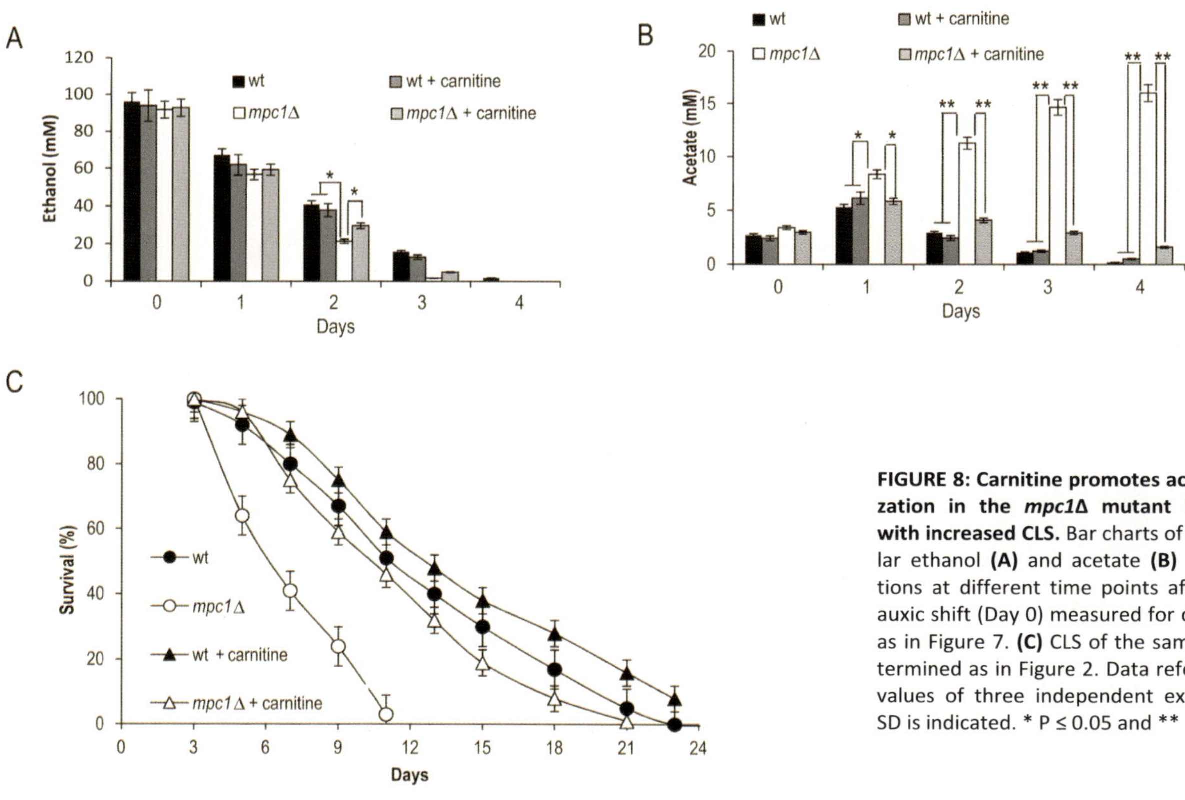

FIGURE 8: Carnitine promotes acetate utilization in the *mpc1Δ* mutant in concert with increased CLS. Bar charts of extracellular ethanol **(A)** and acetate **(B)** concentrations at different time points after the diauxic shift (Day 0) measured for cells grown as in Figure 7. **(C)** CLS of the same cells determined as in Figure 2. Data refer to mean values of three independent experiments. SD is indicated. * $P \leq 0.05$ and ** $P \leq 0.01$.

restored chronological longevity (Fig. 8C).

In conclusion, these data collectively indicate that the lack of the Mpc1 transporter brings about a chain of metabolic events which, in order to counteract the decrease of the pyruvate supply in the mitochondria, by influencing the global acetyl-CoA metabolism ultimately restrict cell survival during chronological aging. In particular, after the diauxic shift when cells utilize the earlier produced ethanol/acetate and increase their respiration demand, one of the metabolic traits of the *mpc1Δ* mutant is a TCA cycle operating in a "branched" fashion with a propensity to shunt intermediates towards pyruvate generation via the malic enzyme. This kind of not-complete cyclic functioning of the TCA cycle by depleting it of intermediates influences not only the respiration, which is reduced in the mutant, but also might reduce mitochondrial acetyl-CoA pool. In fact, a TCA cycle characterized by low levels of intermediates (Fig. 4B-D) generates less succinyl-CoA. This is the substrate for the CoA-transferase's reaction from succinyl-CoA to acetate, catalyzed by Ach1 in cells released from glucose repression [20]. In the mitochondria, this reaction allows the production of acetyl-CoA [8]. Moreover, the CoA-transferase's reaction is using acetate as acceptor, which implies that Ach1 is also important for acetate detoxification and mitochondrial functionality during chronological aging [47]. Consequently, in a condition where acetate-generating activities of Ald4/Ald5 are increased, as it is the case in the mutant, a reduction in CoA-transferase's enzymatic activity could play a causative role in promoting/enhancing the mitochondrial damage observed in the mutant.

Furthermore, since during the utilization of ethanol and acetate, the sole possible route for the net synthesis of C4 dicarboxylic acids for replenishing the TCA cycle of intermediates is the glyoxylate shunt, it follows that in the *mpc1Δ* cells this anaplerotic shunt is enhanced in order to keep the "branched" TCA cycle functioning. In turn, it follows that the pathway providing the cytosolic acetyl-CoA must be increased to support an enhanced glyoxylate demand. In line with this, the cytosolic Ald6 enzymatic activity is increased (Fig. 3C) and hyperactivation of the Acs2 activity has been detected [8]. Interestingly, this synthetase is responsible not only for supplying acetyl-CoA for carbon metabolism, but also for protein acetylation, particularly of histones [19]. In addition, it has been shown that during chronological aging, upregulation of Acs2 activity culminates in histone H3 hyperacetylation associated with transcriptional downregulation of several autophagy-essential *ATG* genes [48]. In this context, *mpc1Δ* cells display an age-dependent loss of autophagy [8]; this feature, given the reciprocal cross-talk between autophagy and mitochondria, can negatively affect the removal of the damaged mitochondria of the mutant and consequently contribute to its inability to maintain proper cellular homeostasis during the aging process. Notably, Ald6, which is responsible of generating cytosolic acetate, is degraded preferentially by autophagy [49] and the persistence of its enzymatic activity seems to be disadvantageous for the survival during nitrogen starvation [50]. Thus, *mpc1Δ* cells make up for their impairment in mitochondrial pyruvate with a metabolic rewiring in which the pro-aging outcome prevails.

MATERIALS AND METHODS

Yeast strains and growth conditions

The *mcp1Δ* mutant (*mcp1Δ::HIS3*) was generated by PCR-based methods in a BY4741 background (*MATa his3Δ-1 leu2Δ-0 met15Δ-0 ura3Δ-0)* and the accuracy of gene replacement was verified by PCR with flanking and internal primers. At least two different clones were tested for any experiment. Yeast cells were grown in batches at 30°C in minimal medium (Difco Yeast Nitrogen Base without amino acids, 6.7 g/L) with 2% glucose and the required supplements added in excess to a final concentration of 200 mg/L, except for leucine at 500 mg/L to avoid auxotrophy starvation [51, 52]. L-carnitine (Sigma) was supplemented to a concentration of 10 mg/L. Strains were inoculated at the same cellular density (culture volume no more than 20% of the flask volume) and growth was monitored by determining cell number using a Coulter Counter-Particle Count and Size Analyser, as described [53]. Duplication times (Td) were obtained by linear regression of the cell number increase over time on a semi-logarithmic plot.

CLS determination

Survival experiments in expired medium were performed on cells grown in minimal medium/2% glucose and the required supplements as described above. During growth, cell number and extracellular glucose, ethanol and acetic acid were measured in order to define the growth profile (exponential phase, diauxic shift, post-diauxic phase and stationary phase) of the culture (Fig. S1). Cell survival was monitored by harvesting aliquots of cells starting with 72 h (Day 3, first age-point) after the diauxic shift (Day 0). CLS was measured according to [51] by counting colony-forming units (CFUs) every 2-3 days. The number of CFUs on Day 3 was considered the initial survival (100%).

Metabolite measurements and enzymatic assays

At designated time points, aliquots of the yeast cultures were centrifuged and both pellets (washed twice) and supernatants were frozen at −80°C until used. Rapid sampling for intracellular metabolite measurements was performed according to the leakage-free cold methanol quenching method developed by [54] in which pure methanol at ≤ -40°C and a ratio of cell culture to quenching solvent of 1:5 (final methanol concentration ≥ 83%) were used. Metabolites from the cell pellets were extracted in 5 ml of a solution of 75% (v/v) boiling absolute ethanol containing 0.25 M Hepes, pH 7.5, as described in [55]. The concentrations of glucose, ethanol, acetate, pyruvate, citrate, succinate and malate were determined using enzymatic assays (K-HKGLU, K-ETOH, K-ACET, K-PYRUV, K-SUCC, K-CITR and K-LMALR kits from Megazyme). Ethanol specific consumption rate (qEtOH), expressed in mmol·g·DW $^{-1}$·h $^{-1}$, was calculated from measured cell dry weights (DWs) and extracellular ethanol concentrations. DW was measured as described [56].

All the enzymatic activities were assayed immediately after preparation of cell-free extracts. Cells were resuspended in 100 mM potassium phosphate buffer, pH 7.5, containing 2 mM $MgCl_2$ and 1 mM dithiothreitol and broken with acid-washed glass beads by shaking on a vortex for several cycles interspersed with cooling on ice. The activities of cytosolic and mitochondrial aldehyde dehydrogenase (Ald) were measured as described by [57], of alcohol dehydrogenase (Adh) according to [58], of phosphoenolpyruvate carboxykinase (Pck1) and isocitrate lyase (Icl1) as in [34]. Malic enzyme activities were determined according to [31] with either 0.4 mM NAD^+ or $NADP^+$ as the redox cofactor. The enzymatic activity was measured in the decarboxylation direction to avoid interference with pyruvate decarboxylase and Adh. Total protein concentration was estimated using the BCA™ Protein Assay Kit (Pierce).

Oxygen consumption and fluorescence microscopy

The basal oxygen consumption of intact cells was measured at 30°C using a 'Clark-type' oxygen electrode in a thermostatically controlled chamber (Oxygraph System, Hansatech Instruments, Norfolk, UK) as previously reported [25]. Data were recorded at sampling intervals of 1 s (Oxygraph Plus software, Hansatech Instruments, Norfolk, UK). All assays were conducted in biological triplicate.

ROS were detected with dihydroethidium (DHE, Sigma) according to [59]. The mitochondrial membrane potential was assessed by staining with $DiOC_6$ (Molecular Probes, Invitrogen), according to [39]; cells were also counterstained with propidium iodide to discriminate between live and dead cells. A Nikon Eclipse E600 fluorescence microscope equipped with a Leica DC 350F ccd camera was used. Digital images were acquired using FW4000 software (Leica).

Statistical analysis of data

All values are presented as the mean of three independent experiments with the corresponding Standard Deviation (SD). Three technical replicates were analyzed in each independent experiments. Statistical significance was assessed by one-way ANOVA test. P value of ≤ 0.05 was considered statistically significant.

ACKNOWLEDGMENTS

The authors are grateful to Neil Campbell for English editing. We acknowledge funding of the project SysBioNet, Italian Roadmap Research Infrastructures.

CONFLICT OF INTEREST

The authors declare no conflict of interest.

REFERENCES

1. Longo VD, Kennedy BK (**2006**). Sirtuins in aging and age-related disease. **Cell** 126(2): 257-68.

2. MacLean M, Harris N, Piper PW (**2001**). Chronological lifespan of stationary phase yeast cells; a model for investigating the factors that might influence the ageing of postmitotic tissues in higher organisms. **Yeast** 18(6): 499-509.

3. Fabrizio P, Longo VD (**2003**). The chronological life span of *Saccharomyces cerevisiae*. **Aging Cell** 2(2): 73-81.

4. Fontana L, Partridge L, Longo VD (**2010**). Extending healthy life span - from yeast to humans. **Science** 328(5976): 321-6.

5. Swinnen E, Ghillebert R, Wilms T, Winderickx J (**2013**). Molecular mechanisms linking the evolutionary conserved TORC1-Sch9 nutrient signalling branch to lifespan regulation in *Saccharomyces cerevisiae*. **FEMS Yeast Res** 14(1): 17-32.

6. Houtkooper RH, Williams RW, Auwerx J (**2010**). Metabolic networks of longevity. **Cell** 142(1): 9-14.

7. de Cabo R, Carmona-Gutierrez D, Bernier M, Hall MN, Madeo F (**2014**). The search for antiaging interventions: from elixirs to fasting regimens. **Cell** 157(7): 1515-26.

8. Eisenberg T, Schroeder S, Andryushkova A, Pendl T, Kuttner V, Bhukel A, Mariño G, Pietrocola F, Harger A, Zimmermann A, Moustafa T, Sprenger A, Jany E, Buttner S, Carmona-Gutierrez D, Ruckenstuhl C, Ring J, Reichelt W, Schimmel K, Leeb T, Moser C, Schatz S, Kamolz LP, Magnes C, Sinner F, Sedej S, Frohlich KU, Juhasz G, Pieber TR, Dengjel J, Sigrist SJ, Kroemer G, Madeo F (**2014**). Nucleocytosolic depletion of the energy metabolite acetyl-coenzyme A stimulates autophagy and prolongs lifespan. **Cell Metab** 19(3): 431-44.

9. Friis RM, Glaves JP, Huan T, Li L, Sykes BD, Schultz MC (**2014**). Rewiring AMPK and mitochondrial retrograde signaling for metabolic control of aging and histone acetylation in respiratory-defective cells. **Cell Rep** 7(2): 565-74.

10. Mariño G, Pietrocola F, Eisenberg T, Kong Y, Malik SA, Andryushkova A, Schroeder S, Pendl T, Harger A, Niso-Santano M, Zamzami N, Scoazec M, Durand S, Enot DP, Fernandez AF, Martins I, Kepp O, Senovilla L, Bauvy C, Morselli E, Vacchelli E, Bennetzen M, Magnes C, Sinner F, Pieber T, Lopez-Otin C, Maiuri MC, Codogno P, Andersen JS, Hill JA, Madeo F, Kroemer G (**2014**). Regulation of autophagy by cytosolic acetyl-coenzyme A. **Mol Cell** 53(5): 710-25.

11. Henriksen P, Wagner SA, Weinert BT, Sharma S, Bacinskaja G, Rehman M, Juffer AH, Walther TC, Lisby M, Choudhary C (**2012**). Proteome-wide analysis of lysine acetylation suggests its broad regulatory scope in *Saccharomyces cerevisiae*. **Mol Cell Proteomics** 11(11): 1510-22.

12. Kaelin WG Jr., McKnight SL (**2013**). Influence of metabolism on epigenetics and disease. **Cell** 153(1): 56-69.

13. Lu C, Thompson CB (**2012**). Metabolic regulation of epigenetics. **Cell Metab** 16(1): 9-17.

14. Millar CB, Grunstein M (**2006**). Genome-wide patterns of histone modifications in yeast. **Nat Rev Mol Cell Biol** 7(9): 657-66.

15. Eisenberg T, Knauer H, Schauer A, Buttner S, Ruckenstuhl C, Carmona-Gutierrez D, Ring J, Schroeder S, Magnes C, Antonacci L, Fussi H, Deszcz L, Hartl R, Schraml E, Criollo A, Megalou E, Weiskopf D, Laun P, Heeren G, Breitenbach M, Grubeck-Loebenstein B, Herker E, Fahrenkrog B, Frohlich KU, Sinner F, Tavernarakis N, Minois N, Kroemer G, Madeo F (**2009**). Induction of autophagy by spermidine promotes longevity. **Nat Cell Biol** 11(11): 1305-14.

16. Rubinsztein DC, Mariño G, Kroemer G (**2011**). Autophagy and aging. **Cell** 146(5): 682-95.

17. Cuervo AM (**2008**). Autophagy and aging: keeping that old broom working. **Trends Genet** 24(12): 604-12.

18. van den Berg MA, de Jong-Gubbels P, Kortland CJ, van Dijken JP, Pronk JT, Steensma HY (**1996**). The two acetyl-coenzyme A synthetases of *Saccharomyces cerevisiae* differ with respect to kinetic properties and transcriptional regulation. **J Biol Chem** 271(46): 28953-9.

19. Takahashi H, McCaffery JM, Irizarry RA, Boeke JD (**2006**). Nucleocytosolic acetyl-coenzyme A synthetase is required for histone acetylation and global transcription. **Mol Cell** 23(2): 207-17.

20. Fleck CB, Brock M (**2009**). Re-characterisation of *Saccharomyces cerevisiae* Ach1p: fungal CoA-transferases are involved in acetic acid detoxification. **Fungal Genet Biol** 46(6-7): 473-85.

21. Pronk JT, Yde Steensma H, Van Dijken JP (**1996**). Pyruvate metabolism in *Saccharomyces cerevisiae*. **Yeast** 12(16): 1607-33.

22. Bricker DK, Taylor EB, Schell JC, Orsak T, Boutron A, Chen YC, Cox JE, Cardon CM, Van Vranken JG, Dephoure N, Redin C, Boudina S, Gygi SP, Brivet M, Thummel CS, Rutter J (**2012**). A mitochondrial pyruvate carrier required for pyruvate uptake in yeast, *Drosophila*, and humans. **Science** 337(6090): 96-100.

23. Herzig S, Raemy E, Montessuit S, Veuthey JL, Zamboni N, Westermann B, Kunji ER, Martinou JC (**2012**). Identification and functional expression of the mitochondrial pyruvate carrier. **Science** 337(6090): 93-6.

24. Casal M, Paiva S, Queiros O, Soares-Silva I (**2008**). Transport of carboxylic acids in yeasts. **FEMS Microbiol Rev** 32(6): 974-94.

25. Orlandi I, Ronzulli R, Casatta N, Vai M (**2013**). Ethanol and acetate acting as carbon/energy sources negatively affect yeast chronological aging. **Oxid Med Cell Longev** Article ID:802870.

26. de Smidt O, du Preez JC, Albertyn J (**2008**). The alcohol dehydrogenases of *Saccharomyces cerevisiae*: a comprehensive review. **FEMS Yeast Res** 8(7): 967-78.

27. Saint-Prix F, Bonquist L, Dequin S (**2004**). Functional analysis of the ALD gene family of *Saccharomyces cerevisiae* during anaerobic growth on glucose: the $NADP^+$-dependent Ald6p and Ald5p isoforms play a major role in acetate formation. **Microbiology** 150(7): 2209-20.

28. de Jong-Gubbels P, van den Berg MA, Luttik MA, Steensma HY, van Dijken JP, Pronk JT (**1998**). Overproduction of acetyl-coenzyme A synthetase isoenzymes in respiring *Saccharomyces cerevisiae* cells does not reduce acetate production after exposure to glucose excess. **FEMS Microbiol Lett** 165(1): 15-20.

29. Remize F, Andrieu E, Dequin S (**2000**). Engineering of the pyruvate dehydrogenase bypass in *Saccharomyces cerevisiae*: role of the cytosolic $Mg(2^+)$ and mitochondrial $K(^+)$ acetaldehyde dehydrogenases Ald6p and Ald4p in acetate formation during alcoholic fermentation. **Appl Environ Microbiol** 66(8): 3151-9.

30. Lee YJ, Jang JW, Kim KJ, Maeng PJ (**2011**). TCA cycle-independent acetate metabolism via the glyoxylate cycle in *Saccharomyces cerevisiae*. **Yeast** 28(2): 153-66.

31. Boles E, de Jong-Gubbels P, Pronk JT (**1998**). Identification and characterization of *MAE1*, the *Saccharomyces cerevisiae* structural gene encoding mitochondrial malic enzyme. **J Bacteriol** 180(11): 2875-82.

32. dos Santos MM, Gombert AK, Christensen B, Olsson L, Nielsen J (**2003**). Identification of *in vivo* enzyme activities in the cometabolism of glucose and acetate by *Saccharomyces cerevisiae* by using ^{13}C-labeled substrates. **Eukaryot Cell** 2(3): 599-608.

33. Palmieri L, Lasorsa FM, De Palma A, Palmieri F, Runswick MJ, Walker JE (**1997**). Identification of the yeast *ACR1* gene product as a succinate-fumarate transporter essential for growth on ethanol or acetate. **FEBS Lett** 417(1): 114-8.

34. Casatta N, Porro A, Orlandi I, Brambilla L, Vai M (**2013**). Lack of Sir2 increases acetate consumption and decreases extracellular pro-aging factors. **Biochim Biophys Acta** 1833(3): 593-601.

35. Lin YY, Lu JY, Zhang J, Walter W, Dang W, Wan J, Tao SC, Qian J, Zhao Y, Boeke JD, Berger SL, Zhu H (**2009**). Protein acetylation microarray reveals that NuA4 controls key metabolic target regulating gluconeogenesis. **Cell** 136(6): 1073-84.

36. Guan KL, Xiong Y (**2011**). Regulation of intermediary metabolism by protein acetylation. **Trends Biochem Sci** 36(2): 108-16.

37. Choudhary C, Weinert BT, Nishida Y, Verdin E, Mann M (**2014**). The growing landscape of lysine acetylation links metabolism and cell signalling. **Nat Rev Mol Cell Biol** 15(8): 536-50.

38. Cai L, Sutter BM, Li B, Tu BP (**2011**). Acetyl-CoA induces cell growth and proliferation by promoting the acetylation of histones at growth genes. **Mol Cell** 42(4): 426-37.

39. Koning AJ, Lum PY, Williams JM, Wright R (**1993**). $DiOC_6$ staining reveals organelle structure and dynamics in living yeast cells. **Cell Motil Cytoskel** 25(2): 111-28.

40. Knorre DA, Popadin KY, Sokolov SS, Severin FF (**2013**). Roles of mitochondrial dynamics under stressful and normal conditions in yeast cells. **Oxid Med Cell Longev** Article ID:139491.

41. Merz S, Hammermeister M, Altmann K, Durr M, Westermann B (**2007**). Molecular machinery of mitochondrial dynamics in yeast. **Biol Chem** 388(9): 917-26.

42. Breitenbach M, Rinnerthaler M, Hartl J, Stincone A, Vowinckel J, Breitenbach-Koller H, Ralser M (**2014**). Mitochondria in ageing: there is metabolism beyond the ROS. **FEMS Yeast Res** 14(1): 198-212.

43. Demir AB, Koc A (**2010**). Assessment of chronological lifespan dependent molecular damages in yeast lacking mitochondrial antioxidant genes. **Biochem Biophys Res Commun** 400(1): 106-10.

44. van Roermund CW, Hettema EH, van den Berg M, Tabak HF, Wanders RJ (**1999**). Molecular characterization of carnitine-dependent transport of acetyl-CoA from peroxisomes to mitochondria in *Saccharomyces cerevisiae* and identification of a plasma membrane carnitine transporter, Agp2p. **EMBO J** 18(21): 5843-52.

45. Swiegers JH, Dippenaar N, Pretorius IS, Bauer FF (**2001**). Carnitine-dependent metabolic activities in *Saccharomyces cerevisiae*: three carnitine acetyltransferases are essential in a carnitine-dependent strain. **Yeast** 18(7): 585-95.

46. Palmieri L, Lasorsa FM, Iacobazzi V, Runswick MJ, Palmieri F, Walker JE (**1999**). Identification of the mitochondrial carnitine carrier in *Saccharomyces cerevisiae*. **FEBS Lett** 462(3): 472-6.

47. Orlandi I, Casatta N, Vai M (**2012**). Lack of Ach1 coa-transferase triggers apoptosis and decreases chronological lifespan in yeast. **Front Oncol** 2: 67.

48. Eisenberg T, Schroeder S, Buttner S, Carmona-Gutierrez D, Pendl T, Andryushkova A, Mariño G, Pietrocola F, Harger A, Zimmermann A, Magnes C, Sinner F, Sedej S, Pieber TR, Dengjel J, Sigrist S, Kroemer G, Madeo F (**2014**). A histone point mutation that switches on autophagy. **Autophagy** 10(6): 1143-45.

49. Suzuki K, Nakamura S, Morimoto M, Fujii K, Noda NN, Inagaki F, Ohsumi Y (**2014**). Proteomic profiling of autophagosome cargo in *Saccharomyces cerevisiae*. **PLoS One** 9(3): e91651.

50. Onodera J, Ohsumi Y (**2004**). Ald6p is a preferred target for autophagy in yeast, *Saccharomyces cerevisiae*. **J Biol Chem** 279(16): 16071-6.

51. Fabrizio P, Gattazzo C, Battistella L, Wei M, Cheng C, McGrew K, Longo VD (**2005**). Sir2 blocks extreme life-span extension. **Cell** 123(4): 655-67.

52. Boer VM, Amini S, Botstein D (**2008**). Influence of genotype and nutrition on survival and metabolism of starving yeast. **Proc Natl Acad Sci USA** 105(19): 6930-5.

53. Vanoni M, Vai M, Popolo L, Alberghina L (**1983**). Structural heterogeneity in populations of the budding yeast *Saccharomyces cerevisiae*. **J Bacteriol** 156(3): 1282-91.

54. Canelas A, Ras C, ten Pierick A, van Dam J, Heijnen J, van Gulik W (**2008**). Leakage-free rapid quenching technique for yeast metabolomics. **Metabolomics** 4(3): 226-39.

55. Canelas AB, ten Pierick A, Ras C, Seifar RM, van Dam JC, van Gulik WM, Heijnen JJ (**2009**). Quantitative evaluation of intracellular metabolite extraction techniques for yeast metabolomics. **Anal Chem** 81(17): 7379-89.

16

At neutral pH the chronological lifespan of *Hansenula polymorpha* increases upon enhancing the carbon source concentrations

Adam Kawałek and Ida J. van der Klei*

Molecular Cell Biology, Groningen Biomolecular Sciences and Biotechnology Institute, Systems Biology Centre for Metabolism and Ageing, University of Groningen, the Netherlands.

* Corresponding Author: Prof. Dr. Ida J. van der Klei, P.O. Box 11103; 9700CC Groningen, the Netherlands

Email: i.j.van.der.klei@rug.nl

ABSTRACT Dietary restriction is generally assumed to increase the lifespan in most eukaryotes, including the simple model organism *Saccharomyces cerevisiae*. However, recent data questioned whether this phenomenon is indeed true for yeast. We studied the effect of reduction of the carbon source concentration on the chronological lifespan of the yeast *Hansenula polymorpha* using four different carbon sources. Our data indicate that reduction of the carbon source concentration has a negative (glucose, ethanol, methanol) or positive (glycerol) effect on the chronological lifespan. We show that the actual effect of carbon source concentrations largely depends on extracellular factor(s). We provide evidence that *H. polymorpha* acidifies the medium and that a low pH of the medium alone is sufficient to significantly decrease the chronological lifespan. However, glucose-grown cells are less sensitive to low pH compared to glycerol-grown cells, explaining why only the reduction of the glycerol-concentration (which leads to less medium acidification) has a positive effect on the chronological lifespan. Instead, the positive effect of enhancing the glucose concentrations is much larger than the negative effect of the medium acidification at these conditions, explaining the increased lifespan with increasing glucose concentrations. Importantly, at neutral pH, the chronological lifespan also decreases with a reduction in glycerol concentrations. We show that for glycerol cultures this effect is related to acidification independent changes in the composition of the spent medium. Altogether, our data indicate that in *H. polymorpha* at neutral pH the chronological lifespan invariably extends upon increasing the carbon source concentration.

***Keywords**: Hansenula polymorpha, chronological lifespan, ageing, acidification, dietary restriction.*

***Abbreviations**:*

CLS - chronological lifespan,
DR - dietary restriction,
MM - mineral medium,
OD - optical density,
WT (wild type).

INTRODUCTION

Dietary restriction (DR) is defined as reduction in nutrient availability without malnutrition [1,2]. DR has been proposed to be a general intervention to prevent ageing in a large variety of organisms, ranging from simple model organisms like yeast to higher eukaryotes, such as rodents [3–5].

The reason as to why DR enhances yeast lifespan is currently debated. The chronological lifespan (CLS) of yeast is defined as the time non-dividing cells remain viable after exit from the growth phase [6,7]. In *Saccharomyces cerevisiae* DR is typically defined as the reduction of the glucose concentration in the batch medium from 2% to 0.5%. *S. cerevisiae* is a Crabtree-positive yeast, meaning that mitochondrial oxidative metabolism is repressed in media with high concentrations of glucose. In 2% glucose containing media, the carbon source is initially fermented to ethanol, which is subsequently utilized when glucose is depleted. This results in the so called diauxic shift [8]. During glucose fermentation *S. cerevisiae* also secretes acetate. Is has been argued that acetate secretion together with the acidification of the medium to values below pH 4.5 is the major cause of the reduced lifespan of *S. cerevisiae* cells grown on 2% glucose [8–10]. Indeed the pH of *S. cerevisiae* cultures containing synthetic complete medium supplemented with 2% glucose can drop to values of 2.5- 2.8 [9,11], whereas cultures containing 0.5% glucose do not acidify [12]. Also, buffering the medium to pH 6.0 or resuspension of stationary cells in water has been shown to extend the CLS of yeast cells grown on media containing 2% glucose [9].

So far, the effects of DR, acidification and acetate on yeast CLS were mainly studied using *S. cerevisiae* and glucose as a carbon source. Here we study the effects of these parameters using the Crabtree negative yeast *Hansenula polymorpha*. This yeast is unable to inhibit respiration in the presence of high levels of glucose in favour of fermentation and is generally assumed not to secrete acetate during growth on glucose. In addition to the effect of different glucose concentrations on the CLS, we studied the effects of different carbon source concentrations and acidification when cells were grown on alternative carbon sources, namely glycerol, ethanol and methanol. Our data indicate that reduction of the carbon source concentration can have a negative (glucose, ethanol, methanol) or positive (glycerol) effect on the CLS of *H. polymorpha*. However, at neutral pH the CLS invariably increases upon enhancing the carbon source concentration. This indicates that reducing the carbon source concentration is not a common intervention to enhance yeast lifespan. Also, we show that a low pH especially reduces the CLS of *H. polymorpha* after exit from the growth phase and acts independent of the presence of compounds (like acetate) secreted in the medium during the growth phase.

RESULTS

The effect of carbon source concentration on the chronological lifespan is carbon source dependent

Recent reports indicated that medium composition strongly influences yeast CLS [2,11,13]. The type of nutrient limitation as well as the concentration of amino acids required for auxotrophic laboratory strains strongly affect the survival of the cells in the stationary phase [14–16]. To avoid these issues we performed our studies using a prototrophic *H. polymorpha* strain and mineral media (MM) in which the carbon source is the only limiting factor for growth (meaning that the cells exit the growth phase solely due to depletion of the carbon source). We therefore defined DR in our current study as a reduction in carbon source concentration under conditions that all other medium components are present in excess. At these conditions a reduction in carbon source concentration results in a proportional reduction in the growth yield (final optical density). We used methylamine as N-source in all our studies, as we recently showed that this results in a longer CLS of *H. polymorpha* relative to the use of ammonium sulphate [17,18].

To determine suitable low and high carbon source concentrations for the four different carbon sources used in this study (glycerol, glucose, methanol, ethanol), we grew wild-type (WT) *H. polymorpha* strain at various concentrations of these compounds and determined the optical densities (OD) of the stationary cultures. A linear increase in final OD with increasing carbon source concentrations was

TABLE 1. Mean and maximum lifespan of *H. polymorpha* WT cells upon growth on different concentrations of glycerol and glucose.

	Glycerol		Glucose	
Concentration (%)	**Mean lifespan (days)**	**Max. lifespan (days)**	**Mean lifespan (days)**	**Max. lifespan (days)**
0.1	5.72 ± 0.69	7.64 ± 0.46	1.08 ± 0.20	2.02 ± 0.29
0.15	4.73 ± 0.51	6.42 ± 0.25	1.40 ± 0.03	3.19 ± 0.03
0.2	4.61 ± 0.39	6.43 ± 0.20	2.02 ± 0.28	3.60 ± 0.60
0.25	4.19 ± 0.44	6.31 ± 0.03	3.44 ± 0.15	5.04 ± 0.18
0.3	3.78 ± 0.17	5.31 ± 0.17	3.77 ± 0.03	5.06 ± 0.03
0.35	3.69 ± 0.09	5.29 ± 0.06	4.17 ± 0.18	5.85 ± 0.27
0.4	3.67 ± 0.36	5.35 ± 0.43	4.38 ± 0.06	5.94 ± 0.62
0.45	3.26 ± 0.13	4.44 ± 0.25	4.31 ± 0.15	6.81 ± 0.15
0.5	2.94 ± 0.18	4.08 ± 0.44	4.29 ± 0.29	6.10 ± 0.27
0.6	ND	ND	5.15 ± 0.51	7.9 ± 0.47
0.7	ND	ND	5.75 ± 0.12	7.75 ± 0.18
0.8	ND	ND	5.67 ± 0.24	8.02 ± 0.15

The mean lifespan was calculated as a time point when cultures reached 50% viability; the maximum lifespan was calculated as a time point when cultures reached 10% viability. The data represents mean ± SD from at least 4 independent cultures. ND – not determined.

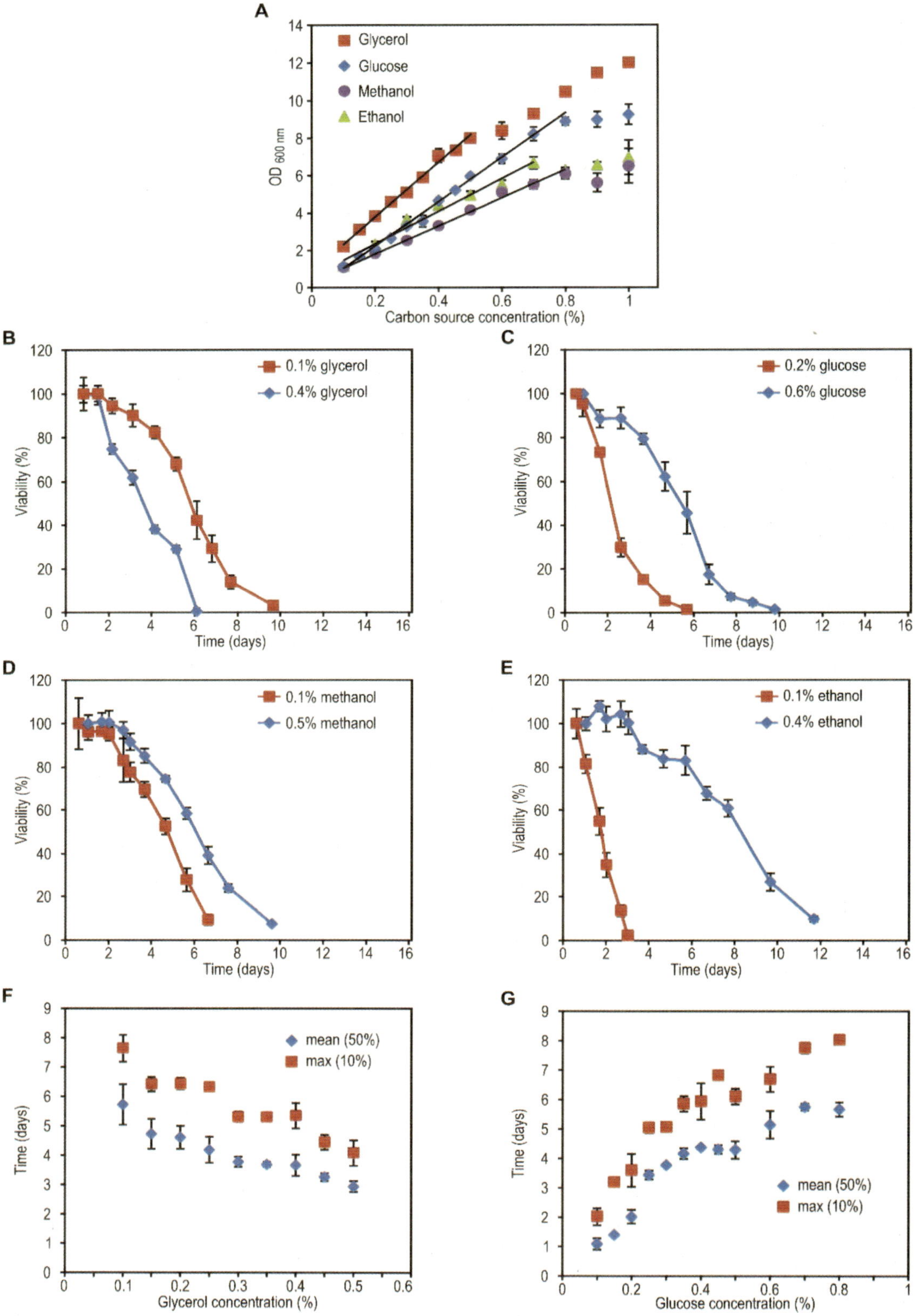

FIGURE 1: The effect of carbon source concentration on yeast chronological lifespan. *H. polymorpha* cells were grown on various concentrations of the indicated carbon sources and 0.25% methylamine as nitrogen source. The $OD_{600\,nm}$ was measured when cultures did exit the growth phase **(A)**. Lines indicate ranges in which a linear correlation (R>0.98) between final OD and carbon source concentration was observed. Data represent mean ± SD (n=3). Chronological lifespan of cells grown at a low and a high concentration of glycerol **(B)**, glucose **(C)**, methanol **(D)**, ethanol **(E)**. Mean lifespan (mean) and maximum lifespan (max) were calculated as the time when cultures reach 50% and 10% viability, respectively, for cultures grown on different concentrations of glycerol **(F)** or glucose **(G)**. Data represent mean ± SD from 4 to 12 independent cultures.

observed up to 0.5% glycerol, 0.8% glucose, 0.8% methanol and 0.7% ethanol (Fig. 1A). Based on these data we selected the low concentrations as those that resulted in final ODs between 1 and 2, whereas for the high concentrations we chose those that resulted in ODs of 1 – 2 OD units below the maximal OD obtained with the highest concentrations of the linear range (Fig. 1A).

The lifespan of cells grown on the high glycerol concentration was shorter compared to cells grown on a low concentration of glycerol (Fig. 1B). However, when glucose, methanol or ethanol was used, the opposite was observed, namely an increase in CLS with increasing carbon source concentration (Fig. 1C-E).

To further analyse the observed opposite effects of reducing the carbon source concentrations on CLS for different carbon sources, we confined our further studies using glycerol, as an example of a carbon source for which reduction in the concentration enhanced the lifespan, and glucose, as representative of a carbon source for which the opposite was observed.

Analysis of the residual glucose and glycerol concentrations confirmed that indeed at the chosen concentrations for these two carbon sources, the cultures did exit the growth phase due to carbon source depletion (Fig. 2A-B). At maximum concentration analysed for glucose, re-addition of glucose to the spent medium allowed growth of the cells, however, the doubling time and final OD were not as high as for fresh mineral medium (Fig. S1). Moreover when higher concentrations, namely 1 or 2% of glycerol or glucose, were used, these carbon sources were not depleted when growth ceased (Fig. S2A-B) and hence cells did exit the growth phase because of other, yet unknown, reasons.

Systematic analysis of the CLS of glycerol and glucose cultures using the previously defined ranges of carbon source concentrations revealed that the CLS gradually decreased when the glycerol concentrations increased from 0.1% to 0.5% (Fig. 1F, Table 1). Using higher concentrations of glycerol (1 or 2%) prolonged the lifespan in comparison to 0.5% glycerol (Fig. S2C). The CLS of glucose grown cells gradually increased with increase of glucose concentrations in the range of 0.2 – 0.8% (Fig. 1G, Table 1). The lifespan of the cells grown on 1% or 2% glucose was comparable to the lifespan of cells grown on 0.5% glucose (Fig. S2D).

To measure the viability in CLS experiments, we routinely determine the colony forming units (CFU) using YPD plates. To exclude that the use of glucose plates for both glucose and glycerol cultures affected the results, we also determined the viability using YP-glycerol plates. Plating on YPD and YP-glycerol plates resulted in similar amount of colonies (Fig. S3A). Also very similar CLS curves were obtained when YPD or YP-glycerol plates were used (Fig. S3B-C). Based on this observation we continued our studies using YPD plates.

Extracellular factor(s) are involved in carbon source concentration-dependent lifespan changes

To test whether the composition of the spent medium differentially affects the CLS in glycerol and glucose media, we performed spent medium swap experiments. Because glycerol and glucose were completely depleted after 20h of growth (Fig. 2A-B), this time point was used to swap the media.

Incubating cells grown on 0.1% glycerol in spent medium of cultures grown on 0.4% glycerol strongly reduced their lifespan to values observed for cultures grown on 0.4% glycerol and kept in this medium (Fig. 2C, Table 2). However, upon incubation of cells grown on 0.2% glucose in spent medium of cultures containing 0.6% glucose, there was no major reduction in lifespan relative to cells grown on 0.2% glucose and kept in their spent medium (Fig. 2D).

Placing cells grown on 0.4% glycerol into spent medium of cells grown on 0.1% glycerol strongly extended their CLS (Fig. 2C). The mean and maximum lifespan of these cells increased almost three times compared to cells grown on 0.4% glycerol and kept in their spent medium (Table 2 and Table 1). Incubating cells grown in 0.6% glucose in spent medium from cultures grown on 0.2% glucose also extended their lifespan, but the difference was smaller (less than 1.5 fold increase) (Fig. 2D, Table 2 and Table 1).

This data indicates that composition of the spent medium strongly affect the chronological lifespan both for glycerol and glucose containing cultures. Furthermore, the CLS of glycerol-grown cells changed much more (either positively or negatively) upon the medium swap relative to the glucose-grown cells.

TABLE 2. Mean and maximum lifespan of *H. polymorpha* cells grown on low and high concentrations of glycerol and glucose and placed in the spent medium from cultures containing other concentration of same carbon source.

	Concentration used for growth of cells (%)	Concentration used to obtain spent medium (%)	Mean lifespan (days)	Max lifespan (days)
Glycerol	0.1	0.4	3.46 ± 0.06	4.58 ± 0.59
	0.4	0.1	10.15 ± 0.23	14.76 ± 0.52
Glucose	0.2	0.6	2.02 ± 0.09	3.08 ± 0.12
	0.6	0.2	7.08 ± 0.59	10.21 ± 0.29

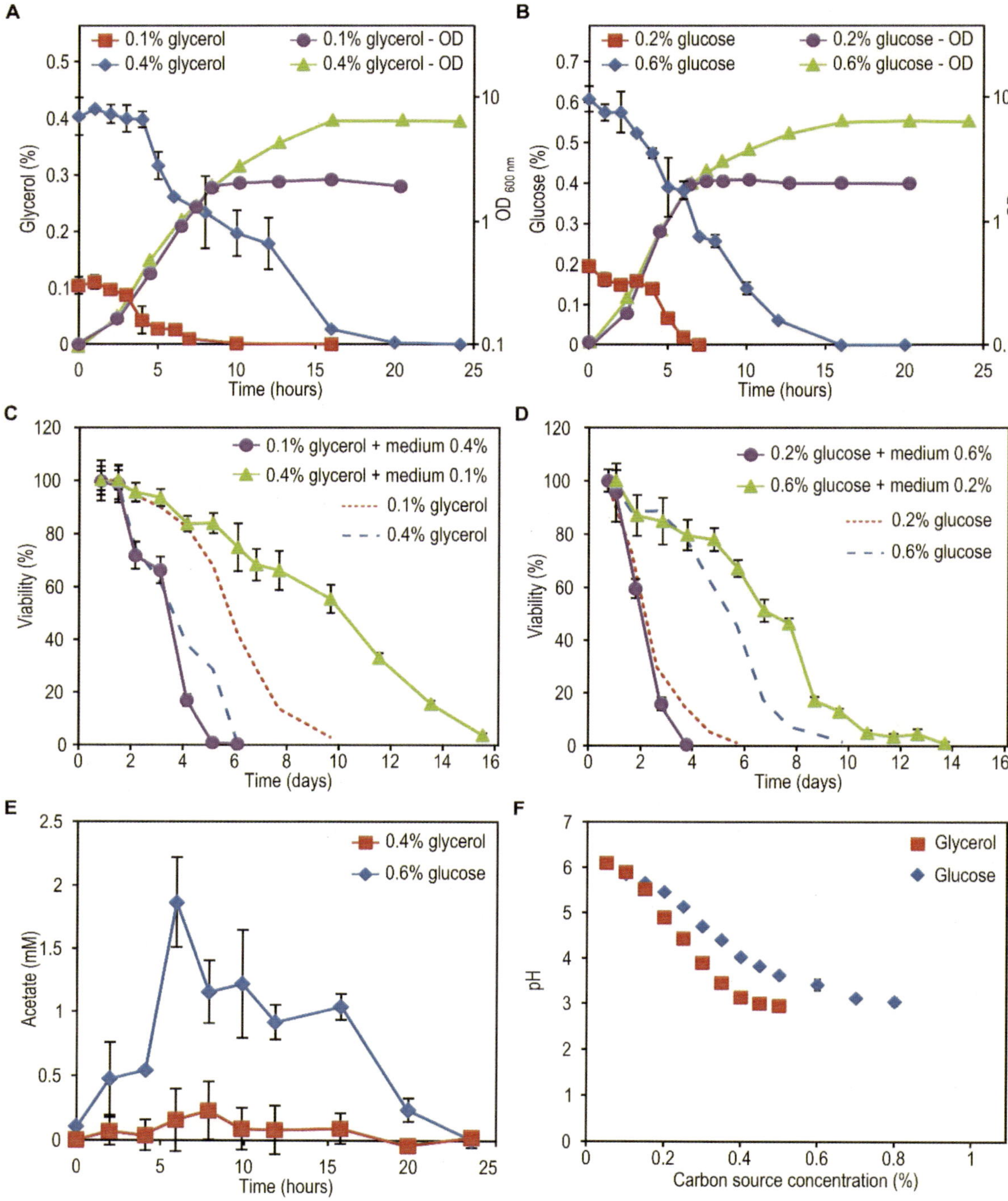

FIGURE 2: The impact of extracellular factors on DR mediated lifespan changes. Cells were grown on 0.1% and 0.4% glycerol **(A)** as well as 0.2% and 0.6% glucose **(B)**. Growth and carbon source depletion were monitored in time. Data represent mean glycerol or glucose concentration ± SD (n=3) and mean OD_{600nm} ± SD. Cells were grown on 0.1% and 0.4% glycerol **(C)** or 0.2% and 0.6% glucose **(D)**. Spent medium of cultures grown on one concentration of carbon source was replaced by the spent medium originating from cultures grown on the second concentration of the same carbon source and viability of the cultures was measured in time. Data represent mean viability ± SD from 3 independent cultures. Lines indicating viability of cultures left in its own medium were redrawn from Figure 1B and 1C. **(E)** The concentration of acetate was measured in clarified medium at different time points upon shifting the cells to medium containing 0.4% glycerol or 0.6% glucose and 0.25% methylamine. Data represent mean ± SD (n=3). **(F)** Cells were grown on different concentrations of glycerol or glucose. The pH of the cultures was measured at the beginning of the stationary phase. Data represent mean ± SD (n=3).

Acetic acid is not a major factor reducing *H. polymorpha* CLS

Despite the fact that *H. polymorpha* is a Crabtree negative yeast, it has been reported that this yeast may secrete acetic acid as a consequence of the overflow metabolism [19]. Being a weak organic acid, acetic acid was shown to directly reduce viability of chronologically ageing *S. cerevisiae* cells [9]. As shown in Fig. 2E, external acetic acid concentrations up to 2 mM were detected during the growth phase of 0.6% glucose containing cultures, but this compound was subsequently depleted from the medium within 24h (i.e. the beginning of the CLS measurements). When cells were grown on 0.4% glycerol, no significant amounts of external acetic acid were detected. These data indicate that the acetic acid, which is secreted during growth, is not a toxic compound in the spent medium of high glucose or glycerol cultures, that affects the CLS as it is either not present (glycerol) or quickly depleted from the spent medium (glucose).

Medium acidification was previously shown to affect the lifespan of *S. cerevisiae* at a variety of growth conditions [9,11,20]. As shown in Fig. 2F, the pH of the cultures grown on 0.4% glycerol or 0.6% glucose significantly decreased to 3.1 and 3.4, respectively, whereas the pH of cultures grown at low carbon source concentrations remained above 5.0. Hence, the reduced CLS of cells grown on low glycerol or glucose media upon incubation in spent medium of high glycerol or glucose cultures could (partially) be explained by the low pH of these solutions. However, it does not explain why the reduction of the CLS of the glycerol-grown cells is much stronger compared to the glucose-grown cells. Possibly, other compounds than acetic acid are present in the medium and toxic at low pH in glycerol-grown cultures, but not or less in glucose cultures. We therefore set out to experiments to further dissect the role of low pH and putative secreted toxic compounds on *H. polymorpha* CLS.

Low pH of the milieu shortens the chronological lifespan

For a further detailed analysis of the effect of pH on CLS, we first confined our studies to glycerol containing cultures and one (intermediate) carbon source concentration (0.2%). Cells were grown at different, constant pH values using pH controlled batch fermenters. CLS measurements revealed a sudden increase in mean and maximum lifespan when cells were grown at pH values equal or higher than 6.0, relative to lower pH values (Fig. 3A). These differences are not related to differences in growth rates, because in media with pH values in the range of pH 3.5 to 6.5 the growth curves (doubling time, final OD) were similar (data not shown). Hence, these data confirm that a low pH reduces the CLS of glycerol-grown *H. polymorpha*.

To investigate whether a low pH affects the CLS during the growth, after the growth phase or both, we performed pH swap experiments. Cells were grown on 0.2% glycerol at a constant pH of 6.5 or 3.5 and upon exit from the growth phase, the pH was reduced or increased by the addition of H_3PO_4 or NaOH, respectively. Neutralizing the medium of cells grown at pH 3.5 to 6.5 resulted in a similar CLS compared to cells grown and kept at pH 6.5 during the stationary phase (Fig. 3B). Lowering the pH of cultures grown at pH 6.5 to 3.5 resulted in a strong decrease in CLS (Fig. 3B), but the CLS was not shortened to the values observed for cells that were grown and subsequently kept at pH 3.5. These data indicate that a low pH after exit from the growth phase strongly reduces the CLS of glycerol containing cultures.

Next, we asked whether only the low pH is responsible for lifespan shortening or whether it reduced the CLS in combination with other factors in the spent media. For instance, the toxicity of weak organic acids depends on the concentration of the acid and increases with decreasing pH [9,21]. Cells were grown in a batch fermenter at a constant pH of 6.5 or 3.5 on 0.2% glycerol. After exit from the growth phase cells were collected by centrifugation and resuspended in fresh mineral medium without carbon source (MM) at a pH of 3.5 or 6.5. Shifting the cells grown at pH 3.5 and pH 6.5 to fresh MM with the same pH did not alter the lifespan in comparison to cells left in spent medium (Fig. 3C, compare with Fig. 3B). Resuspension of the cells grown at pH 3.5 in fresh MM with a pH of 6.5 prolonged the CLS to the same extent as cells grown at pH 6.5 and shifted to mineral medium pH 6.5. Conversely, resuspension of cells grown at pH 6.5 in MM with a pH value of 3.5 shortened the CLS to a similar extent as the pH swap from pH 6.5 to 3.5 in spent medium (Fig. 3C, compare with Fig. 3B).

These data indicate that for cells grown on 0.2% glycerol, a low pH after the growth phase is the major factor reducing the CLS and not toxic components present in the spent medium. This conclusion is furthermore supported by the outcome of experiments in which non-growing cells were transferred to 25 mM phosphate buffer or 50 mM MES buffer with a pH of 3.5 or 6.5 instead of MM (Fig. 3D). Also, the same effect was observed when methylamine was replaced by other nitrogen sources (ammonium sulphate (AS) or urea) (Fig. 3E), indicating that the effect was not specific for cells utilizing methylamine. Importantly, the observed negative effect of a low pH after the growth phase was also observed when cells were grown on 0.4% glucose (Fig. 3F). Hence, the CLS of both glycerol and glucose-grown cells decreases when cells are incubated at a low pH after exit from the growth phase.

Acidification-independent effect of carbon source concentration on lifespan

Although a low pH affects the CLS of both glucose and glycerol-grown cells and a similar acidification is observed for both carbon sources (Fig. 2F), it remained unclear why DR has a positive effect when glycerol is the carbon source and a negative effect when glucose is present in the medium. One explanation may be that the glycerol-grown cells are more sensitive to a low pH. To investigate this, we analysed the effect of neutralizing the pH after exit from the growth phase. Cells were grown in media containing 0.1% to 0.5% glycerol or 0.1% to 0.8% glucose and the pH of the cultures was adjusted to 6.5 when cells stopped to grow. Neutralizing the pH significantly increased the CLS of cells

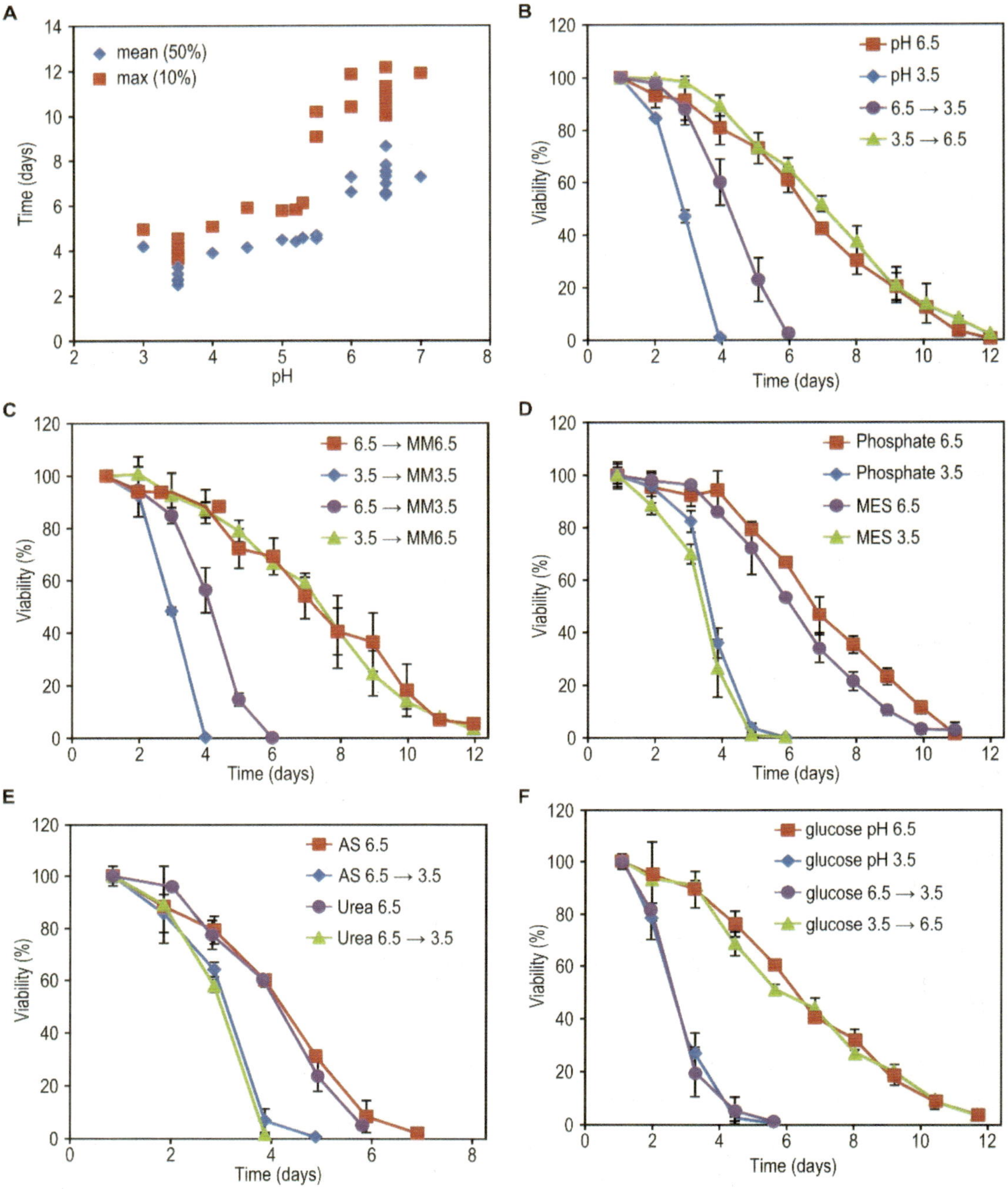

FIGURE 3: The effect of medium pH on yeast chronological lifespan. (A) Cells were grown on 0.2% glycerol and 0.25% methylamine in batch fermenters at different pH values. After 20h cells were shifted to flasks and the viability of the cultures was measured over time. Data represent time when viability of the cells reached 50% (mean) and 10% (max) for all individual fermenters. **(B)** Cells were grown on 0.2% glycerol / 0.25% methylamine at pH 6.5 or 3.5. After 20h of cultivation in batch fermenters, the pH of the cultures was changed by the addition of 1 M H_3PO_4 (6.5 → 3.5) or 1 M NaOH (3.5 → 6.5). Cells before the pH change were kept as the controls. Data represent mean ± SD (n=3). **(C)** Cells were grown on 0.2% glycerol / 0.25% methylamine at pH 6.5 or 3.5. After 20h, spent medium of the cultures was replaced by fresh medium (without carbon source) with a pH of 6.5 (MM6.5) or 3.5 (MM3.5). Data represent mean ± SD (n=3). **(D)** Cells were grown on 0.2% glycerol / 0.25% methylamine at pH 6.5 or 3.5. After 20h, spent medium of the cultures was replaced by potassium phosphate buffer pH 6.5 (phosphate 6.5), pH 3.5 (phosphate 3.5) or MES buffer pH 6.5 (MES 6.5) or pH 3.5 (MES 3.5). Data represent mean ± SD (n=3). **(E)** Cells were grown on 0.2% glycerol and 0.25% ammonium sulphate (AS) or 0.2% urea at pH 6.5. After 20h of cultivation in batch fermenters, the pH of the cultures was changed by addition of concentrated H_3PO_4 (6.5 → 3.5). Cells before the pH change were kept as the controls. Data represent mean ± SD (n=2). **(F)** Cells were grown on 0.4% glucose / 0.25% methylamine at pH 6.5 (glucose pH 6.5) or 3.5 (glucose pH 3.5) in batch fermenters. After 20h the pH of the cultures was changed by addition of H_3PO_4 (glucose 6.5 → 3.5) or NaOH (glucose 3.5 → 6.5). Cells before the pH change were kept as the controls. Data represent mean viability ± SD (n=2).

grown on 0.15 to 0.5% glycerol (Fig. 4A, compare with Fig. 1F, Table 1 and 3). Notably, under these conditions, the CLS increased with increasing glycerol concentrations. Neutralizing the medium of glucose-grown cultures also extended the CLS, which increased with increasing carbon source concentrations, like in the non-neutralized cultures (Fig. 4B, compare with Fig. 1G, Table 1 and 3). The CLS extending effect of neutralizing the pH was much less pronounced in glucose-grown cultures in comparison to glycerol-grown cultures (Fig. 4C).

We next asked whether the glucose-grown cells are only more resistant to low pH or also against acetic acid. As shown in Fig. 4D, cells grown on 0.6% glucose are also more resistant to short exposure to acetic acid than cells grown on 0.4% glycerol (Fig. 4D) suggesting that an adaptation may have occurred during the growth phase in cultures containing 0.6% glucose.

Exposure to weak organic acids at sublethal concentrations triggers cell cycle exit and prolonged cell stasis rather than cell death. Notably, an efficient cell cycle arrest at G_0/G_1 is crucial for longevity of chronologically ageing *S. cerevisiae* [20,22–24]. To assess the efficiency of cell cycle arrest we counted the percentage of cells in cultures grown on a low and a high concentration of glycerol and glucose after exit from the growth phase. Cultures grown on 0.6% glucose showed almost complete lack of budding cells, whereas more than 8% of the cells grown on 0.2% glucose and 0.1% or 0.4% glycerol contained buds (Fig. 4E).

The above data indicate that buffering has a weaker lifespan extending effect on cells grown on a high concentration of glucose. This effect is accompanied by elevated resistance to acetic acid and a more efficient cell cycle arrest.

Acidification is not the only extracellular factor in spent medium that affects the chronological lifespan

High carbon source concentrations (glucose or glycerol) are positive for the CLS of *H. polymorpha* cultures when the pH is neutral in the stationary phase. This could be either related to a direct positive effect of the high carbon source levels on the viability of the cells or due to an altered composition of the spent medium (the secretion of higher amounts of compounds that stimulate longevity or depletion of medium component negatively affecting the viability of the cells).

To investigate the acidification-independent effect of spent medium composition we performed buffered spent medium swap experiments. Cells were grown in media containing low or high carbon source concentrations (glycerol or glucose), collected by centrifugation and resuspended in spent medium from cultures containing either the low or the high concentration of carbon source. The pH was adjusted to 6.5 at the onset of the CLS experiment.

Replacing the medium of cells grown on 0.1% glycerol with buffered spent medium from cultures containing 0.4% glycerol strongly extended the CLS to values obtained for

TABLE 3. Mean and maximum lifespan of *H. polymorpha* cells upon growth on different concentrations of glycerol and glucose in medium buffered to pH 6.5.

	Glycerol		Glucose	
Concentration (%)	**Mean lifespan (days)**	**Max. lifespan (days)**	**Mean lifespan (days)**	**Max. lifespan (days)**
0.1	5.92 ± 0.48	8.57 ± 0.40	1.04 ± 0.06	1.79 ± 0.06
0.15	6.68 ± 0.29	10.06 ± 0.59	1.88 ± 0.18	3.73 ± 0.56
0.2	8.19 ± 1.15	12.82 ± 1.83	2.99 ± 0.17	5.50 ± 0.90
0.25	10.75 ± 0.15	15.71 ± 0.98	6.38 ± 0.47	9.50 ± 0.71
0.3	12.06 ± 0.47	17.01 ± 1.19	7.23 ± 0.56	11.42 ± 0.35
0.35	12.08 ± 1.30	16.15 ± 1.92	7.46 ± 0.12	11.49 ± 0.69
0.4	10.64 ± 0.94	14.35 ± 1.92	7.99 ± 0.23	11.81 ± 0.23
0.45	12.06 ± 0.82	16.03 ± 1.18	8.00 ± 0.47	12.83 ± 1.41
0.5	10.14 ± 0.67	16.25 ± 1.50	8.50 ± 0.40	13.25 ± 0.25
0.6	ND	ND	9.77 ± 0.49	14.85 ± 0.63
0.7	ND	ND	9.75 ± 1.06	15.31 ± 0.56
0.8	ND	ND	11.63 ± 0.40	16.36 ± 0.61

The mean lifespan was calculated as a time point when cultures reached 50% viability; the maximum lifespan was calculated as a time point when cultures reached 10% viability. The data represents mean ± SD from at least 4 independent cultures. ND – not determined.

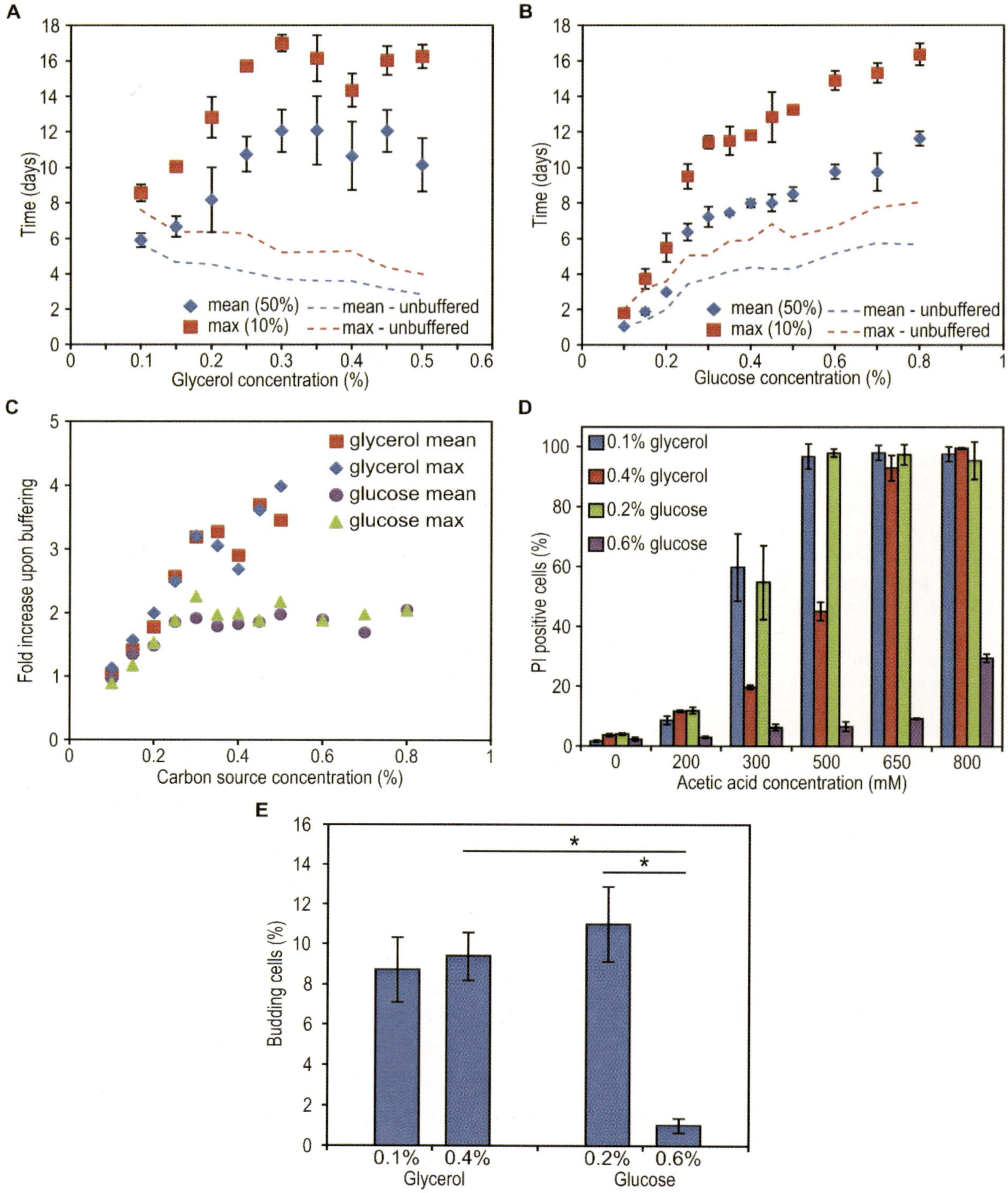

FIGURE 4: The effect of carbon source concentration on yeast lifespan upon buffering of spent medium.
Cells were grown on different concentrations of glycerol **(A)** or glucose **(B)** and 0.25% methylamine as the nitrogen source. The pH of the cultures was adjusted to pH 6.5. Data represent mean ± SD from 4 to 8 independent cultures. Trend lines indicating changes in mean and maximum lifespan of cells in non-buffered media were redrawn from Figure 1F and 1G. **(C)** Fold increase of mean and maximum lifespan upon buffering the cultures grown on different concentrations of glycerol and glucose calculated as lifespan after buffering divided by lifespan before buffering. **(D)** The resistance of cells to acetic acid treatment. Upon exit of the growth phase, cells were treated with increasing concentration of acetic acid followed by PI staining and FACS analysis. Data represent mean number of PI positive cells ± SD from 4 cultures. **(E)** The number of budding cells upon exit from the growth phase in unbuffered medium. Data represent mean percentage of cells containing a bud from 6 cultures ± SD.

TABLE 4. Mean and maximum lifespan of *H. polymorpha* WT cells grown on low and high concentrations of glycerol and glucose and placed in the buffered spent medium from cultures containing other concentration of same carbon source.

	Concentration used for growth of cells (%)	Concentration used to obtain spent medium (%)	Mean lifespan (days)	Max lifespan (days)
Glycerol	0.1	0.4	11.40 ± 0.35	16.04 ± 0.56
	0.4	0.1	10.00 ± 0.42	13.28 ± 0.46
Glucose	0.2	0.6	3.15 ± 0.15	6.67 ± 0.59
	0.6	0.2	7.25 ± 0.94	12.56 ± 0.50

cells grown in 0.4% glycerol and buffered upon exit from the growth phase (Fig. 5A, Table 4). Conversely, replacing the medium of cells grown in 0.4% glycerol with buffered medium originating from cultures containing 0.1% glycerol only had a slight reducing effect on the CLS (Fig. 5A). This observation suggests that the spent medium from cultures grown on a low concentration of glycerol possibly contains a factor negatively affecting the lifespan and/or that the spent medium of cells grown on high glycerol contains a factor positively affecting the CLS.

Essentially, similar experiments performed with cultures containing glucose as a carbon source, indicate that in this case extracellular factors only slightly affect the CLS. Replacing the medium with buffered spent medium of cells grown at higher or lower glucose concentrations did not strongly affect the mean and maximum lifespan values (Fig. 5B and Table 4). Based on these observations we conclude that the increased CLS with enhancing carbon source concentration is in part related to the composition of the medium in case of glycerol, but not for glucose.

DISCUSSION

Dietary restriction is the gold standard intervention that increases the lifespan of many organisms. Reduction of nutrient availability also affects the lifespan of *S. cerevisiae*. The relatively short lifespan and the possibility to use media with various nutrient compositions favour this organism as a model to unravel the mechanism of DR. A disadvantage however is that *S. cerevisiae* can ferment part of the glucose to acetic acid, which, together with the low pH of the spent medium is thought to be the primary cause of life span reduction in 2% glucose containing cultures. The exact mechanisms remain however controversial [9–12].

Here we analysed the effect of carbon source concentration on the chronological lifespan of the Crabtree negative yeast *H. polymorpha*. Because the effect of glucose concentration on CLS of *S. cerevisiae* was shown to largely depend on nutrient balance and amino acid concentrations in the medium [2], we used in our study mineral media in which the carbon source is the only growth limiting medium compound as well as a prototrophic strain.

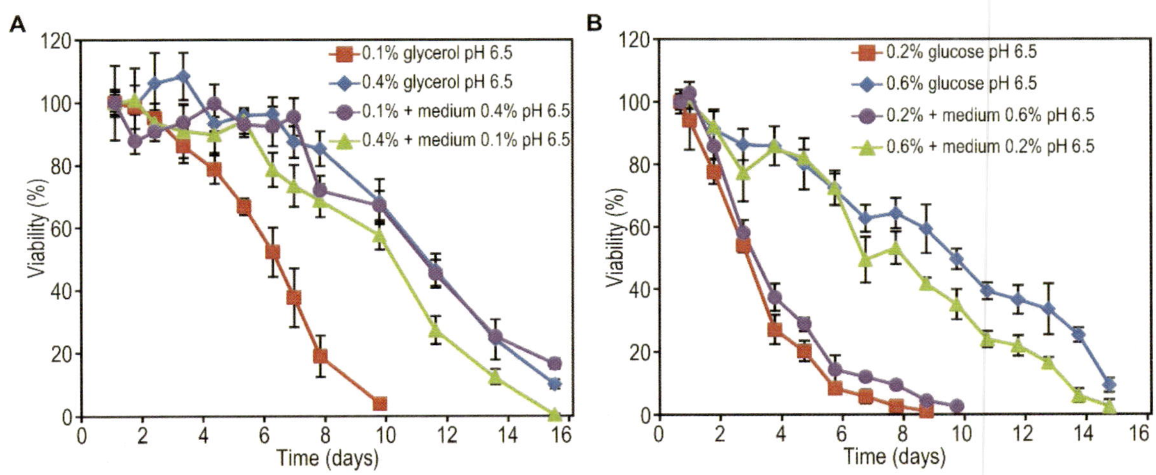

FIGURE 5: Acidification-independent impact of spent medium composition on lifespan. Cells were grown on 0.1% and 0.4% glycerol **(A)** as well as 0.2% and 0.6% glucose **(B)** for 20 hours. The spent medium from cultures grown on one concentration of carbon source was replaced by the spent medium originating from cultures containing other concentration of that carbon source with pH adjusted to 6.5. Data represent mean viability ± SD from 3 independent cultures. Representative data of two experiments with similar results are presented.

Our data indicate that the reduction of carbon source concentration has a negative effect on the CLS when glucose is used, whereas a positive effect was observed for glycerol. Reduction of glucose concentration was previously also shown to drastically reduce the lifespan of another Crabtree negative yeast, namely *Kluyveromyces lactis* [25]. Spent medium swap experiments revealed that for glycerol, and to the lesser extent for glucose, the effect of changing the carbon source concentrations is dependent on extracellular factors.

Our data indicate that unlike for glucose-grown *S. cerevisiae* [9], acetic acid secretion is not a major factor in CLS reduction for glucose-grown *H. polymorpha*. Instead we provide evidence that a low pH alone (below pH 5.5) is sufficient to reduce the lifespan. In *S. cerevisiae* the cytosolic pH is kept around 7.0 when cells are grown on glucose even when the external pH is as low as 3.0 [26]. However, when glucose is depleted a low extracellular pH promotes intracellular acidification to the minimum values of 5.0 - 5.5 [27–30]. This process can be stimulated by the presence of low molecular weak organic acids like acetic acid [21,28]. As lowering the pH of the medium from 6.5 to 3.5 upon exit from the growth phase decreased the lifespan of cells grown on 0.2% glycerol in a similar way as transfer to new medium / buffer with 2 distinct pH values (Fig. 3B-D), compounds secreted during growth are unlikely to influence the CLS. Instead, we cannot rule out that compounds acting like weak organic acids would be released from dead / lysed cells after the growth phase to accelerate cytosol acidification in the remaining cells.

A weaker impact of buffering on cells grown on 0.6% glucose in comparison to cells grown on glycerol correlates with their elevated resistance to acetic acid (Fig. 4D). It is possible that in cultures containing 0.6% glucose low concentrations of acetic acid, produced as a consequence of the overflow metabolism [19], trigger acid adaptation and cross protection against the effect of low pH later on in the stationary phase. Remarkably, low pH pre-treatment of *S. cerevisiae* cells elevates their resistance to subsequent treatment with acetic acid [31]. Such hormetic adaptation and higher initial resistance of 0.6% glucose grown cells to the impact of low pH would explain the limited lifespan extension upon placing these cells into spent medium from cultures containing initially 0.2% glucose (Fig. 2D) and the lower impact of medium buffering on the lifespan of glucose containing cultures (Fig. 4C). We speculate that the adaptation process requires the presence of a weak organic acid (likely acetic acid) in the growth phase and a low pH of the medium. Such adaptation effect could also explain the differences in lifespan observed initially in cultures containing a low and a high concentration of ethanol (which can be converted into acetic acid) and methanol (where formic acid is produced) [32]. Consequently, such adaptation may be weak or not occurring in cells grown on 0.4% glycerol rendering these cells more fragile to subsequent exposure to low pH.

The intracellular pH is a parameter that affects a whole range of cellular functions [33,34]. The process of cytosolic pH maintenance and weak acid extrusion is energy demanding. In yeast, intracellular acidification activates the accumulation of cAMP [35–37] followed by activation of protein kinase A (PKA) targets. Through mobilization of trehalose and glycogen this process can help the cells to overcome ATP shortage [38–40]. However, low intracellular pH-mediated increase in Ras signalling could also promote chronological ageing via induction of replication stress [22,23]. Remarkably, cultures grown on 0.6% glucose also display less budded cells than cultures grown on 0.4% glycerol which is accompanied by less impact of low medium pH on the lifespan of these cells. A stronger arrest in G0/G1 phase after exit from the growth phase observed in cultures containing 0.6% glucose, possibly resulting from the presence of acetic acid in the cultures during the growth phase, could be beneficial in counteracting the induction of growth signalling by the low pH. Consequently, such adaptation is not occurring in the cells grown on high concentration of glycerol, thus these cells are more fragile to the impact of low pH. The fact that a high budding index of the 0.4% glycerol grown cultures is not a problem in buffered medium suggests that the arrest in G0/G1 phase is important for survival only in medium with a low pH. Consequently, when a low concentration of glucose (0.2%) or glycerol (0.1%) is used, the acidification is minor (Fig. 2F) and cells are simply not exposed to low pH.

Altogether our data indicate that the pH of the medium is an important factor determining *H. polymorpha* chronological lifespan. Similarly, the pH was recently shown to affect the chronological senescence in cultured mammalian cells [41,42] suggesting that indeed the mechanism of cellular response to the acidification could be conserved.

The actual effect of dietary restriction strongly depends on the organism and environmental conditions [3]. We have shown that acidification impacts the effect of carbon source concentration on lifespan in a carbon source dependent manner. Upon buffering, the chronological lifespan invariably increases with increasing glycerol and glucose concentrations. The differences in lifespan between high and low concentrations of carbon source are mediated by a combination of extracellular and intracellular factors. The actual effect of DR in *S. cerevisiae* depends not only on carbon source concentration, but also on the nutrient composition of the initial medium [2]. This yeast also secretes a variety of compounds to the medium [8,9], with further impact on chronological lifespan. Shifting the cells to buffer or water should rule out the impact of extracellular factors on the lifespan of *H. polymorpha*, like previously demonstrated in *S. cerevisiae* [7,43].

In *S. cerevisiae* and other yeast species DR is routinely obtained by reduction of carbon source concentrations from 1%, 2% or higher to 0.5% or less [2,25,44]. When *H. polymorpha* cells are grown on 1 or 2% glycerol the CLS is enhanced in comparison to cultures grown on 0.5% glycerol (Fig. S2C). Growth of cells in 1 or 2% glucose resulted in a similar CLS as obtained with 0.5% glucose (Fig. S2D). This data indicate that also at the carbon source concentrations generally used for *S. cerevisiae*, we also see no positive effect of reducing the carbon source concentration on CLS.

Summarizing, our data demonstrate that decreasing the carbon source concentrations in yeast cultures is not a general intervention that leads to an increase in lifespan.

MATERIALS AND METHODS

Strains and growth conditions

A wild-type prototrophic strain was obtained by complementation of *H. polymorpha* NCYC495 leu1.1 [45] by multicopy integration of pHIPX7 [46], containing *S. cerevisiae LEU2* gene under its own promoter in the *H. polymorpha TEF1* promoter region. Cells were grown in mineral medium [47] containing the indicated carbon sources and 0.25% methylamine as nitrogen source unless stated otherwise. Where indicated, the cells were grown on media containing 0.25% ammonium sulphate or 0.2% urea as nitrogen source. 6 mM K_2SO_4 was added when methylamine or urea were used as nitrogen sources. In all experiments the cells were intensively precultivated in MM containing 0.25% glucose and 0.25% ammonium sulphate. When the OD_{600nm} of the precultures reached 1.5-2.0, cells were diluted to OD_{600nm} = 0.1 in the final medium. Culturing was performed in flasks closed with a cotton plug at a medium to flask volume ratio of 1:5, at 37°C and with shaking at 200 rpm. When pH control was needed, cells were grown in batch fermenters (culture volume 1 L) in a 2 L fermenter (Applikon, The Netherlands) at 37°C, 300 rpm stirring and aeration rate of 0.4 L/min. pH was controlled by the addition of 1 M NaOH.

Medium swap and buffering experiments

In the spent medium swap experiments cells were spun down for 5 min at 3000 g at 37°C, spent medium was collected and clarified by another centrifugation step for 5 min at 3000 g, 37°C. The cells were washed once with warm (37°C) sterile water and resuspended in the desired spent medium. Cells were similarly transferred to 25 mM phosphate buffer pH 3.5 or 6.5, 50 mM MES buffer pH 3.5 or 6.5 or fresh MM supplemented with 0.25% methylamine.

In buffering experiments spent medium from part of the culture was clarified by centrifugation and the pH was adjusted by addition of 1 M NaOH. Cells from equal volumes of the cultures were recovered by centrifugation and resuspended in filtered and pre-warmed medium with adjusted pH. The same approach was used for buffered spent medium swap experiments.

Chronological lifespan measurements

The viability of the cultures was assessed essentially as described before [18]. Briefly, the number of cells per ml was measured using a CASY Model TT (Roche) and 500 cells were plated on YPD agar plates (1% yeast extract, 1% peptone, 1% glucose, 2% agar) or where indicated YP-glycerol plates (1% yeast extract, 1% peptone, 1% glycerol, 2% agar). After 36-48 hours of incubation at 37°C the plates were photographed and colonies were counted using an ImageJ plugin. The number of colonies obtained at the first time point was set as 100%.

Glycerol, glucose and acetate measurements

Glycerol, glucose and acetate were assayed in clarified medium collected at different time points. For acetate determination the pH of clarified media was adjusted to 7.0 by the addition of NaOH before analysis. Glycerol concentrations were assayed with a Glycerol GK Assay Kit (Megazyme, Ireland), glucose with a D-Glucose HK assay kit (Megazyme, Ireland). Acetate was measured with an Acetic acid assay kit (Acetate kinase analyser format, Megazyme, Ireland). All measurements were performed according to the manufacturer's protocols.

Acetic acid resistance and flow cytometry

Cells grown for 20 hours were harvested and resuspended in 50 mM potassium phosphate buffer pH 3.0 and treated with different concentrations of acetic acid (at $OD_{600\ nm}$ = 0.7) in a 96 well microtiter plate for 1 h at 37°C with shaking (900 rpm). The cells were washed once with 50 mM potassium phosphate buffer pH 7.0 and stained for 10 min with 10 μg/ml propidium iodide in the same buffer. After subsequent washing, the fluorescence of 10000 cells was analysed using a FACS Aria II Cell sorter (BD Biosciences) using a 488 nm laser, a 550 nm long pass mirror and a 575/25 nm band-pass filter. Data were recorded and analysed using FACSDiva software (ver. 6.1.2). Stained non-treated and boiled cells were used to set the gates.

Analysis of budding index

The number of budding cells was determined in cultures grown for 20 hours. Bright field mosaic images were made using a Zeiss Observer Z1 microscope. The number of cells containing a not separated bud was counted manually in at least 500 cells per culture.

CONFLICT OF INTEREST

The authors declare no conflict of interest.

REFERENCES

1. Kaeberlein M, Burtner CR, and Kennedy BK (**2007**). Recent developments in yeast aging. **PLoS Genet** 3(5): e84.

2. Wu Z, Liu SQ, and Huang D (**2013**). Dietary restriction depends on nutrient composition to extend chronological lifespan in budding yeast *Saccharomyces cerevisiae*. **PLoS ONE** 8(5): e64448.

3. Skinner C and Lin S-J (**2010**). Effects of calorie restriction on life span of microorganisms. **Appl Microbiol Biotechnol** 88(4): 817–828.

4. Masoro EJ (**2005**). Overview of caloric restriction and ageing. **Mech Ageing Dev** 126(9): 913–922.

5. Mair W and Dillin A (**2008**). Aging and survival: the genetics of life span extension by dietary restriction. **Annu Rev Biochem** 77(1): 727–754.

6. Fabrizio P and Longo VD (**2003**). The chronological life span of *Saccharomyces cerevisiae*. **Aging Cell** 2(2): 73–81.

7. MacLean M, Harris N, and Piper PW (**2001**). Chronological lifespan of stationary phase yeast cells; a model for investigating the factors that might influence the ageing of postmitotic tissues in higher organisms. **Yeast** 18(6): 499–509.

8. Orlandi I, Ronzulli R, Casatta N, and Vai M (**2013**). Ethanol and acetate acting as carbon/energy sources negatively affect yeast chronological aging. **Oxid Med Cell Longev** 2013.

9. Burtner CR, Murakami CJ, Kennedy BK, and Kaeberlein M (**2009**). A molecular mechanism of chronological aging in yeast. **Cell Cycle** 8(8): 1256–1270.

10. Longo VD, Shadel GS, Kaeberlein M, and Kennedy B (**2012**). Replicative and chronological aging in *Saccharomyces cerevisiae*. **Cell Metab** 16(1): 18–31.

11. Murakami CJ, Wall V, Basisty N, and Kaeberlein M (**2011**). Composition and acidification of the culture medium influences chronological aging similarly in vineyard and laboratory yeast. **PLoS ONE** 6(9): e24530.

12. Tahara EB, Cunha FM, Basso TO, Della Bianca BE, Gombert AK, and Kowaltowski AJ (**2013**). Calorie restriction hysteretically primes aging *Saccharomyces cerevisiae* toward more effective oxidative metabolism. **PLoS ONE** 8(2): e56388.

13. Piper PW (**2012**). Maximising the yeast chronological lifespan. In: Breitenbach M, Jazwinski SM, Laun P, editors. Aging Research in Yeast. Springer Netherlands; pp. 145–159.

14. Alvers AL, Fishwick LK, Wood MS, Hu D, Chung HS, Dunn Jr WA, and Aris JP (**2009**). Autophagy and amino acid homeostasis are required for chronological longevity in *Saccharomyces cerevisiae*. **Aging Cell** 8(4): 353–369.

15. Gomes P, Sampaio-Marques B, Ludovico P, Rodrigues F, and Leão C (**2007**). Low auxotrophy-complementing amino acid concentrations reduce yeast chronological life span. **Mech Ageing Dev** 128(5–6): 383–391.

16. Wu Z, Song L, Liu SQ, and Huang D (**2013**). Independent and additive effects of glutamic acid and methionine on yeast longevity. **PLoS ONE** 8(11): e79319.

17. Kumar S, Lefevre SD, Veenhuis M, and van der Klei IJ (**2012**). Extension of yeast chronological lifespan by methylamine. **PLoS ONE** 7(11): e48982.

18. Kawalek A, Lefevre SD, Veenhuis M, and van der Klei IJ (**2013**). Peroxisomal catalase deficiency modulates yeast lifespan depending on growth conditions. **Aging** 5(1): 67–83.

19. Stöckmann C, Losen M, Dahlems U, Knocke C, Gellissen G, and Büchs J (**2003**). Effect of oxygen supply on passaging, stabilising and screening of recombinant *Hansenula polymorpha* production strains in test tube cultures. **FEMS Yeast Res** 4(2): 195–205.

20. Murakami C, Delaney JR, Chou A, Carr D, Schleit J, Sutphin GL, An EH, Castanza AS, Fletcher M, Goswami S, Higgins S, Holmberg M, Hui J, Jelic M, Jeong K-S, Kim JR, Klum S, Liao E, Lin MS, Lo W, Miller H, Moller R, Peng ZJ, Pollard T, Pradeep P, Pruett D, Rai D, Ros V, Schuster A, Singh M, Spector BL, Wende HV, Wang AM, Wasko BM, Olsen B, and Kaeberlein M (**2012**). pH neutralization protects against reduction in replicative lifespan following chronological aging in yeast. **Cell Cycle** 11(16): 3087–3096.

21. Piper P, Calderon CO, Hatzixanthis K, and Mollapour M (**2001**). Weak acid adaptation: the stress response that confers yeasts with resistance to organic acid food preservatives. **Microbiology** 147(10): 2635–2642.

22. Weinberger M, Feng L, Paul A, Smith DL Jr, Hontz RD, Smith JS, Vujcic M, Singh KK, Huberman JA, and Burhans WC (**2007**). DNA replication stress is a determinant of chronological lifespan in budding yeast. **PLoS ONE** 2(8): e748.

23. Weinberger M, Mesquita A, Carroll T, Marks L, Yang H, Zhang Z, Ludovico P, and Burhans WC (**2010**). Growth signaling promotes chronological aging in budding yeast by inducing superoxide anions that inhibit quiescence. **Aging** 2(10): 709–726.

24. Burhans WC and Weinberger M (**2009**). Acetic acid effects on aging in budding yeast: are they relevant to aging in higher eukaryotes? **Cell Cycle** 8(14): 2300–2302.

25. Oliveira GA, Tahara EB, Gombert AK, Barros MH, and Kowaltowski AJ (**2008**). Increased aerobic metabolism is essential for the beneficial effects of caloric restriction on yeast life span. **J Bioenerg Biomembr** 40(4): 381–388.

26. Orij R, Postmus J, Beek AT, Brul S, and Smits GJ (**2009**). In vivo measurement of cytosolic and mitochondrial pH using a pH-sensitive GFP derivative in *Saccharomyces cerevisiae* reveals a relation between intracellular pH and growth. **Microbiology** 155(1): 268–278.

27. Valli M, Sauer M, Branduardi P, Borth N, Porro D, and Mattanovich D (**2005**). Intracellular pH distribution in *Saccharomyces cerevisiae* cell populations, analyzed by flow cytometry. **Appl Environ Microbiol** 71(3): 1515–1521.

28. Carmelo V, Santos H, and Sá-Correia I (**1997**). Effect of extracellular acidification on the activity of plasma membrane ATPase and on the cytosolic and vacuolar pH of *Saccharomyces cerevisiae*. **Biochim Biophys Acta BBA - Biomembr** 1325(1): 63–70.

29. Imai T and Ohno T (**1995**). Measurement of yeast intracellular pH by image processing and the change it undergoes during growth phase. **J Biotechnol** 38(2): 165–172.

30. Dechant R, Binda M, Lee SS, Pelet S, Winderickx J, and Peter M (**2010**). Cytosolic pH is a second messenger for glucose and regulates the PKA pathway through V-ATPase. **EMBO J** 29(15): 2515–2526.

31. Giannattasio S, Guaragnella N, Corte-Real M, Passarella S, and Marra E (**2005**). Acid stress adaptation protects *Saccharomyces cerevisiae* from acetic acid-induced programmed cell death. **Gene** 354: 93–98.

32. Klei IJ van der, Harder W, and Veenhuis M (**1991**). Methanol metabolism in a peroxisome-deficient mutant of *Hansenula polymorpha*: a physiological study. **Arch Microbiol** 156(1): 15–23.

33. Orij R, Brul S, and Smits GJ (**2011**). Intracellular pH is a tightly controlled signal in yeast. **Biochim Biophys Acta BBA - Gen Subj** 1810(10): 933–944.

34. Orij R, Urbanus ML, Vizeacoumar FJ, Giaever G, Boone C, Nislow C, Brul S, and Smits GJ (**2012**). Genome-wide analysis of intracellular pH reveals quantitative control of cell division rate by pHc in *Saccharomyces cerevisiae*. **Genome Biol** 13(9): R80.

35. Trevillyan JM and Pall ML (**1979**). Control of cyclic adenosine 3′,5′-monophosphate levels by depolarizing agents in fungi. **J Bacteriol** 138(2): 397–403.

36. Thevelein JM, Beullens M, Honshoven F, Hoebeeck G, Detremerie K, Hollander J a. D, and Jans AWH (**1987**). Regulation of the cAMP level in the yeast *Saccharomyces cerevisiae*: intracellular pH and the effect of membrane depolarizing compounds. **J Gen Microbiol** 133(8): 2191–2196.

37. Colombo S, Ma P, Cauwenberg L, Winderickx J, Crauwels M, Teunissen A, Nauwelaers D, de Winde JH, Gorwa M-F, Colavizza D, and Thevelein JM (**1998**). Involvement of distinct G-proteins, Gpa2 and Ras, in glucose- and intracellular acidification-induced cAMP signalling in the yeast *Saccharomyces cerevisiae*. **EMBO J** 17(12): 3326–3341.

38. Thevelein JM and De Winde JH (**1999**). Novel sensing mechanisms and targets for the cAMP–protein kinase A pathway in the yeast *Saccharomyces cerevisiae*. **Mol Microbiol** 33(5): 904–918.

39. Lillie SH and Pringle JR (**1980**). Reserve carbohydrate metabolism in *Saccharomyces cerevisiae*: responses to nutrient limitation. **J Bacteriol** 143(3): 1384–1394.

40. Thevelein JM (**1984**). Activation of trehalase by membrane-depolarizing agents in yeast vegetative cells and ascospores. **J Bacteriol** 158(1): 337–339.

41. Leontieva OV and Blagosklonny MV (**2011**). Yeast-like chronological senescence in mammalian cells: phenomenon, mechanism and pharmacological suppression. **Aging** 3(11): 1078–1091.

42. Fabrizio P and Wei M (**2011**). Conserved role of medium acidification in chronological senescence of yeast and mammalian cells. **Aging** 3(12): 1127–1129.

43. Fabrizio P, Battistella L, Vardavas R, Gattazzo C, Liou L-L, Diaspro A, Dossen JW, Gralla EB, and Longo VD (**2004**). Superoxide is a mediator of an altruistic aging program in *Saccharomyces cerevisiae*. **J Cell Biol** 166(7): 1055–1067.

44. Chen B-R and Runge KW (**2009**). A new *Schizosaccharomyces pombe* chronological lifespan assay reveals that caloric restriction promotes efficient cell cycle exit and extends longevity. **Exp Gerontol** 44(8): 493–502.

45. Gleeson MA and Sudbery PE (**1988**). Genetic analysis in the methylotrophic yeast *Hansenula polymorpha*. **Yeast** 4(4): 293–303.

46. Baerends RJS, Salomons FA, Faber KN, Kiel JAKW, Van Der Klei IJ, and Veenhuis M (**1997**). Deviant Pex3p levels affect normal peroxisome formation in *Hansenula polymorpha*: high steady-state levels of the protein fully abolish matrix protein import. **Yeast** 13(15): 1437–1448.

47. Dijken LPV, Otto R, and Harder W (**1976**). Growth of *Hansenula polymorpha* in a methanol-limited chemostat. **Arch Microbiol** 111(1-2): 137–144.

17

Extracellular calcium triggers unique transcriptional programs and modulates staurosporine-induced cell death in *Neurospora crassa*

A. Pedro Gonçalves[1,2,*], João Monteiro[2], Chiara Lucchi[2], David J. Kowbel[3], J. Miguel Cordeiro[1,4], Paulo Correia-de-Sá[1,4], Daniel J. Rigden[5], N. Louise Glass[3], Arnaldo Videira[1,2,*]

[1] ICBAS-Instituto de Ciências Biomédicas de Abel Salazar, Universidade do Porto, Rua de Jorge Viterbo Ferreira 228, 4050-313 Porto, Portugal.

[2] IBMC-Instituto de Biologia Molecular e Celular - Universidade do Porto, Rua do Campo Alegre 823, 4150-180 Porto, Portugal.

[3] Plant and Microbial Biology Department, The University of California, Berkeley, CA 94720, USA.

[4] UMIB-Unidade Multidisciplinar de Investigação Biomédica, Universidade do Porto, Rua de Jorge Viterbo Ferreira 228, 4050-313 Porto, Portugal.

[5] Institute of Integrative Biology, University of Liverpool, Liverpool, L69 7ZB, United Kingdom.

* Corresponding Authors: A. Pedro Gonçalves, ICBAS-Instituto de Ciências Biomédicas de Abel Salazar, Universidade do Porto, Rua de Jorge Viterbo Ferreira 228; 4050-313 Porto, Portugal; E-mail: apgoncalves@ibmc.up.pt AND Arnaldo Videira, ICBAS-Instituto de Ciências Biomédicas de Abel Salazar, Universidade do Porto, Rua de Jorge Viterbo Ferreira 228; 4050-313 Porto, Portugal; E-mail: avideira@ibmc.up.pt

ABSTRACT Alterations in the intracellular levels of calcium are a common response to cell death stimuli in animals and fungi and, particularly, in the *Neurospora crassa* response to staurosporine. We highlight the importance of the extracellular availability of Ca^{2+} for this response. Limitation of the ion in the culture medium further sensitizes cells to the drug and results in increased accumulation of reactive oxygen species (ROS). Conversely, an approximately 30-fold excess of external Ca^{2+} leads to increased drug tolerance and lower ROS generation. In line with this, distinct staurosporine-induced cytosolic Ca^{2+} signaling profiles were observed in the absence or presence of excessive external Ca^{2+}. High-throughput RNA sequencing revealed that different concentrations of extracellular Ca^{2+} define distinct transcriptional programs. Our transcriptional profiling also pointed to two putative novel Ca^{2+}-binding proteins, encoded by the NCU08524 and NCU06607 genes, and provides a reference dataset for future investigations on the role of Ca^{2+} in fungal biology.

***Keywords**: calcium, cell death, ROS, Ca2+-binding motif, Neurospora crassa.*

INTRODUCTION

Calcium (Ca^{2+}) is crucial for diverse processes in fungi, including tip growth, virulence, circadian rhythms, nutrient sensing, cell wall regeneration, chemotropic interaction and cell death [1]. Ca^{2+} and cell death are intimately connected, especially because of the vital role of Ca^{2+} as an intracellular messenger affecting numerous signaling pathways that determine cell fate. Therefore, it is not surprising that cells commonly respond to cell death stimuli with rises in the intracellular levels of this divalent ion. On one hand, these transient or stable modifications in the levels of Ca^{2+} can operate as survival signals, leading to the activity of anti-death proteins or stimulating transcriptional defence responses. On the other hand, the loss of Ca^{2+} homeostasis can behave as a pro-death signal, since in some systems overload or a simple perturbation of the distribution of the ion within storage organelles is sufficient to trigger cell death [2]. In yeast species like *Saccharomyces cerevisiae*, *Candida albicans* and *Cryptococcus neoformans*, cytosolic Ca^{2+} increases are a common feature of the cellular response to a number of antifungal stresses, including pheromone [3], the plant essential oils carvacrol [4] and eugenol [5], amiodarone [3, 6], and ER stress-inducing agents like tunicamycin and azole drugs [7, 8]. In *N. crassa*, a rise in the cytosolic levels of Ca^{2+} was associated with cell death induced by chitosan [9], *Penicillium chrysogenum* protein PAF [10], PAF26 [11] and staurosporine [12].

We have been using *N. crassa* to investigate the molecular mechanisms underlying the cell death process induced by the bacterial alkaloid staurosporine. Treatment with staurosporine leads to a complex response that includes the efflux of reduced glutathione (GSH), with a concomitant change in the intracellular redox state to a more oxidative environment and the consequent accumulation of reactive oxygen species (ROS) [13, 14]. The drug also activates a drug resistance pathway that comprises the zinc binuclear cluster transcription factor CZT-1 [15] and ABC-3, an ATP-binding cassette (ABC) transporter [16]. A combined genetic and pharmacological strategy coupled to measurements of cytosolic alterations in Ca^{2+} levels with the photoreporter aequorin revealed a three-step signature response to the drug that involves mobilization of the ion from intracellular stores as well as uptake from the external medium. *N. crassa* is a good model for the study of Ca^{2+} dynamics as sequencing of its genome revealed a rich and versatile assortment of molecules involved in Ca^{2+} homeostasis, including channels, pumps and signaling transducers [17-19]. A substantial fraction of the genome, however, remains to be characterized and additional unknown members of Ca^{2+} machinery may exist. The best-studied Ca^{2+}-binding moiety is the classical EF-hand, in which a Ca^{2+}-interacting loop is flanked by two helices [20]. However, there are variations on this motif, especially at the structural level, and it is now accepted that more than a dozen different types of Ca^{2+}-binding domains exist [21]. In particular, several proteins classes were shown to carry convergently evolved Ca^{2+}-binding Dx[DN]xDG motifs. The distinct evolutionary origins of the motif are evident from the variability of structural contexts flanking the Ca^{2+}-binding region [22, 23]. Though initially identified in bacteria [24], it was later perceived that this motif is widespread across all kingdoms [22, 23], including in a fungal caleosin [23, 25] and lectin from *Psathyrella velutina* [22, 26].

We have recently shown that staurosporine-induced cell death in *N. crassa* involves a precise sequence of Ca^{2+} signaling events which are completely abolished by the inhibition of extracellular Ca^{2+} uptake [12]. Thus, we investigated the effects of the drug in *N. crassa* cells growing in limited or excessive amounts of the ion. The results presented here revealed that extracellular Ca^{2+} modulates cell death and the transcriptional alterations induced by staurosporine, and led to the identification of two novel putative Ca^{2+}-binding proteins, encoded by the NCU08524 and NCU06607 genes.

RESULTS

The availability of extracellular Ca^{2+} impacts staurosporine-induced cell death and intracellular Ca^{2+} signaling

Intracellular Ca^{2+} dynamics play an important role during staurosporine-induced cell death in *N. crassa*, eliciting a complex response that includes Ca^{2+} release from internal stores as well as uptake from the extracellular space [12]. Thus, we tested the response to staurosporine of cells growing with different Ca^{2+} concentrations. We prepared a culture medium, designated 'no Ca^{2+}', by removing $CaCl_2$ from Vogel's minimal medium (MM) [27]. Small amounts of Ca^{2+} as impurities from the other reagents in the solution allowed *N. crassa* to grow well under such conditions (Fig. 1A, middle panel). In contrasting experiments, we supplemented medium with 20 mM $CaCl_2$ ('20 mM $CaCl_2$' medium), which represents an approximately 30 fold (but not toxic) increase as compared with the 0.68 mM $CaCl_2$ present in standard Vogel's medium (Fig. 1A, lower panel). We inoculated wild type cells on the centre of Petri dishes containing standard, no Ca^{2+} or 20 mM $CaCl_2$ solid Vogel's MM supplemented with staurosporine and measured radial growth during a 104 hour time course. The drug inhibited growth in all media (Fig. 1A-C). However, the staurosporine inhibitory effect was amplified in the absence of Ca^{2+} (e.g., ~39% of inhibition after 32 hours of treatment with 1 µM staurosporine in no Ca^{2+} versus ~14% in standard MM). On the other hand, growth inhibition was partially overcome in 20 mM $CaCl_2$ medium (e.g., ~37% of inhibition after 32 hours of treatment with 2.5 µM staurosporine in 20 mM $CaCl_2$ versus ~65% in standard MM). In GFS medium, which contains sorbose that promotes colonial growth [28], the outcome of modifying the extracellular concentration of Ca^{2+} was similar: the effects of staurosporine were exacerbated in no Ca^{2+} medium whereas growth inhibition was partially suppressed in the presence of 20 mM $CaCl_2$, (Fig. 1D). Enhancement and suppression of inhibition of growth by absence of Ca^{2+} and 20 mM $CaCl_2$, respectively, was not reproduced when cells are treated with phytosphingosine (data not shown), another cell death inducer in *N. crassa*, supporting previous observations that staurosporine and phytosphingosine act by distinct mechanisms [14, 29, 30].

In order to check cell death, we analysed cell staining with YOPRO-1 (early apoptosis marker) or PI (late apoptosis/necrosis marker) by flow cytometry. A 2-hour treatment with staurosporine in standard MM resulted in ~22% and ~17% of YOPRO-1 and PI positive cells, respectively (Fig. 1E-F). In no Ca^{2+} medium, these levels were augmented to ~40% (p-value ~ 0) and ~34% (p-value = 0.029), respectively (Fig. 1E-F). In contrast, the percentage of YOPRO-1 and PI-positive cells was reduced to ~15% (p-value = 0.042) and ~6% (p-value = 0.050), respectively, in 20 mM $CaCl_2$ medium.

The production of ROS is an essential event during the response of *N. crassa* to staurosporine, as the addition of antioxidants blocks cell death [13]. We asked if the availability of Ca^{2+} in the culture medium influenced cellular ROS production induced by staurosporine. While in standard MM cells stressed for 30 minutes with staurosporine displayed a ~4.2 fold-increase in ROS accumulation, the equivalent accumulation was ~10.4 (p-value = 0.001) and ~1.3 (p-value = 0.014) in no Ca^{2+} and 20 mM $CaCl_2$ media, respectively (Fig. 1G). Given the importance of ROS formation for cell death provoked by staurosporine, this Ca^{2+}-ROS dependence is likely associated with the distinct cell death phenotypes in the different culture media.

To further stress the modulatory effect of altering the levels of Ca^{2+} in staurosporine-induced cell death, we combined the drug with the Ca^{2+} chelator BAPTA and the Ca^{2+} ionophore A23187 (calcimycin) in GFS plates with standard MM. As expected, extracellular Ca^{2+} blockage with BAPTA

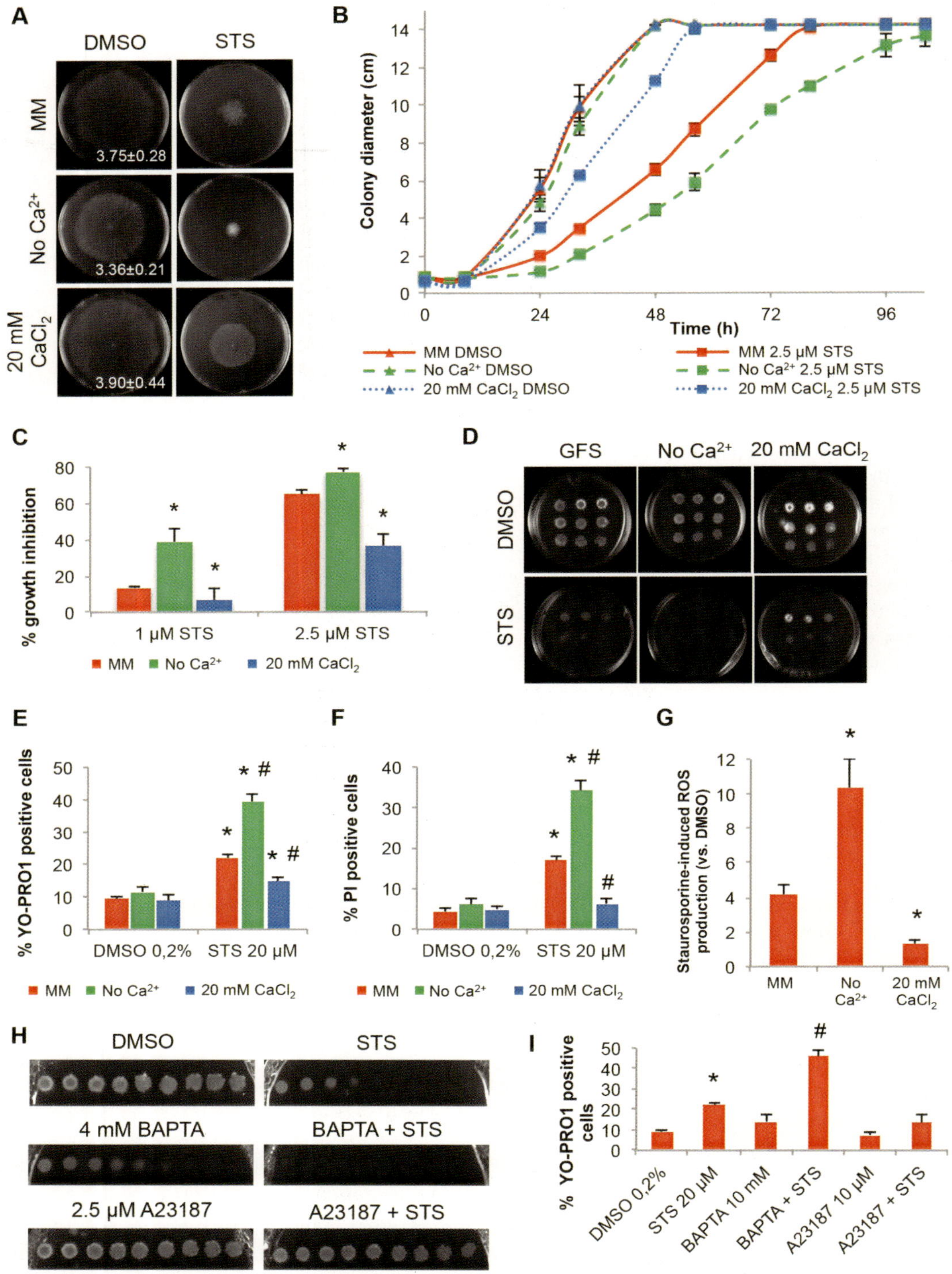

FIGURE 1: Extracellular Ca^{2+} modulates the *N. crassa* sensitivity to staurosporine. (A-C) Conidia were inoculated on the centre of a Petri dish containing Vogel's MM with 2.5 µM staurosporine (STS) and different concentrations of Ca^{2+}. Growth at 32h **(A)**, over time **(B)** and the percentage of growth in STS-treated cells versus control **(C)** are shown. Growth rates (mm/h) are indicated in the left panels in **(A)**. *, p-value ≤ 0.05. **(D)** Serial dilutions (from left to right) of conidia spotted in GFS agar medium supplemented with 5 µM STS and different concentrations of Ca^{2+} were incubated for 3 days. **(E-F)** Cell death following treatment of the cells with 20 µM staurosporine in the indicated liquid culture media was evaluated by flow cytometry quantification of positive cells for YOPRO-1 **(E)** or PI **(F)**. *, p-value ≤ 0.05 for the comparison STS versus DMSO in each media; #, p-value ≤ 0.05 for the comparison between media after the treatment with STS. **(G)** The fold increase in cellular ROS accumulation following treatment with 20 µM staurosporine in the indicated liquid culture media was evaluated by staining with DHR123. *, p-value ≤ 0.05. **(H-I)** Growth inhibition and cell death in 2.5 µM STS-treated cells in the presence of BAPTA or A23187 was assessed by spots in GFS medium **(H)** or cell growth in liquid medium followed by staining with YOPRO-1 **(I)** *, p-value ≤ 0.05 for the comparison STS versus DMSO; #, p-value ≤ 0.05 for the comparison between STS alone and BAPTA+STS.

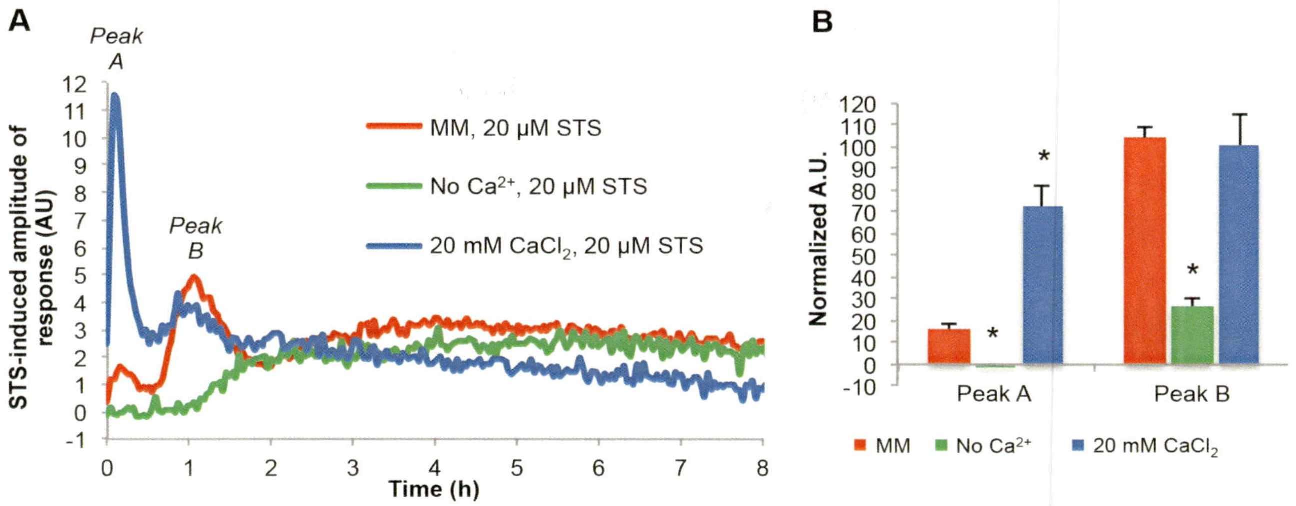

FIGURE 2: Extracellular Ca^{2+} availability affects intracellular Ca^{2+} dynamics in response to staurosporine. (A) Aequorin-expressing wild type cells were cultured for 6 hours in the indicated media and after the addition of 20 µM staurosporine (STS), luminescence was followed over time. The STS-induced amplitude of response is shown. **(B)** Quantification (in arbitrary units) of the cytosolic Ca^{2+} peaks "A" and "B" (indicated in A). *, p-value ≤ 0.05.

synergized whereas elevation of intracellular Ca^{2+} with A23187 antagonized the effects of staurosporine (Fig. 1H). In agreement, pre-incubation with BAPTA significantly enhanced apoptotic levels induced by staurosporine, whereas they were partially inhibited by A23187, as measured by YOPRO-1 staining (Fig. 1I). Altogether, these data consistently indicate that the absence of Ca^{2+} enhances staurosporine-induced cell death, whereas excess (but not toxic) concentrations of Ca^{2+} partially prevent it.

Using cells expressing the cytosolic Ca^{2+} reporter gene aequorin, we described recently that staurosporine induces a three-step Ca^{2+} response comprising three main signals called "A", "B" and "C". Peak "A" occurs immediately after the addition of the drug and lasts approximately 20 minutes; peak "B" appears after 35-40 minutes and lasts approximately 80 minutes; phase "C" is a prolonged and continuous elevation of cytosolic Ca^{2+} [12]. In the absence of extracellular Ca^{2+}, the Ca^{2+} response to staurosporine was compromised, with peaks "A" and "B" showing substantial reduction (Fig. 2A-B). In medium with 20 mM $CaCl_2$ there was a strong boost in peak "A", while peak "B" remained similar to standard MM. These alterations in intracellular Ca^{2+} dynamics are in agreement with our findings that extracellular Ca^{2+} uptake is important for the Ca^{2+} signaling response to staurosporine [12].

Distinct transcriptional programs in staurosporine-treated cells depend on the extracellular Ca^{2+} availability

Treatment with 20 µM staurosporine led to significant alterations in the expression of ~28% of the *N. crassa* genes (1921 genes) [15]. The extent of this transcriptional response was amplified in the absence of Ca^{2+} (~36% of genes (2749 genes) were altered by the drug), and markedly suppressed by 20 mM of $CaCl_2$ (~6% genes (454 genes) showed altered profiles) (Fig. 3A-B). In standard medium, approximately half of the genes were induced and approximately half were repressed by staurosporine [15]. In no Ca^{2+} medium, 43.1% of the altered genes were upregulated, whereas 56.9% where downregulated. Remarkably, in the presence of excessive Ca^{2+} most of the altered genes were repressed (~68%). The complete dataset obtained from these RNA-seq experiments is presented as File S1.

We analysed the distribution of the individual genes induced and repressed by staurosporine in the different culture media. In the absence of Ca^{2+}, 792 (66.8%) of the induced genes were specific to this condition while in 20 mM $CaCl_2$ this corresponded to 67 (46.2%) genes (Fig. 3C and File S1). Among repressed genes, 875 (56.0%) and 157 (50.8%) genes were specific to no Ca^{2+} and 20 mM $CaCl_2$ media, respectively (Fig. 3D and File S1). Only minor fractions of the altered genes (30 and 74 genes, induced and repressed, respectively) were common to all conditions.

FunCat was used to examine the enrichment of functional categories in the different gene sets (File S2; Table S1-S2). The most interesting Ca^{2+} concentration-specific differences were found among the downregulated genes. In the absence of Ca^{2+}, staurosporine led to the downregulation of genes included in numerous categories. These encompass various groups within the 'Metabolism' supercategory such as 'metabolism of the aspartate family', 'purine nucleotide/nucleoside/nucleobase metabolism', 'phosphate metabolism' and 'tetracyclic and pentacyclic triterpenes metabolism'. Also, there was a specific repression of genes involved in 'Cell cycle and DNA processing' and 'Transcription', namely 'mRNA synthesis' and 'mRNA processing'. Categories related to stress responses like 'Protein fate', 'unfolded protein response', 'cellular sensing and response to external stimulus', 'cell growth', 'anti-apoptosis' were also repressed. Finally, while in standard MM there was a downregulation of 'aerobic respiration'

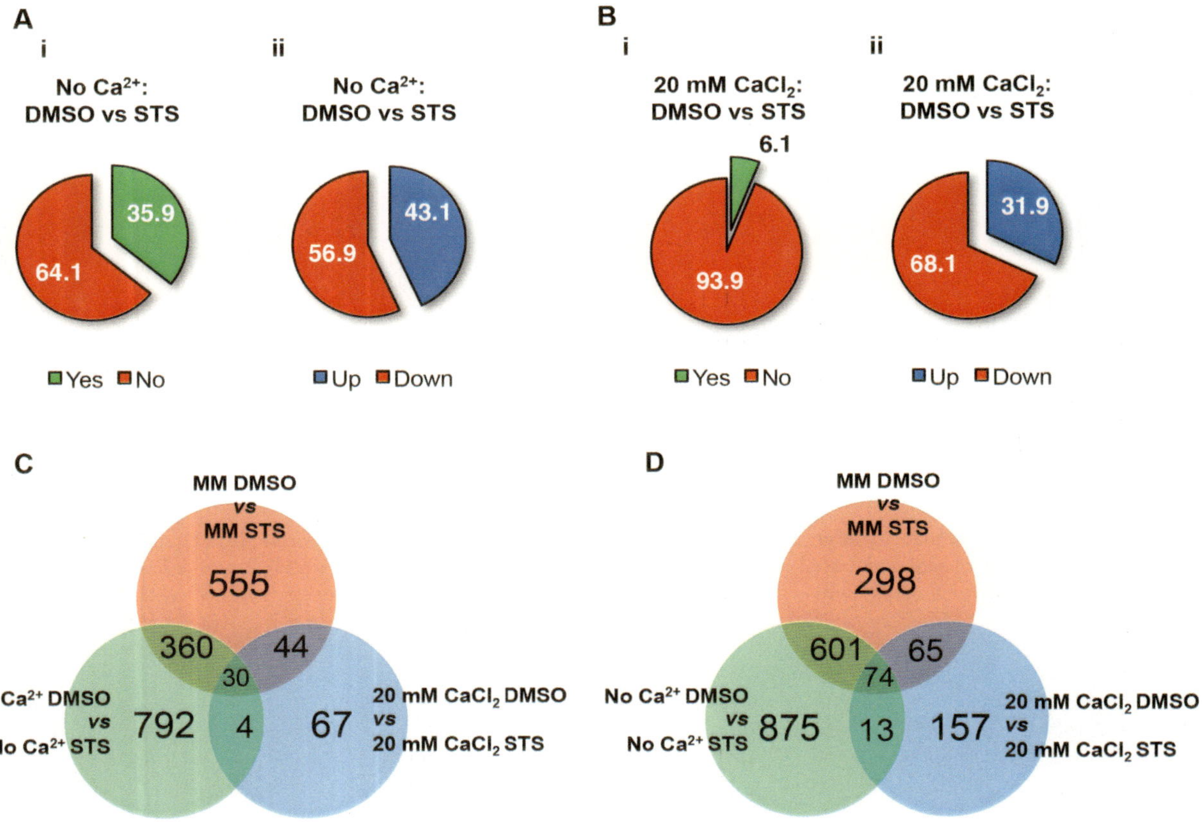

FIGURE 3: Overview of the *N. crassa* transcriptional response to staurosporine in minimal media containing different concentrations of Ca^{2+}. (A-B) The percentages of genes with altered expression upon treatment with staurosporine (i) and the fraction of induced and repressed genes (ii) were calculated for culture medium with no Ca^{2+} **(A)** or with 20 mM $CaCl_2$ **(B)**. **(C-D)** Venn diagrams were used to assess the amount of Ca^{2+}-specific staurosporine-induced **(C)** and -repressed genes **(D)**. General statistics for Ca^{2+}-specific transcriptional responses are included.

genes, in no Ca^{2+} medium, other bioenergetic pathways were repressed: 'glycolysis and gluconeogenesis' and 'pentose-phosphate pathway'. These data indicates that a highly complex transcriptional program is established by staurosporine-treated cells growing in the absence of Ca^{2+} resulting in the repression of several genes related with an anti-stress response. It seems that, when Ca^{2+} is limited, staurosporine causes the downregulation of cellular defences and this may be related to the high susceptibility of these cells to the drug.

RNA-seq analysis of the fungal reaction to staurosporine revealed that the drug can trigger very distinct transcriptional responses depending on the concentrations of Ca^{2+} in the culture medium. This is probably associated with the observed differences in the levels of staurosporine-induced cell death in the three media.

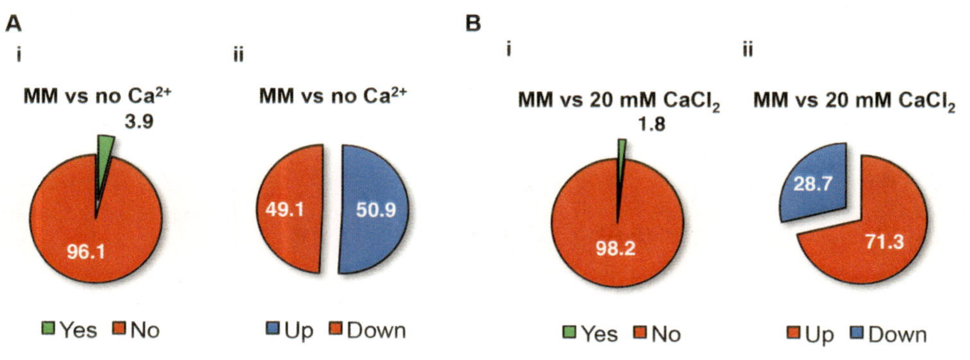

FIGURE 4: Overview of the *N. crassa* transcriptional response to limited or excess Ca^{2+}. (A-B) The percentages of genes with altered expression (i) and the fraction of induced and repressed genes (ii) were calculated for no Ca^{2+} **(A)** and 20 mM $CaCl_2$ medium **(B)** in comparison with standard MM.

A

No Ca^{2+} upregulated genes (139)

ID	Category	P-value
01.01	amino acid metabolism	3.52E-06
01.02	nitrogen, sulfur and selenium metabolism	3.95E-08
02.13.01	anaerobic respiration	0.002958
20.01.01.01	cation transport	2.19E-05
20.01.01.07	anion transport	4.29E-05
20.03	transport facilities	1.59E-06
20.09.18.07	non-vesicular cellular import	3.39E-07
32.01.04	pH stress response	0.001205
34.01.01	homeostasis of cations	0.000272
34.01.03	homeostasis of anions	4.92E-05

20 mM $CaCl_2$ upregulated genes (88)

ID	Category	P-value
02.11	electron transport and membrane-associated energy conservation	0.000179
02.11.05	accessory proteins of electron transport and membrane-associated energy conservation	0.002529
02.13	respiration	0.000695
20.09.18.07	non-vesicular cellular import	9.84E-07

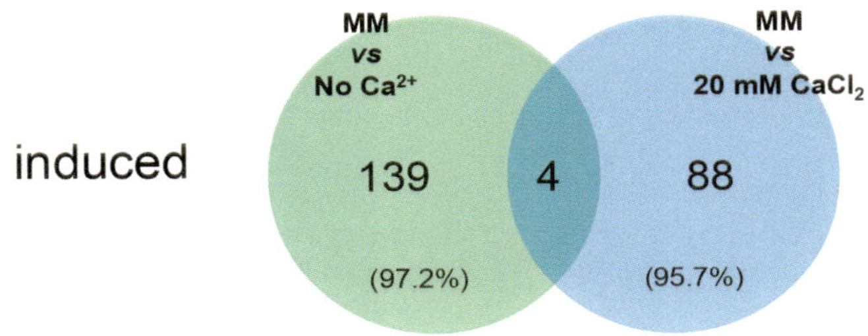

B

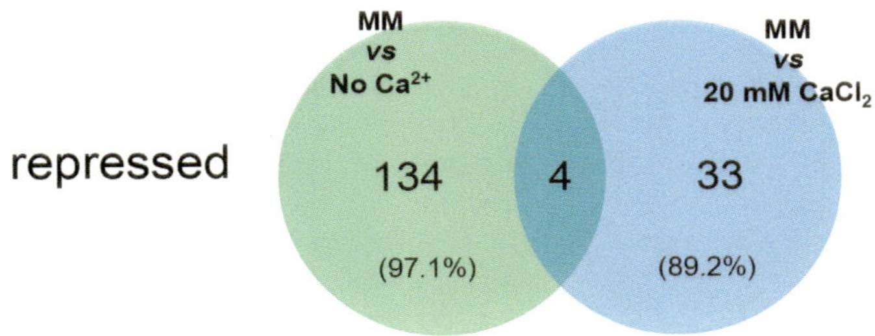

No Ca^{2+} downregulated genes (134)

ID	Category	P-value
20.03.02.03.01	proton driven antiporter	0.001853

20 mM $CaCl_2$ downregulated genes (33)

ID	Category	P-value
20.01.01.01	cation transport	3.50E-05
20.01.01.07	anion transport	2.61E-06
32.01.04	pH stress response	1.66E-07
34.01.01	homeostasis of cations	7.25E-07
34.01.03	homeostasis of anions	1.80E-07

FIGURE 5: Functional enrichment analysis of the transcriptional alterations caused by limited or excess Ca^{2+}. (A-B) The transcriptional response to the absence or excess of Ca^{2+} was investigated by building Venn diagrams showing the distribution of genes with induced **(A)** or repressed **(B)** expression in no Ca^{2+} and 20 mM $CaCl_2$ medium. Lists of enriched categories are included.

Limited and excess of Ca^{2+} alters basal gene expression

The transcriptional response to the lack or excess of Ca^{2+} was analysed by comparing basal gene expression in no Ca^{2+}, 20 mM $CaCl_2$ and standard MM (File S3). The fraction of genes with altered expression in no Ca^{2+} and 20 mM $CaCl_2$ compared with standard MM was 3.9% and 1.8%, respectively (Fig. 4A-B), corresponding to 281 and 129 genes, respectively. While half of the genes altered in no Ca^{2+} were induced and half were repressed, in 20 mM $CaCl_2$ most of the altered genes were upregulated (71.3%). Most of the genes were specifically induced or repressed depending on the particular culture medium: 97.2% and 97.1%, respectively, for no Ca^{2+}, and 95.7% and 89.2%, respectively, for 20 mM $CaCl_2$ (Fig. 5A-B). Not surprisingly, categories related to 'ion transport', 'homeostasis of cations' and 'homeostasis of anions' were enriched in the set of genes induced in the absence of Ca^{2+} (Fig. 5A and File S4), which is perhaps linked to a cellular attempt to recover proper conditions of osmolarity. Consistently, these categories were also significantly enriched among the genes repressed in the presence of 20 mM $CaCl_2$ (Fig. 5B). Lack of Ca^{2+} also seems to be associated to a metabolic adaptation of the cells, since there was enrichment in the induction of genes falling on the 'metabolism' super-category, like 'amino acid metabolism', 'nitrogen, sulfur and selenium metabolism' and 'anaerobic respiration'.

We looked for alterations in the expression of known Ca^{2+} channels, Ca^{2+}-ATPases, Ca^{2+}-exchangers, Ca^{2+}-dependent signaling molecules and other Ca^{2+} binding proteins [18, 19, 31, 32] (Fig. S1 and File S5). NCU11680, NCU04736 and NCU07075, encoding the TRP channel YVC-1, the Ca^{2+}-ATPase NCA-2 and the Ca^{2+}/H^{+} antiporter CAX, respectively, were repressed in the absence of Ca^{2+}. NCU04265, encoding the invertase enzyme (INV) was induced in the absence of Ca^{2+}. The induction of INV, an enzyme that hydrolyses extracellular sucrose into glucose and fructose, is a marker of the process of carbon catabolite repression [33, 34]. NCU08147, encoding the stress-related Ca^{2+}-ATPase ENA-2 [35] was repressed in the presence of 20 mM $CaCl_2$. NCU05046 and NCU07966, encoding ENA-1 and TRM-1, respectively, were induced by no Ca^{2+} and repressed in 20 mM $CaCl_2$ medium. These results further suggest an intracellular metabolic remodeling in response to Ca^{2+} availability.

Two putative novel components of the *N. crassa* Ca^{2+} homeostasis machinery

We found that 188 genes with unknown function ("hypothetical proteins") had altered expression in no Ca^{2+} or in 20 mM $CaCl_2$ medium. We screened the respective protein sequences for the presence of Ca^{2+}-binding patterns using CaPS [20] and identified 6 candidates (NCU01697, NCU08524, NCU06116, NCU07582, NCU06607 and NCU03647). We further considered NCU08524 and NCU06607, because the others did not display convincing similarities with known proteins, even around the predicted Ca^{2+}-binding motif. NCU08524 was repressed in Ca^{2+}-free medium whereas NCU06607 was induced by 20 mM $CaCl_2$ (Fig. 6A). Fig. 6B shows the alignment of the predicted Ca^{2+}-binding motifs of NCU08524 and NCU06607 with those of proteins containing the Dx[DN]xDG motif [22, 23]. The NCU08524 motif sequence resembles that of the rhamnogalacturonan lyase YesW from *Bacillus subtilis*. YesW exemplifies β-propeller structures in which the Dx[DN]xDG motif can be embedded in one of the propeller blades, forming a 'calcium blade' structure [22]. Such a calcium blade was recently described in a fungal lectin from *Psathyrella velutina* [22]. NCU08524 is similar to a fucose-specific lectin from *Macrophomina phaseolina* (Table S3) and matches from HHpred suggest that it contains a 6-bladed propeller structure in which β-strands flank the putative Ca^{2+}-binding motif. Taking advantage of the similarity between fucose-specific lectins and NCU08524, molecular models were built for the latter in a situation of Ca^{2+} binding (Fig. S2A-C). In proteins containing the Dx[DN]xDG motif, interactions with bound Ca^{2+} ions are supplemented by the interaction of side chains from at least one downstream acidic residue [22]. Therefore, three molecular models were prepared, considering Glu^{502}, Glu^{503} or Asp^{504} as the possible additional acidic residue. The models show that NCU08524 can accommodate a Ca^{2+}-binding geometry by combining the Dx[DN]xDG motif with any of these downstream residues (Fig. S2A-C).

The predicted Ca^{2+}-binding motif of NCU06607 (Fig. 6C) is comparable to the Dx[DN]xDG motif of a calmodulin from *Paramecium tetraurelia* [23]. Molecular modeling of NCU06607 was not feasible because the Dx[DN]xDG motif is not located within a folded domain of known structure, but at the beginning of a predicted intrinsically disordered region. Anyway, the Dx[DN]xDG motif is sometimes found in highly disordered regions [36]. NCU06607 has a predicted extracellular localization and there is a strong bias in the localization of proteins possessing the Dx[DN]xDG motif towards the cell surface or secretion [22]. Plasma membrane localization together with a transmembrane domain was predicted for NCU08524 (Table S3).

The growth of ΔNCU08524 and ΔNCU06607 was compared with the wild type strain under different conditions of Ca^{2+} availability. The growth of both mutants was significantly reduced in no Ca^{2+} medium and in medium containing the Ca^{2+} chelator BAPTA (Fig. 6D). ΔNCU06607 also showed a significant reduction in growth in the presence of 20 mM $CaCl_2$ as compared with wild type. The observed differences in growth were not merely due to non-specific osmotic stress, since all strains were similarly affected by the addition of 4% NaCl. ΔNCU06607 was slightly more sensitive to staurosporine than wild type, suggesting that the respective protein may be involved in the response to the drug (Fig. 6E). In summary, the RNA-seq dataset and further analysis revealed two putative novel components of the Ca^{2+}-handling machinery in *N. crassa*.

Several members of the Ca^{2+} homeostasis machinery are involved in the fungal response to staurosporine

To further substantiate the crucial role of Ca^{2+} during the *N. crassa* response to staurosporine and to identify mediators of the process, we assayed the drug sensitivity profile of approximately 50 deletion strains encoding proteins involved in Ca^{2+} homeostasis. We showed recently that deletion of genes encoding phospholipase C enzymes, especially PLC-2, results in mutants resistant to staurosporine, while strains lacking the Ca^{2+} channels CCH-1 and MID-1 are hypersensitive [12]. Several other deletion strains that lack Ca^{2+}-permeable channels, Ca^{2+}-ATPases or other Ca^{2+} binding molecules involved in cellular signaling showed a resistance phenotype to staurosporine that was dissimilar from the wild type strain (Fig. 7). A resistance phenotype was particularly strong in deletion mutants for genes encoding the Ca^{2+}-ATPases NCA-2 and PMR-1, and deletion of other Ca^{2+}-ATPases, namely ENA-1, ENA-2 and NCU10143 also led to increased resistance. The disruption of the signaling molecules calmodulin, CAMK-4, and CSE-1 aug-

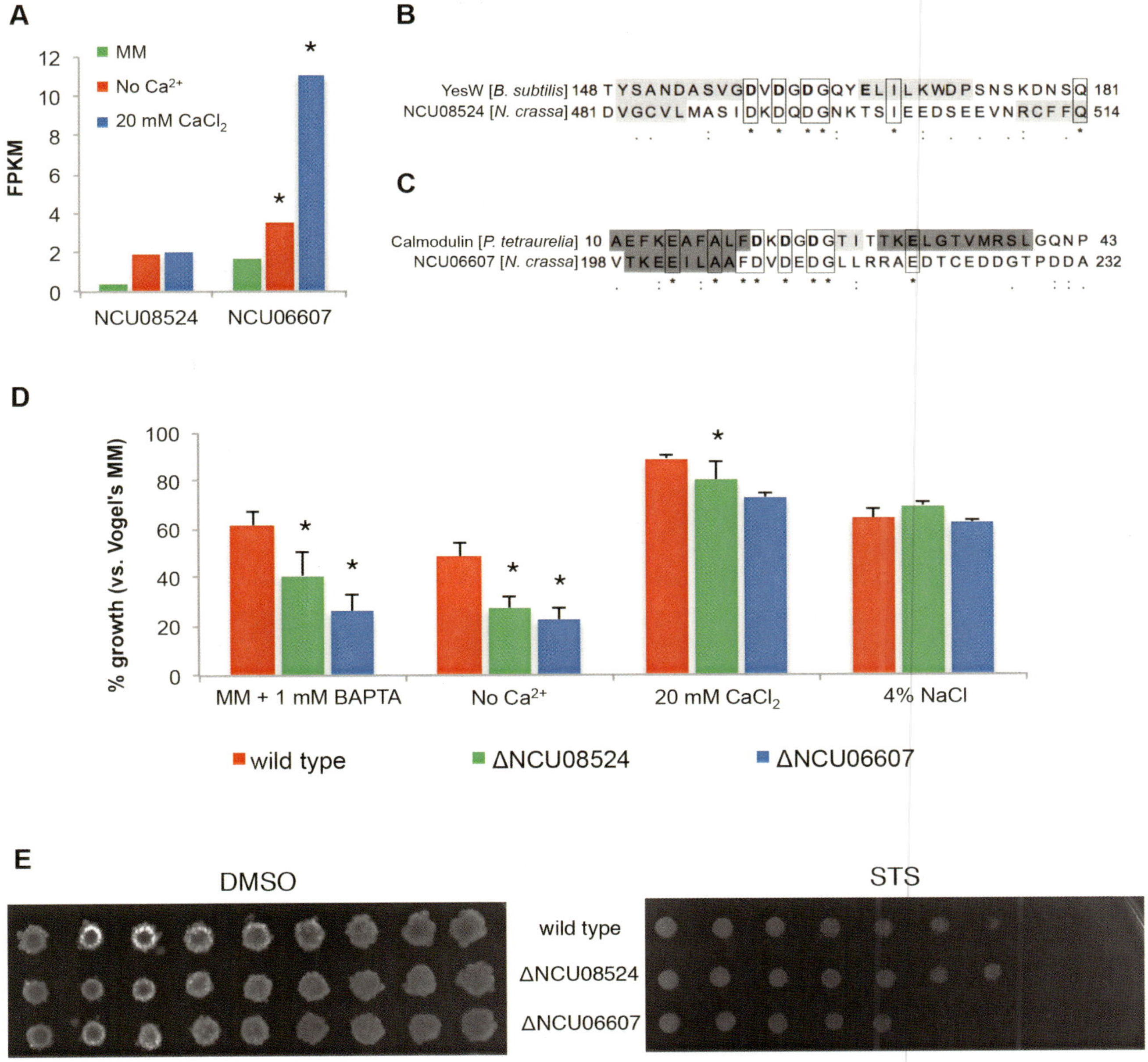

FIGURE 6: Identification of putative novel Ca^{2+}-binding proteins. (A) Expression levels of NCU08524 and NCU06607 in no Ca^{2+}, standard MM and 20 mM $CaCl_2$ media. *, p-value ≤ 0.05. **(B-C)** Sequence alignments of the EF-hand-like domain of NCU08524 and YesW from *B. subtilis* (accession number: O31526; PDB code: 2z8r) **(B),** and NCU06607 and calmodulin from *P. tetraurelia* (accession number: P07463; PDB code: 1exr) **(C)**. Residues binding Ca^{2+} in YesW and calmodulin are in bold; β-strands and α-helices are shaded in light and dark grey, respectively (confirmed by crystallography for YesW [61] and calmodulin [62] and predicted with MEMSAT3 for NCU08524 and NCU06607). **(D)** The percentages of growth of the different strains in the indicated media versus standard MM were determined by measuring absorbance of the respective liquid cultures at 450 nm. *, p-value ≤ 0.05. **(E)** The spot assay was used to examine the sensitivity of ΔNCU08524 and ΔNCU06607 to 5 μM staurosporine (STS).

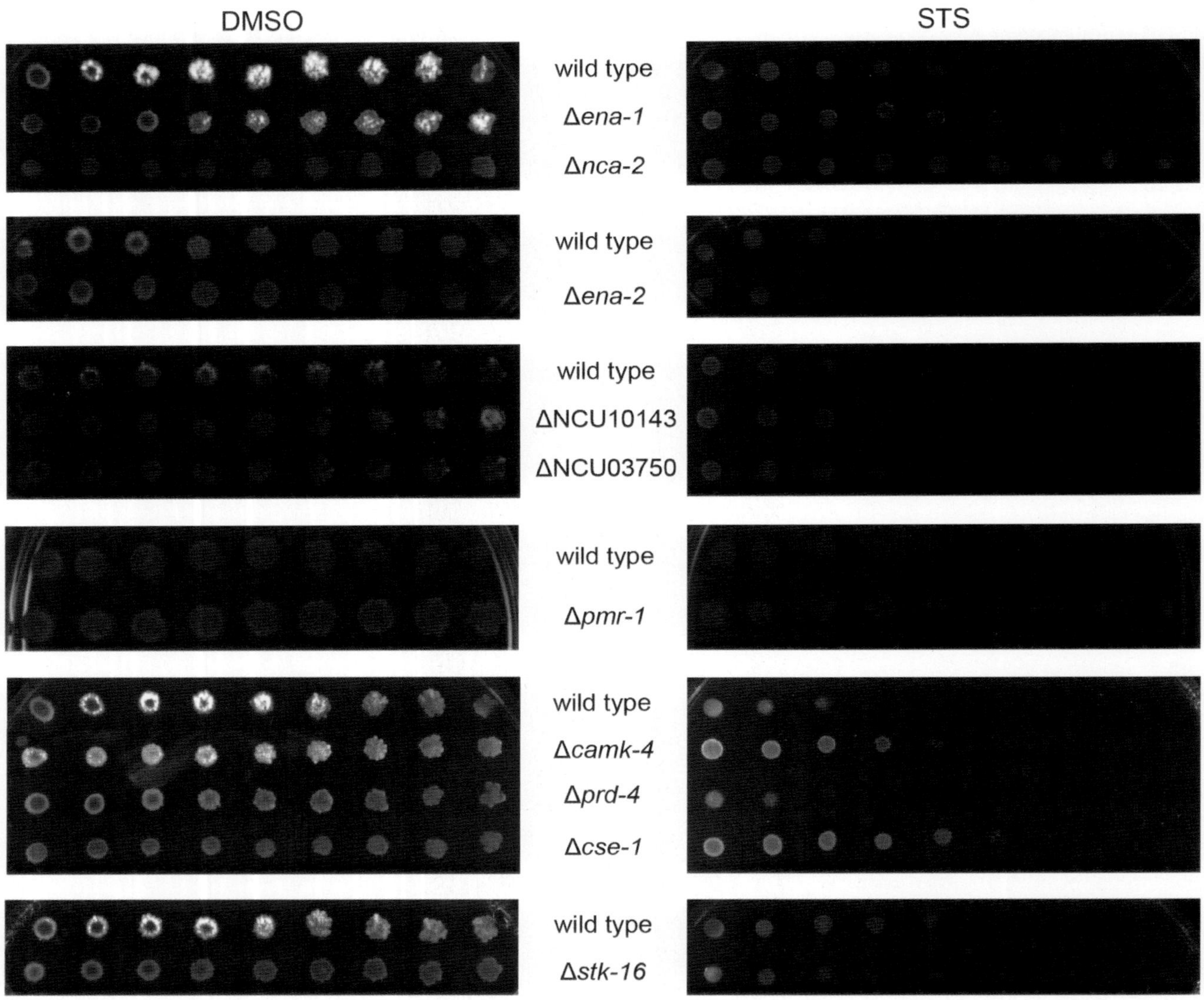

FIGURE 7: Several members of the Ca^{2+} handling machinery are involved in the fungal response to staurosporine. Serial dilutions (from left to right) of conidia from the indicated strains were spotted on GFS medium supplemented with 5 μM staurosporine (STS) and incubated for 3 days.

mented resistance, whereas strains carrying deletions in genes encoding STK-16 and PRD-4 resulted in increased susceptibility to staurosporine. Table 1 lists strains with altered sensitivity to staurosporine and Table S4 lists all strains tested. The deletion strains for NCU05360, NCU03804, NCU03833, NCU09871, NCU06347, NCU06617, NCU02411 and NCU09265 were not tested due to strain unavailability or their heterokaryotic nature. This data further stresses the importance and complexity of the fungal response to staurosporine, implicating numerous Ca^{2+}-related proteins of different function and subcellular localization in the process.

DISCUSSION

Here we show that controlling the amount of Ca^{2+} available in the extracellular milieu is enough for a robust modulation of the effects of staurosporine in *N. crassa*. Cell death is strongly enhanced by Ca^{2+} limitation whereas it is partially suppressed by a ~30-fold excess of external Ca^{2+}. A similar behavior was observed in *S. cerevisiae* cells treated with miconazole or terbinafine, inhibitors of ergosterol biosynthesis [37], suggesting common stress-responsive pathways in the two fungi, even when the stimuli have apparently distinct targets. Supplementation of the medium with high Ca^{2+} concentrations (20 mM $CaCl_2$, as used here) protected *N. crassa* cells from toxicity induced by PAF, an antifungal protein from *Penicillium chrysogenum* that also disturbs intracellular Ca^{2+} homeostasis [10]. Our results substantiate the view that Ca^{2+} is an important player during cell death.

The effects of staurosporine in media containing different amounts of Ca^{2+} are remarkably paralleled by distinct

TABLE 1. Strains with Ca^{2+}-related deleted genes that exhibit altered sensitivity to staurosporine when compared with wild type *N. crassa*.

Strain	Gene name	Classification [17-19]	Phenotype
ΔNCU02762	*cch-1*	Ca^{2+}-permeable channel	SS [12]
ΔNCU06703	*mid-1*	Ca^{2+}-permeable channel	SS [12]
ΔNCU04736	*nca-2*	Ca^{2+}-ATPase	RR
ΔNCU05046	*ena-1*	Ca^{2+}-ATPase	R
ΔNCU08147	*ena-2; ph-7*	Ca^{2+}-ATPase	S
ΔNCU10143		Ca^{2+}-ATPase	R
ΔNCU03292	*pmr-1*	Ca^{2+}-ATPase	RR
ΔNCU06245	*plc-1*	Phospholipase C (δ-type)	R [12]
ΔNCU01266	*plc-2*	Phospholipase C (δ-type)	RR [12]
ΔNCU09655	*plc-3*	Phospholipase C (δ-type)	R [12]
ΔNCU03750		Calmodulin	R
ΔNCU04379	*cse-1*	Ca^{2+} and/or CaM binding protein	R
ΔNCU09212	*camk-4*	Ca^{2+} and/or CaM binding protein	R
ΔNCU00914	*stk-16*	Ca^{2+} and/or CaM binding protein	S
ΔNCU02814	*prd-4*	Ca^{2+} and/or CaM binding protein	S

SS: much more sensitive; S: slightly more sensitive; RR: much more resistant; R: slightly more resistant.

cytosolic Ca^{2+} dynamics, in line with our previous findings that extracellular Ca^{2+} uptake is required for the development of the cytosolic Ca^{2+} signature induced by staurosporine [12]. We have recently demonstrated that during the staurosporine-induced Ca^{2+} signature, peaks "A" and "B" precede cell death, suggesting that they represent signaling events. In the presence of 20 mM $CaCl_2$ the peak "A" is enhanced, whereas in medium lacking Ca^{2+} peaks "A" and "B" of the staurosporine-induced cytosolic Ca^{2+} response are abolished. Thus, intracellular Ca^{2+} signaling that is triggered by staurosporine is markedly modulated by the concentration of Ca^{2+} in the external medium and might have a role in determining cell survival. However, the intracellular response to staurosporine is very complex and direct conclusions on the susceptibility of the cells based solely on the cytosolic Ca^{2+} profile, although tempting, are not possible. For instance, a Δ*plc-2* mutant shows increased resistance to the drug despite that its cytosolic Ca^{2+} response to staurosporine is nearly obliterated [12]. Thus, it seems that events downstream of the cytosolic changes in Ca^{2+} levels ultimately define cell fate in response to staurosporine.

The screening for staurosporine sensitivity of Ca^{2+}-handling mutant strains revealed a number of proteins likely involved in the cell death response to the drug. Further studies are needed to contextualize them functionally in the mechanism of action of the drug. NCA-2 is a Ca^{2+}-ATPase which pumps excess of cytosolic Ca^{2+} to the vacuoles or extracellular space [38, 39], and the respective knockout cells are very resistant to staurosporine. The knockout strain for PMR-1, a Golgi-localized Ca^{2+}-ATPase [40], is also very resistant to staurosporine and yeast Δ*PMR1* knockout cells accumulate much less Ca^{2+} in the ER [41]. Lack of these functions in Δ*nca-2* and Δ*pmr-1* cells is likely related to the phenotype of the strains upon treatment with staurosporine, as both the ER and the vacuoles play an important role during the fungal response to staurosporine [12]. It is plausible that the lack of NCA-2 or PMR-1 results in an increased accumulation of Ca^{2+} in the cytosol and this is consistent with the observations that high levels of Ca^{2+} (below a toxicity threshold) protect against the effects of staurosporine. The lack of NCA-2 has been recently associated with increased sensitivity to UV exposure, further implicating the molecule in cell survival [42]. Other Ca^{2+}-related deletion strains present altered resistance to staurosporine, including Δ*cse-1*, Δ*camk-4* and Δ*stk-16*. Knockout of *cse-1* has been shown to lead to impaired germling communication and fusion [43], as well as increased sensitivity to UV irradiation and Ca^{2+} [44]. A previous study has shown that disruption of *camk-4* results in benomyl resistance, while disruption of *stk-16* leads to increased sensitivity to fludioxonil, sodium chloride and benomyl [45].

The *N. crassa* transcriptional response to staurosporine is robustly reduced in the presence of 20 mM $CaCl_2$ when compared with standard MM, whereas limited extracellular Ca^{2+} causes alterations in an increased number of genes. Thus, different concentrations of extracellular Ca^{2+} not only lead to distinct intracellular Ca^{2+} dynamics, ROS accumulation and cell death levels, but also seem to be coupled to unique transcriptional profiles. It appears that, in the absence of Ca^{2+}, cells treated with staurosporine turn off a diverse array of biological processes. We discovered several enriched functional categories in the set of no Ca^{2+} medium-specific repressed genes, including the repression of genes involved in cell cycle, signal transduction, cell growth, anti-apoptosis, cellular polarization, protein fate and the unfolded protein response. It is possible that the decreased activity of various intracellular pathways explains the inability of cells to cope with the staurosporine insult and the consequent increased cell death.

N. crassa clearly translates changes in extracellular Ca^{2+} into a transcriptional response, since alterations in gene expression are caused alone by extracellular Ca^{2+} limitation or overload. This likely results from the homeostatic adaptation to the abnormal Ca^{2+} environment, i.e., the observed differences in gene expression might be triggered after the uptake or release of Ca^{2+} to balance the intracellular levels of the ion. Alternatively, these alterations may be secondary to the activity of a surface-localized Ca^{2+} sensor, although a sequence homologue of the extracellular Ca^{2+}-sensing receptor of animals and plants [46, 47] has not been found so far in *N. crassa*. Given the predicted cell surface localization of the two putative novel Ca^{2+}-binding proteins, NCU08524 and NCU06607, and the growth phenotype of the respective deletion strains under Ca^{2+} stress, it is tempting to speculate that they could play a role in Ca^{2+} sensing. The chemotropic interaction during conidial anastomosis tubes during cell fusion is Ca^{2+}-dependent [43] and both NCU08524 and NCU06607 genes have been recently shown to be under the control of the PP-1 transcription factor, a regulator of cell fusion [48, 49].

In summary, we show that the effects of the antifungal [13-16, 50] and anticancer [51] agent staurosporine are robustly modulated by the accessibility to Ca^{2+} in the culture medium. The amount of extracellular Ca^{2+} affects staurosporine-triggered intracellular events like Ca^{2+} signaling, the production of ROS and cell death. Our results indicate that these differences are linked to the stimulation of unique transcriptional programs and further highlight the importance of Ca^{2+} during fungal cell death. In addition, our RNA-seq dataset will be a useful resource for investigations on the role of Ca^{2+} on different aspects of fungal biology.

MATERIALS AND METHODS

Strains, culture media and chemicals

Wild type 74-OR23-1VA strain and deletion mutants were obtained from the Fungal Genetics Stock Center [52] and handled with standard procedures [28]. Minimal medium refers to Vogel's medium containing 1.5% (w/v) sucrose [27] plus 1.5% (w/v) agar to obtain solid medium. The concentration of KH_2PO_4 in the 50x Vogel's stock solution was limited to 10 mM to avoid precipitation with supplemental calcium [10]. The following chemicals were used: staurosporine (from LC Laboratories), dimethyl sulfoxide (DMSO), A23187, calcium chloride and sodium chloride (Sigma-Aldrich), phytosphingosine (Avanti Polar Lipids), 1,2-bis(ortho-aminophenoxy)ethane-N,N,N',N'-tetrasodium (BAPTA) (Merck Biochemicals).

Growth assays

Hyphal growth in solid medium at 26°C was obtained by measuring colony elongation after the inoculation of 20 µl containing $5x10^4$ conidia on the centre of large Petri dishes (φ 14.2 cm). Growth in liquid medium was examined by incubating $1x10^4$ conidia/ml at 26°C, 100 rpm, under constant light in 96-well plates (200 µl/well) and following absorbance at 450 nm [53] during 24 hours. The % growth versus the respective control and the growth rate were calculated for each condition. For the spot assays, 3-fold serial dilutions up to $1x10^4$ cells/ml were prepared for each strain. 5 µl from each dilution were spotted on plates containing glucose-fructose-sorbose (GFS) [28] plus the relevant chemicals and incubated for 3 days at 26°C.

Flow cytometry measurements of cell death and reactive oxygen species (ROS)

For each experiment and strain, a control without staining was prepared to define autofluorescence. Samples were read in a BD FACS Calibur and data analyzed with FlowJo (Tree Star). The levels of cell death were determined by staining cells with YOPRO-1 (Life Technologies) or propidium iodide (PI; Sigma-Aldrich). An inoculum of 10^6 conidia/ml in the appropriate medium was incubated for 4 hours at 26°C (140 rpm, constant light). Staurosporine (20 µM) was added and growth resumed for further 2 hours. The cells were harvested by centrifugation, washed twice with PBS and incubated either with 0.1 µM YOPRO-1 (samples were kept 20 min on ice before reading) or 2 µg/ml PI (samples were immediately ready for cytometry). The production of ROS was measured using the fluorescent probe dihydrorhodamine 123 (Sigma-Aldrich). After growing 10^6 conidia/ml for 4 hours in the indicated medium at 26°C, 20 µg/ml dihydrorhodamine 123 and staurosporine were added for further 30 minutes. Samples were harvested by centrifugation and washed twice with PBS before being resuspended in PBS and read in the cytometer.

Cytosolic Ca^{2+} measurements

A bioluminescent method based on the reaction of cytosolic Ca^{2+} with the genetic probe aequorin in wild type cells was employed as previously described [12]. Briefly, 100 µl of conidia at a concentration of $2x10^6$ cells/ml, 5 µM coelenterazine (Santa Cruz Biotechnology) and minimal medium were added to white opaque 96-well plates and incubated for 6 hours at 26°C, in the dark, without agitation. Luminescence (in RLU, relative light units) was captured over time on a Bio-Tek Synergy HT microplate reader. Maximum levels of luminescence, determined in extra wells by measuring luminescence during 3 mins after the injection of 100 µl of 3M $CaCl_2$ in 20% ethanol, were used to normalize each experiment (cytosolic Ca^{2+} levels (arbitrary units) = experimental RLU values / total emitted luminescence). Quantification was performed by summing the normalized values and data are expressed as mean ± SEM. The values from control DMSO-treated samples were subtracted from staurosporine-treated samples to obtain the "staurosporine-induced amplitude of response".

Statistical and protein sequence analyses

The non-parametric Mann-Whitney test was used for comparisons between two groups using SPSS 20 (SPSS Inc.) and p-values ≤ 0.05 were considered statistically significant. ClustalW2 [54] was used to align protein sequences and PSI-BLAST [55] was used for iterative searches. Secondary structure was predicted with MEMSAT3 [56], conserved domains with InterProScan [57], transmembrane domains with TMHMM 2.0 [58], subcellular localization with WoLF PSORT [59] and Ca^{2+}-binding motifs with CaPS [20].

High-throughput RNA sequencing (RNA-seq)

Conidial suspensions of 1x10^6 cells/ml were incubated in culture medium with the indicated concentration of Ca^{2+} for 6 hours (26°C, 140 rpm, constant light), staurosporine (or DMSO alone) was added and growth resumed for 1 more hour. Cells were harvested and immediately frozen in liquid nitrogen. RNA isolation, mRNA purification and fragmentation and cDNA synthesis was performed as previously described [15]. The cDNA libraries were generated using an Illumina TruSeq kit and sequenced in an Illumina HiSeq2000 (single reads of 50 bp were obtained). Sequencing data was handled with Tophat, Cufflinks and Cuffdiff and expression levels are presented as Fragments Per Kilobase of transcript per Million mapped reads (FPKM). Functional enrichment of sets of genes was assessed with FunCat [60]. The resulting dataset is available at the NCBI GEO database (http://www.ncbi.nlm.nih.gov/geo/; series record: GSE53013).

ACKNOWLEDGMENTS

We would like to thank Dr. Nick D Read (Manchester Fungal Infection Group, Institute of Inflammation and Repair, University of Manchester, UK) for providing the aequorin-containing plasmid. A.P.G. was recipient of a fellowship from Fundação Calouste Gulbenkian (104210) and a short-term fellowship from EMBO (329-2012). J.M.C. was hired under the scope of FCT Portugal CIÊNCIA 2008 Programme (FSE-POPH-QREN). This work was supported by FCT Portugal (PEst-C/SAU/LA0002/2013 and FCOMP-01-0124-FEDER-037277 to A.V. and PEst-OE/SAU/UI0215/2014 to P.C.S.), the European POCI program of QCAIII co-participated by FEDER (NORTE-07-0124-FEDER-000003), a grant from the University of Porto (PP_IJUP2011-20 to A.V.) and a grant from the National Institutes of Health (NIH R24 GM081597 to N.L.G. with Drs. JW Taylor and RB Brem).

CONFLICT OF INTEREST

The authors declare no conflict of interest.

REFERENCES

1. Shaw BD, Hoch HC (**2001**). Biology of the fungal cell. In: Howard RJ, Gow NAR, editors. The Mycota VIII. Springer-Verlag KG, Berlin; pp 73-89.

2. Zhivotovsky B, Orrenius S (**2011**). Calcium and cell death mechanisms: a perspective from the cell death community. **Cell Calcium** 50(3): 211-221.

3. Pozniakovsky AI, Knorre DA, Markova OV, Hyman AA, Skulachev VP, Severin FF (**2005**). Role of mitochondria in the pheromone- and amiodarone-induced programmed death of yeast. **J Cell Biol** 168(2): 257-269.

4. Rao A, Zhang Y, Muend S, Rao R (**2010**). Mechanism of antifungal activity of terpenoid phenols resembles calcium stress and inhibition of the TOR pathway. **Antimicrob Agents Chemother** 54(12): 5062-5069.

5. Roberts SK, McAinsh M, Widdicks L (**2012**). Cch1p mediates Ca^{2+} influx to protect *Saccharomyces cerevisiae* against eugenol toxicity. **PLoS One** 7(9): e43989.

6. Gupta SS, Ton VK, Beaudry V, Rulli S, Cunningham K, Rao R (**2003**). Antifungal activity of amiodarone is mediated by disruption of calcium homeostasis. **J Biol Chem** 278(31): 28831-28839.

7. Bonilla M, Nastase KK, Cunningham KW (**2002**). Essential role of calcineurin in response to endoplasmic reticulum stress. **EMBO J** 21(10): 2343-2353.

8. Martin DC, Kim H, Mackin NA, Maldonado-Baez L, Evangelista CC, Jr., Beaudry VG, Dudgeon DD, Naiman DQ, Erdman SE, Cunningham KW (**2011**). New regulators of a high affinity Ca^{2+} influx system revealed through a genome-wide screen in yeast. **J Biol Chem** 286(12): 10744-10754.

9. Palma-Guerrero J, Huang IC, Jansson HB, Salinas J, Lopez-Llorca LV, Read ND (**2009**). Chitosan permeabilizes the plasma membrane and kills cells of *Neurospora crassa* in an energy dependent manner. **Fungal Genet Biol** 46(8): 585-594.

10. Binder U, Chu M, Read ND, Marx F (**2010**). The antifungal activity of the Penicillium chrysogenum protein PAF disrupts calcium homeostasis in *Neurospora crassa*. **Eukaryot Cell** 9(9): 1374-1382.

11. Munoz A, Marcos JF, Read ND (**2012**). Concentration-dependent mechanisms of cell penetration and killing by the de novo designed antifungal hexapeptide PAF26. **Mol Microbiol** 85(1): 89-106.

12. Gonçalves AP, Cordeiro JM, Monteiro J, Muñoz A, Correia-de-Sá P, Read ND, Videira A (**2014**). Activation of a TRP-like channel and intracellular calcium dynamics during phospholipase C-mediated cell death. **J Cell Sci.** (In press)

13. Castro A, Lemos C, Falcao A, Fernandes AS, Glass NL, Videira A (**2010**). Rotenone enhances the antifungal properties of staurosporine. **Eukaryot Cell** 9(6): 906-914.

14. Fernandes AS, Castro A, Videira A (**2013**). Reduced glutathione export during programmed cell death of *Neurospora crassa*. **Apoptosis** 18(8): 940-948.

15. Goncalves AP, Hall C, Kowbel DJ, Glass NL, Videira A (**2014**). CZT-1 Is a Novel Transcription Factor Controlling Cell Death and Natural Drug Resistance in *Neurospora crassa*. **G3 (Bethesda)** 4(6): 1091-1102.

16. Fernandes AS, Goncalves AP, Castro A, Lopes TA, Gardner R, Glass NL, Videira A (**2011**). Modulation of fungal sensitivity to staurosporine by targeting proteins identified by transcriptional profiling. **Fungal Genet Biol** 48(12): 1130-1138.

17. Galagan JE, Calvo SE, Borkovich KA, Selker EU, Read ND, Jaffe D, FitzHugh W, Ma LJ, Smirnov S, Purcell S, Rehman B, Elkins T, Engels R, Wang S, Nielsen CB, Butler J, Endrizzi M, Qui D, Ianakiev P, Bell-Pedersen D, Nelson MA, Werner-Washburne M, Selitrennikoff CP, Kinsey JA, Braun EL, Zelter A, Schulte U, Kothe GO, Jedd G, Mewes W, et al. (**2003**). The genome sequence of the filamentous fungus *Neurospora crassa*. **Nature** 422(6934): 859-868.

18. Borkovich KA, Alex LA, Yarden O, Freitag M, Turner GE, Read ND, Seiler S, Bell-Pedersen D, Paietta J, Plesofsky N, Plamann M, Goodrich-Tanrikulu M, Schulte U, Mannhaupt G, Nargang FE, Radford A, Selitrennikoff C, Galagan JE, Dunlap JC, Loros JJ, Catcheside D, Inoue H, Aramayo R, Polymenis M, Selker EU, Sachs MS, Marzluf GA, Paulsen I, Davis R, Ebbole DJ, et al. (**2004**). Lessons from the genome sequence of *Neurospora crassa*: tracing the path from genomic blueprint to multicellular organism. **Microbiol Mol Biol Rev** 68(1): 1-108.

19. Zelter A, Bencina M, Bowman BJ, Yarden O, Read ND (**2004**). A comparative genomic analysis of the calcium signaling machinery in *Neurospora crassa*, *Magnaporthe grisea*, and *Saccharomyces cerevisiae*. **Fungal Genet Biol** 41(9): 827-841.

20. Zhou Y, Yang W, Kirberger M, Lee HW, Ayalasomayajula G, Yang JJ (**2006**). Prediction of EF-hand calcium-binding proteins and analysis of bacterial EF-hand proteins. **Proteins** 65(3): 643-655.

21. Santamaria-Hernando S, Krell T, Ramos-Gonzalez MI (**2012**). Identification of a novel calcium binding motif based on the detection of sequence insertions in the animal peroxidase domain of bacterial proteins. **PLoS One** 7(7): e40698.

22. Rigden DJ, Woodhead DD, Wong PW, Galperin MY (**2011**). New structural and functional contexts of the Dx[DN]xDG linear motif: insights into evolution of calcium-binding proteins. **PLoS One** 6(6): e21507.

23. Rigden DJ, Galperin MY (**2004**). The DxDxDG motif for calcium binding: multiple structural contexts and implications for evolution. **J Mol Biol** 343(4): 971-984.

24. Rigden DJ, Jedrzejas MJ, Moroz OV, Galperin MY (**2003**). Structural diversity of calcium-binding proteins in bacteria: single-handed EF-hands? **Trends Microbiol** 11(7): 295-297.

25. Naested H, Frandsen GI, Jauh GY, Hernandez-Pinzon I, Nielsen HB, Murphy DJ, Rogers JC, Mundy J (**2000**). Caleosins: Ca^{2+}-binding proteins associated with lipid bodies. **Plant Mol Biol** 44(4): 463-476.

26. Cioci G, Mitchell EP, Chazalet V, Debray H, Oscarson S, Lahmann M, Gautier C, Breton C, Perez S, Imberty A (**2006**). Beta-propeller crystal structure of *Psathyrella velutina* lectin: an integrin-like fungal protein interacting with monosaccharides and calcium. **J Mol Biol** 357(5): 1575-1591.

27. Vogel HJ (**1956**). A convenient growth medium for Neurospora (Medium N). **Microbial Genet Bull** 13: 42-43.

28. Davis RH, de Serres FJ (**1970**). Genetic and microbiological research techniques for *Neurospora crassa*. **Meth Enzymol** 17:79-143.

29. Castro A, Lemos C, Falcao A, Glass NL, Videira A (**2008**). Increased resistance of complex I mutants to phytosphingosine-induced programmed cell death. **J Biol Chem** 283(28): 19314-19321.

30. Videira A, Kasuga T, Tian C, Lemos C, Castro A, Glass NL (**2009**). Transcriptional analysis of the response of *Neurospora crassa* to phytosphingosine reveals links to mitochondrial function. **Microbiology** 155(Pt 9): 3134-3141.

31. Bick AG, Calvo SE, Mootha VK (**2012**). Evolutionary diversity of the mitochondrial calcium uniporter. **Science** 336(6083): 886.

32. Cavinder B, Trail F (**2012**). Role of Fig1, a component of the low-affinity calcium uptake system, in growth and sexual development of filamentous fungi. **Eukaryot Cell** 11(8): 978-988.

33. Herwig C, Doerries C, Marison I, von Stockar U (**2001**). Quantitative analysis of the regulation scheme of invertase expression in *Saccharomyces cerevisiae*. **Biotechnol Bioeng** 76(3): 247-258.

34. Ziv C, Gorovits R, Yarden O (**2008**). Carbon source affects PKA-dependent polarity of *Neurospora crassa* in a CRE-1-dependent and independent manner. **Fungal Genet Biol** 45(2): 103-116.

35. Benito B, Garciadeblas B, Rodriguez-Navarro A (**2000**). Molecular cloning of the calcium and sodium ATPases in *Neurospora crassa*. **Mol Microbiol** 35(5): 1079-1088.

36. Rodrigues JJ, Ferreira HB, Farias SE, Zaha A (**1997**). A protein with a novel calcium-binding domain associated with calcareous corpuscles in *Echinococcus granulosus*. **Biochem Biophys Res Commun** 237(2): 451-456.

37. Edlind T, Smith L, Henry K, Katiyar S, Nickels J (**2002**). Antifungal activity in *Saccharomyces cerevisiae* is modulated by calcium signalling. **Mol Microbiol** 46(1): 257-268.

38. Bowman BJ, Abreu S, Margolles-Clark E, Draskovic M, Bowman EJ (**2011**). Role of four calcium transport proteins, encoded by nca-1, nca-2, nca-3, and cax, in maintaining intracellular calcium levels in *Neurospora crassa*. **Eukaryot Cell** 10(5): 654-661.

39. Bowman BJ, Draskovic M, Freitag M, Bowman EJ (**2009**). Structure and distribution of organelles and cellular location of calcium transporters in *Neurospora crassa*. **Eukaryot Cell** 8(12): 1845-1855.

40. Bowman BJ, Abreu S, Johl JK, Bowman EJ (**2012**). The pmr gene, encoding a Ca^{2+}-ATPase, is required for calcium and manganese homeostasis and normal development of hyphae and conidia in *Neurospora crassa*. **Eukaryot Cell** 11(11): 1362-1370.

41. Strayle J, Pozzan T, Rudolph HK (**1999**). Steady-state free Ca^{2+} in the yeast endoplasmic reticulum reaches only 10 microM and is mainly controlled by the secretory pathway pump pmr1. **EMBO J** 18(17): 4733-4743.

42. Deka R, Tamuli R (**2013**). *Neurospora crassa* ncs-1, mid-1 and nca-2 double-mutant phenotypes suggest diverse interaction among three Ca^{2+}-regulating gene products. **J Genet** 92(3): 559-563.

43. Palma-Guerrero J, Hall CR, Kowbel D, Welch J, Taylor JW, Brem RB, Glass NL (**2013**). Genome wide association identifies novel loci involved in fungal communication. **PLoS Genet** 9(8): e1003669.

44. Deka R, Kumar R, Tamuli R (**2011**). *Neurospora crassa* homologue of Neuronal Calcium Sensor-1 has a role in growth, calcium stress tolerance, and ultraviolet survival. **Genetica** 139(7): 885-894.

45. Park G, Servin JA, Turner GE, Altamirano L, Colot HV, Collopy P, Litvinkova L, Li L, Jones CA, Diala FG, Dunlap JC, Borkovich KA (**2011**). Global analysis of serine-threonine protein kinase genes in *Neurospora crassa*. **Eukaryot Cell** 10(11): 1553-1564.

46. Brown EM, Gamba G, Riccardi D, Lombardi M, Butters R, Kifor O, Sun A, Hediger MA, Lytton J, Hebert SC (**1993**). Cloning and characterization of an extracellular Ca^{2+}-sensing receptor from bovine parathyroid. **Nature** 366(6455): 575-580.

47. Han S, Tang R, Anderson LK, Woerner TE, Pei ZM (**2003**). A cell surface receptor mediates extracellular Ca^{2+} sensing in guard cells. **Nature** 425(6954): 196-200.

48. Leeder AC, Jonkers W, Li J, Glass NL (**2013**). Early colony establishment in *Neurospora crassa* requires a MAP kinase regulatory network. **Genetics** 195(3): 883-898.

49. Palma-Guerrero J, Leeder AC, Welch J, Glass NL (**2014**). Identification and characterization of LFD1, a novel protein involved in membrane merger during cell fusion in *Neurospora crassa*. **Mol Microbiol** 92(1): 164-182.

50. Gonçalves AP, Videira A (**2014**). Programmed cell death in *Neurospora crassa*. **New J Sci** 2014. Article ID 479015.

51. Gescher A (**2000**). Staurosporine analogues - pharmacological toys or useful antitumour agents? **Crit Rev Oncol Hematol** 34(2): 127-135.

52. McCluskey K, Wiest A, Plamann M (**2010**). The Fungal Genetics Stock Center: a repository for 50 years of fungal genetics research. **J Biosci** 35(1): 119-126.

53. Alberghina FA (**1973**). Growth regulation in *Neurospora crassa*. Effects of nutrients and of temperature. **Arch Mikrobiol** 89(2): 83-94.

54. Larkin MA, Blackshields G, Brown NP, Chenna R, McGettigan PA, McWilliam H, Valentin F, Wallace IM, Wilm A, Lopez R, Thompson JD, Gibson TJ, Higgins DG (**2007**). Clustal W and Clustal X version 2.0. **Bioinformatics** 23(21): 2947-2948.

55. Altschul SF, Madden TL, Schaffer AA, Zhang J, Zhang Z, Miller W, Lipman DJ (**1997**). Gapped BLAST and PSI-BLAST: a new generation of protein database search programs. **Nucleic Acids Res** 25(17): 3389-3402.

56. Jones DT (**2007**). Improving the accuracy of transmembrane protein topology prediction using evolutionary information. **Bioinformatics** 23(5): 538-544.

57. Quevillon E, Silventoinen V, Pillai S, Harte N, Mulder N, Apweiler R, Lopez R (**2005**). InterProScan: protein domains identifier. **Nucleic Acids Res** 33(Web Server issue): W116-120.

58. Krogh A, Larsson B, von Heijne G, Sonnhammer EL (**2001**). Predicting transmembrane protein topology with a hidden Markov model: application to complete genomes. **J Mol Biol** 305(3): 567-580.

59. Horton P, Park KJ, Obayashi T, Fujita N, Harada H, Adams-Collier CJ, Nakai K (**2007**). WoLF PSORT: protein localization predictor. **Nucleic Acids Res** 35(Web Server issue): W585-587.

60. Ruepp A, Zollner A, Maier D, Albermann K, Hani J, Mokrejs M, Tetko I, Guldener U, Mannhaupt G, Munsterkotter M, Mewes HW (**2004**). The FunCat, a functional annotation scheme for systematic classification of proteins from whole genomes. **Nucleic Acids Res** 32(18): 5539-5545.

61. Ochiai A, Itoh T, Maruyama Y, Kawamata A, Mikami B, Hashimoto W, Murata K (**2007**). A novel structural fold in polysaccharide lyases *Bacillus subtilis* family 11 rhamnogalacturonan lyase YesW with an eight-bladed beta-propeller. **J Biol Chem** 282(51): 37134-37145.

62. Wilson MA, Brunger AT (**2000**). The 1.0 A crystal structure of Ca bound calmodulin: an analysis of disorder and implications for functionally relevant plasticity. **J Mol Biol** 301(5): 1237-1256.

Permissions

All chapters in this book were first published in MIC, by Shared Science Publishers OG; hereby published with permission under the Creative Commons Attribution License or equivalent. Every chapter published in this book has been scrutinized by our experts. Their significance has been extensively debated. The topics covered herein carry significant findings which will fuel the growth of the discipline. They may even be implemented as practical applications or may be referred to as a beginning point for another development.

The contributors of this book come from diverse backgrounds, making this book a truly international effort. This book will bring forth new frontiers with its revolutionizing research information and detailed analysis of the nascent developments around the world.

We would like to thank all the contributing authors for lending their expertise to make the book truly unique. They have played a crucial role in the development of this book. Without their invaluable contributions this book wouldn't have been possible. They have made vital efforts to compile up to date information on the varied aspects of this subject to make this book a valuable addition to the collection of many professionals and students.

This book was conceptualized with the vision of imparting up-to-date information and advanced data in this field. To ensure the same, a matchless editorial board was set up. Every individual on the board went through rigorous rounds of assessment to prove their worth. After which they invested a large part of their time researching and compiling the most relevant data for our readers.

The editorial board has been involved in producing this book since its inception. They have spent rigorous hours researching and exploring the diverse topics which have resulted in the successful publishing of this book. They have passed on their knowledge of decades through this book. To expedite this challenging task, the publisher supported the team at every step. A small team of assistant editors was also appointed to further simplify the editing procedure and attain best results for the readers.

Apart from the editorial board, the designing team has also invested a significant amount of their time in understanding the subject and creating the most relevant covers. They scrutinized every image to scout for the most suitable representation of the subject and create an appropriate cover for the book.

The publishing team has been an ardent support to the editorial, designing and production team. Their endless efforts to recruit the best for this project, has resulted in the accomplishment of this book. They are a veteran in the field of academics and their pool of knowledge is as vast as their experience in printing. Their expertise and guidance has proved useful at every step. Their uncompromising quality standards have made this book an exceptional effort. Their encouragement from time to time has been an inspiration for everyone.

The publisher and the editorial board hope that this book will prove to be a valuable piece of knowledge for researchers, students, practitioners and scholars across the globe.

List of Contributors

Christopher Chin
Department of Biology, Providence College, Providence, RI 02918, U.S.A
University of Massachusetts School of Medicine, 55 Lake Ave. N., Worcester, MA 01655, U.S.A

Faith Donaghey, Morgan McCarthy, Stephen Rogers and Nicanor Austriaco
Department of Biology, Providence College, Providence, RI 02918, U.S.A

Katherine Helming
Department of Biology, Providence College, Providence, RI 02918, U.S.A
Dana-Farber Cancer Institute, 44 Binney St., Boston, MA 02115, U.S.A

William R. Proto, Nathaniel G. Jones and Jeremy C. Mottram
Wellcome Trust Centre for Molecular Parasitology, Institute of Infection, Immunity and Inflammation, College of Medical, Veterinary and Life Sciences, University of Glasgow, Glasgow, G12 8TA, UK

Graham H. Coombs
Strathclyde Institute of Pharmacy and Biomedical Sciences, University of Strathclyde, Glasgow, G4 0RE, UK

Daisuke Watanabe, Rie Kikushima, Miho Aitoku, Akira Nishimura, Iwao Ohtsu, Ryo Nasuno and Hiroshi Takagi
Graduate School of Biological Sciences, Nara Institute of Science and Technology, 8916-5 Takayama, Ikoma, Nara 630-0192, Japan

Markus M. M. Bisschops
Department of Biotechnology, Delft University of Technology, Delft, The Netherlands.
Division of Systems and Synthetic Biology, Department of Biology and Biological Engineering & The Novo Nordisk

Tim Vos, Pilar de la Torre Cortés, Jack T. Pronk and Pascale Daran-Lapujade
Department of Biotechnology, Delft University of Technology, Delft, The Netherlands

Rubén Martínez-Moreno
Instituto de Ciencias de la Vid y del Vino, CSIC, Universidad de La Rioja, Gobierno de La Rioja, Logroño, Spain Foundation Center for Biosustainability, Chalmers University of Technology, Gothenburg, Sweden
Quercus Europe S.L., L'Hospitalet de Llobregat, Catalonia, Spain

Michael E. Bond and Rachel Brown
MRC Laboratory for Molecular Cell Biology, University College London, London WC1E 6BT, UK

Charalampos Rallis and Jürg Bähler
Department of Genetics, Evolution and Environment, University College London, London WC1E 6BT,UK
Institute of Healthy Ageing, University College London, London WC1E 6BT, UK

Sara E. Mole
MRC Laboratory for Molecular Cell Biology, University College London, London WC1E 6BT, UK
UCL Institute of Child Health, 30 Guilford Street, London WC1N 1EH, UK
Department of Genetics, Evolution and Environment, University College London, London WC1E 6BT, UK

Chunyan Jin, Stephen K. Kim, Stephen D. Willis and Katrina F. Cooper
Department of Molecular Biology, Rowan University School of Osteopathic Medicine, Stratford, NJ, 08055 USA Shaoxing Women & Children Hospital, 305 East Street, Shaoxing, Zhejiang, China, 312000

Afsaneh Porzoor and Ian G. Macreadie
School of Applied Sciences, RMIT University, Bundoora, Victoria 3083, Australia

Anne Silvestre
Institut Pasteur, Unité Biologie Cellulaire du Parasitisme, F-75015 Paris, France
INSERM U786, F-75015 Paris, France
INRA, UMR1282 Infectiologie et Santé Publique, F-37380 Nouzilly, France
Université de Tours, UMR1282 Infectiologie et Santé Publique, F-37000 Tours, France

Aurélie Plaze, Roman Thibeaux, Nancy Guillen and Elisabeth Labruyère
Institut Pasteur, Unité Biologie Cellulaire du Parasitisme, F-75015 Paris, France
INSERM U786, F-75015 Paris, France

Patricia Berthon
INRA, UMR1282 Infectiologie et Santé Publique, F-37380 Nouzilly, France
Université de Tours, UMR1282 Infectiologie et Santé Publique, F-37000 Tours, France

Chen Sun, Jian Li, Yu Cao, Gongbo Long and Bing Zhou
State Key Laboratory of Biomembrane and Membrane Biotechnology, School of Life Sciences, Tsinghua University, Beijing, China

Samuel Bru, Javier Jiménez and Josep Clotet
Department of Basic Sciences, Faculty of Medicine and Health Sciences. Universitat Internacional de Catalunya. Barcelona, Spain

David Canadell and Joaquín Ariño
Departament de Bioquímica i Biologia Molecular and Institut de Biotecnologia i Biomedicina, Universitat Autònoma de Barcelona. Cerdanyola del Vallès, Spain Cell Signaling Research Group, Departament de Ciències Experimentals i de la Salut, Universitat Pompeu Fabra, Barcelona, Spain

Triana Amen and Daniel Kaganovich
Department of Cell and Developmental Biology, Hebrew University of Jerusalem, Givat Ram, Jerusalem 91904, Israel

Anup Arunrao Deshpande, Muskan Bhatia and Anand Kumar Bachhawat
Department of Biological Sciences, Indian Institute of Science Education and Research (IISER Mohali), S.A.S. Nagar, Punjab 140306, India

Sunil Laxman
Institute for Stem Cell Biology and Regenerative Medicine (inStem), NCBS campus, Bangalore 560065, India

Nkechi E. Egbe
Division of Molecular and Cellular Function, School of Biological Sciences, Faculty of Biology, Medicine and Health, The University of Manchester, Manchester Academic Health Science Centre, Michael Smith Building, Oxford Rd., Manchester, M13 9PT, United Kingdom
Department of Biological Sciences, Nigerian Defence Academy, PMB 2109, Kaduna, Nigeria

Tawni O. Dornelles, Caroline M. Paget and Mark P. Ashe
Division of Molecular and Cellular Function, School of Biological Sciences, Faculty of Biology, Medicine and Health, The University of Manchester, Manchester Academic Health Science Centre, Michael Smith Building, Oxford Rd., Manchester, M13 9PT, United Kingdom

Lydia M. Castelli
Division of Molecular and Cellular Function, School of Biological Sciences, Faculty of Biology, Medicine and Health, The University of Manchester, Manchester Academic Health Science Centre, Michael Smith Building, Oxford Rd., Manchester, M13 9PT, United Kingdom
Sheffield Institute for Translational Neuroscience, University of Sheffield, Sheffield, S10 2HQ, United Kingdom

Ali Ghanem, Ana Kitanovic, Jinda Holzwarth and Stefan Wölfl
Institute of Pharmacy and Molecular Biotechnology, Heidelberg University, Heidelberg, Germany

Petra Gottier, Mauro Serricchio and Peter Bütikofer
Institute for Biochemistry and Molecular Medicine, University of Bern, Bern, Switzerland

Rita Vitale and Angela Corcelli
School of Medicine: Basic Medical Sciences, Neuroscience and Sense Organs, University of Bari Aldo Moro, Bari, Italy

Ivan Orlandi and Marina Vai
SYSBIO Centre for Systems Biology Milano, Italy
Dipartimento di Biotecnologie e Bioscienze, Università di Milano-Bicocca, Piazza della Scienza 2, 20126 Milano, Italy

Damiano Pellegrino Coppola
Dipartimento di Biotecnologie e Bioscienze, Università di Milano-Bicocca, Piazza della Scienza 2, 20126 Milano, Italy

Adam Kawałek and Ida J. van der Klei
Molecular Cell Biology, Groningen Biomolecular Sciences and Biotechnology Institute, Systems Biology Centre for Metabolism and Ageing, University of Groningen, the Netherlands

A. Pedro Gonçalves and Arnaldo Videira
ICBAS-Instituto de Ciências Biomédicas de Abel Salazar, Universidade do Porto, Rua de Jorge Viterbo Ferreira 228, 4050-313 Porto, Portugal

IBMC-Instituto de Biologia Molecular e Celular - Universidade do Porto, Rua do Campo Alegre 823, 4150-180 Porto, Portugal

João Monteiro and Chiara Lucchi
IBMC-Instituto de Biologia Molecular e Celular - Universidade do Porto, Rua do Campo Alegre 823, 4150-180 Porto, Portugal

David J. Kowbel and N. Louise Glass
Plant and Microbial Biology Department, The University of California, Berkeley, CA 94720, USA

J. Miguel Cordeiro and Paulo Correia-de-Sá
ICBAS-Instituto de Ciências Biomédicas de Abel Salazar, Universidade do Porto, Rua de Jorge Viterbo Ferreira 228, 4050-313 Porto, Portugal
UMIB-Unidade Multidisciplinar de Investigação Biomédica, Universidade do Porto, Rua de Jorge Viterbo Ferreira 228, 4050-313 Porto, Portugal

Daniel J. Rigden
Institute of Integrative Biology, University of Liverpool, Liverpool, L69 7ZB, United Kingdom

Index

O

P

R

S

T

V

W

Y